AF411326

Fucoxanthin and Astaxanthin—Production, Biofunction, and Application

Editors

Masashi Hosokawa
Hayato Maeda

MDPI • Basel • Beijing • Wuhan • Barcelona • Belgrade • Manchester • Tokyo • Cluj • Tianjin

Editors
Masashi Hosokawa   Hayato Maeda
Hokkaido University   Hirosaki University
Japan   Japan

Editorial Office
MDPI
St. Alban-Anlage 66
4052 Basel, Switzerland

This is a reprint of articles from the Special Issue published online in the open access journal *Marine Drugs* (ISSN 1660-3397) (available at: https://www.mdpi.com/journal/marinedrugs/special_issues/Fucoxanthin_Astaxanthin).

For citation purposes, cite each article independently as indicated on the article page online and as indicated below:

LastName, A.A.; LastName, B.B.; LastName, C.C. Article Title. *Journal Name* **Year**, *Volume Number*, Page Range.

ISBN 978-3-0365-6324-4 (Hbk)
ISBN 978-3-0365-6325-1 (PDF)

Contents

About the Editors

Masashi Hosokawa

Prof. Masashi Hosokawa has studied marine food chemistry and molecular nutrition of marine chemicals such as carotenoids and n-3 polyunsaturated fatty acids at Hokkaido University. He started his academic career in 1994 as an assistant professor in the Faculty of Fisheries Sciences, Hokkaido University. He became associate professor in 2000 and professor in 2017 at Hokkaido University. From June 2000 to May 2001, he worked on the bioconversion of polyunsaturated fatty acids to hydroxyl and tetrahydrofuranyl fatty acids as a Visiting Scientist at USDA, NCAUR, and ARS. He recently reported that dietary fucoxanthin is metabolized to paracentrone, which is a cleavage product, and exhibits anti-inflammatory effects on activated macrophages and adipocytes.

Hayato Maeda

Hayato Maeda, an associate professor of the Faculty of Agriculture and Life Science, Hirosaki University, Japan, researchers food nutrition and chemistry, specifically investigating bio-active compounds such as lipids and pigments. He has studied the anti-obesity and anti-inflammation effects of carotenoids and unsaturated fatty acids in seaweed and fish. Several of his papers have described the anti-obesity, anti-diabetic, and anti-cancer effects of fucoxanthin, which is contained in brown algae. In addition, he is currently investigating the functional ingredients of agricultural products such as apples, black currants, and garlic. As a member of the Institute of Regional Innovation at Hirosaki University, he is working on joint research with companies to develop functional food and cosmetic products.

 *marine drugs*

Article

Fucoxanthin Suppresses Osteoclastogenesis via Modulation of MAP Kinase and Nrf2 Signaling

You-Jung Ha [1], Yong Seok Choi [2], Ye Rim Oh [2], Eun Ha Kang [1], Gilson Khang [3], Yong-Beom Park [4] and Yun Jong Lee [1,5,*]

1 Division of Rheumatology, Department of Internal Medicine, Seoul National University Bundang Hospital, Seongnam 13620, Korea; hayouya@snubh.org (Y.-J.H.); kangeh@snubh.org (E.H.K.)
2 Medical Science Research Institute, Seoul National University Bundang Hospital, Seongnam 13620, Korea; choicell1@hanmail.net (Y.S.C.); ptojt93@naver.com (Y.R.O.)
3 Department of Bionanotechnology and Bio-Convergence Engineering, Department of PolymerNano Science and Technology and Polymer Materials Fusion Research Center, Chonbuk National University, Jeonju-si 54896, Korea; gskhang@jbnu.ac.kr
4 Division of Rheumatology, Department of Internal Medicine, Yonsei University College of Medicine, Seoul 03722, Korea; yongbpark@yuhs.ac
5 Department of Internal Medicine, Seoul National University College of Medicine, Seoul 03080, Korea
* Correspondence: yn35@snu.ac.kr; Tel.: +82-31-787-7049

Abstract: Fucoxanthin (FX), a natural carotenoid present in edible brown seaweed, is known for its therapeutic potential in various diseases, including bone disease. However, its underlying regulatory mechanisms in osteoclastogenesis remain unclear. In this study, we investigated the effect of FX on osteoclast differentiation and its regulatory signaling pathway. In vitro studies were performed using osteoclast-like RAW264.7 cells stimulated with the soluble receptor activator of nuclear factor-κB ligand or tumor necrosis factor-alpha/interleukin-6. FX treatment significantly inhibited osteoclast differentiation and bone resorption ability, and downregulated the expression of osteoclast-specific markers such as nuclear factor of activated T cells 1, dendritic cell-specific seven transmembrane protein, and matrix metallopeptidase 9. Intracellular signaling pathway analysis revealed that FX specifically decreased the activation of the extracellular signal-regulated kinase and p38 kinase, and increased the nuclear translocation of phosphonuclear factor erythroid 2-related factor 2 (Nrf2). Our results suggest that FX regulates the expression of mitogen-activated protein kinases and Nrf2. Therefore, FX is a potential therapeutic agent for osteoclast-related skeletal disorders including osteoporosis and rheumatoid arthritis.

Keywords: brown seaweed; fucoxanthin; osteoclastogenesis; MAP kinase; Nrf2

Citation: Ha, Y.-J.; Choi, Y.S.; Oh, Y.R.; Kang, E.H.; Khang, G.; Park, Y.-B.; Lee, Y.J. Fucoxanthin Suppresses Osteoclastogenesis via Modulation of MAP Kinase and Nrf2 Signaling. *Mar. Drugs* **2021**, *19*, 132. https://doi.org/10.3390/md19030132

Academic Editors: Masashi Hosokawa and Hayato Maeda

Received: 4 January 2021
Accepted: 23 February 2021
Published: 27 February 2021

Publisher's Note: MDPI stays neutral with regard to jurisdictional claims in published maps and institutional affiliations.

1. Introduction

Bones are a dynamic tissue that undergoes constant renewal and repair through bone remodeling. This process is characterized by the spatiotemporal coupling of osteoclast-induced bone resorption and osteoblast-induced bone formation. An imbalance between bone resorption and bone formation, especially excessive osteoclastic activity, is involved in the pathogenesis of osteoporosis, rheumatoid arthritis, multiple myeloma, and metastatic cancers [1,2]. Osteoclasts are multinucleated cells that are differentiated from hematopoietic precursor cells of monocyte or macrophage lineage by canonical stimulation with macrophage colony-stimulating factor (M-CSF) and receptor activator of nuclear factor-κB (NF-κB) ligand (RANKL) [3,4]. RANKL binds to its receptor on osteoclast precursors to activate the mitogen-activated protein kinase (MAPK) signaling pathway, and downstream transcription factors and osteoclast differentiation markers, including AP-1, NF-κB, and nuclear factor of activated T cells 1 (NFATc1) [2,5,6]. Some cytokines, such as tumor necrosis factor (TNF)-α and interleukin (IL)-6, can serve as noncanonical osteoclastogenic effectors in a RANKL-independent mechanism [7,8].

Fucoxanthin (FX) is an oxygenated carotenoid present in edible brown sea algae such as kombu (*Laminaria japonica*), wakame (*Undaria pinnatifida*), and arame (*Eisenia bicyclis*) [9]. Previous studies demonstrated that FX possesses antiobesity, antidiabetic, anti-inflammatory, anticancer, and hepatoprotective activities, in addition to its cerebrovascular protective effects [10–18]. Fruit carotenoids such as lycopene and cryptoxanthin were reported to inhibit osteoclastogenesis [19,20]. However, the effect and underlying mechanism of FX on osteoclastogenesis remain poorly understood.

There is interest in the effect of FX on osteoclast differentiation. Although FX exhibited a limited antiosteoresorptive effect in a ligature-induced periodontitis mouse model, FX administration significantly reduced the number of RANKL-positive osteoclasts [21]. Das et al. [22] showed that the treatment of osteoclast-like RAW264.7 cells with 2.5 μM FX inhibits RANKL-induced osteoclast differentiation through an induction of apoptosis. Despite this, an extremely high dose of FX did not produce significant side effects in animal models. Moreover, several studies demonstrated that FX inhibits apoptosis or promotes the survival of various nonmalignant cells at concentrations of up to 50 μM [23–26]. Taira et al. [27] reported FX-induced cytotoxicity in RAW264.7 cells at 20 μM. Therefore, the underlying mechanisms of FX effects on the canonical and noncanonical osteoclastogenic signaling pathways could not depend on cellular apoptosis and are yet to be elucidated. In the present study, we investigated the effects of FX on RANKL-dependent and -independent osteoclast differentiation, and identified its molecular regulatory mechanisms.

2. Results

2.1. FX Effect on RAW264.7 Cell Viability

The cytotoxic effect of FX on RAW264.7 cells was determined using the 3-(4,5-dimethylthiazol-2-yl) -2,5-diphenyltetrazolium bromide (MTT) assay (Figure 1A). FX did not affect cell viability at a concentration of ≤5 μM. However, the number of viable cells after treatment with 10 μM of FX was significantly lower than the number of untreated cells was. Additionally, in contrast to the results of Das et al. [22], no cleavage of procaspase-3 and poly ADP ribose polymerase (PARP; Figure 1B) was detected in cells treated with ≤10 μM FX. Therefore, FX was used at a concentration of ≤5 μM in all subsequent experiments.

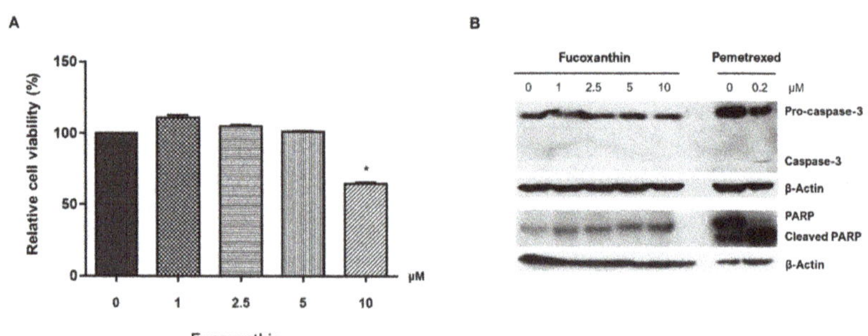

Figure 1. Effect of fucoxanthin (FX) on viability of RAW264.7 cells. (**A**) Cells treated with different concentrations of FX, and cell viability was determined using 3-(4,5-dimethylthiazol-2-yl) -2,5-diphenyltetrazolium bromide (MTT) assay. Data are representative of six independent experiments and are expressed as mean ± standard error of mean (SEM); * $p < 0.05$ versus FX-untreated cells (0 μM). (**B**) Procaspase-3 and poly ADP ribose polymerase (PARP) expression remained uncleaved upon treatment with ≤10 μM FX, unlike in pemetrexed-treated cells of lung-cancer cell line NCI-H3122.

2.2. FX Inhibits Osteoclastogenesis

Tartrate-resistant acid phosphatase (TRAP) is highly expressed in differentiated osteoclasts, and is therefore used as a primary marker of osteoclastogenesis [28]. The treatment of soluble RANKL (sRANKL, 50 ng/mL) or costimulation with TNF-α (50 ng/mL) and IL-6 (50 ng/mL) in RAW264.7 cells increased the number of TRAP-positive and multinucleated cells compared with that of untreated control cells (Figure 2A). Under these conditions, FX treatment decreased the number of TRAP-positive multinucleated cells in a dose-dependent manner (Figure 2B). In experiments using human CD14+ monocytes, 0 to 5 µM of FX dose-dependently suppressed RANKL- and TNF-α/IL-6-induced osteoclast differentiation from osteoclast precursors (Figure 2C). These results suggest that FX inhibits the differentiation of RAW264.7 cells and human CD14+ monocytes to osteoclast-like cells.

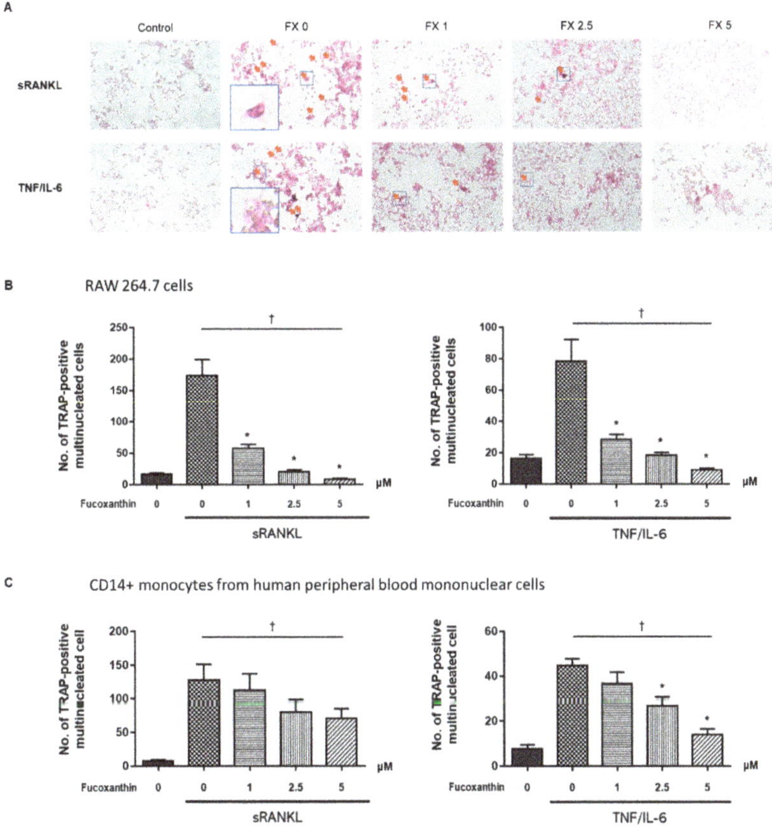

Figure 2. Soluble receptor activator of nuclear factor-κB (NF-κB) ligand (sRANKL)- or tumor necrosis factor (TNF)-α/interleukin (IL)-6-induced differentiation into osteoclast-like cells. (**A**) Representative microscopic images of tartrate-resistant acid phosphatase (TRAP) stained RAW264.7 cells (red arrows; original magnification, 100×). Blue box in bottom corner is a magnified photograph of the smaller boxed area (original magnification, 400×). (**B,C**) Number of TRAP-positive multinucleated cells differentiated from (**B**) RAW264.7 cells and (**C**) human CD14+ monocytes decreased upon treatment with FX in a dose-dependent manner. Data are representative of three independent experiments and are expressed as mean $\pm$ SEM; * $p < 0.05$ versus FX-untreated osteoclast-differentiated cells; † $p < 0.05$ by Jonckheere–Terpstra test. FX, fucoxanthin.

The resorption pit assay was performed to examine osteoclast activity. As shown in Figure 3, the resorption pit area was significantly decreased upon treatment with FX in a dose-dependent manner. These results collectively suggest that FX exerts an inhibitory effect on both sRANKL-dependent and -independent bone-resorbing osteoclast activity.

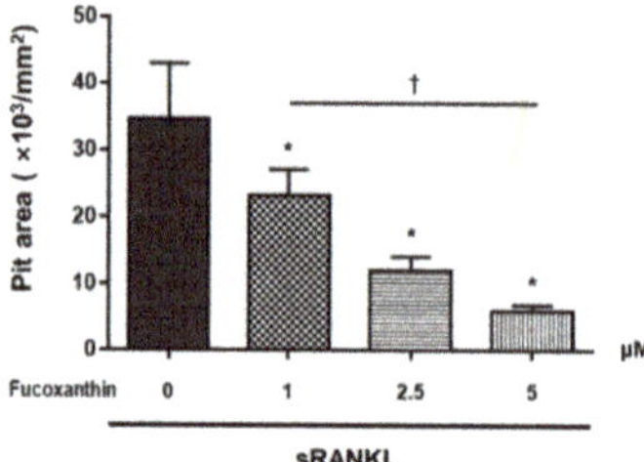
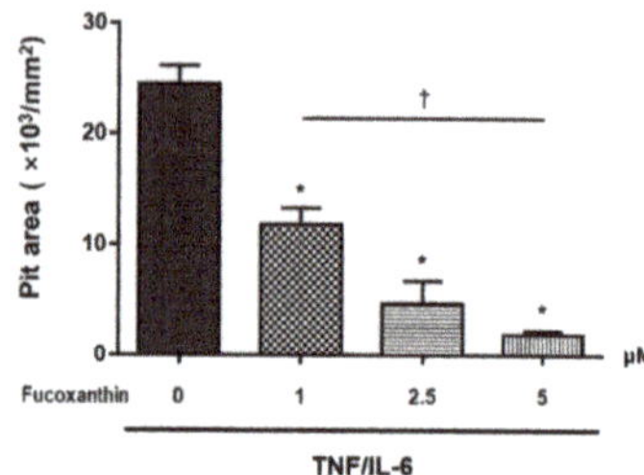

Figure 3. Effect of FX on osteoclast activity. Statistical differences of resorption pit area and trends tests are presented in histograms. Data are representative of three independent experiments and are expressed as mean ± SEM. * $p < 0.05$ versus FX untreated cells; † $p < 0.05$ by Jonckheere–Terpstra test.

2.3. FX Downregulates Osteoclast-Specific Markers and Transcriptional Factors in RAW264.7 Cells

NFATc1 is a master transcription factor for osteoclastogenesis [29]. The expression of NFATc1 in sRANKL- or TNF/IL-6-stimulated RAW264.7 cells decreased upon treatment with FX in a dose-dependent manner (Figure 4A). Dendritic-cell-specific transmembrane protein (DC-STAMP), another essential mediator for osteoclastogenesis, is upregulated upon osteoclastogenic stimulation. Increased DC-STAMP expression, in turn, upregulates the expression of osteoclast-specific markers such as *TRAP* [30]. As shown in Figure 4B, *DC-STAMP* mRNA expression was significantly decreased upon treatment with FX in a dose-dependent manner.

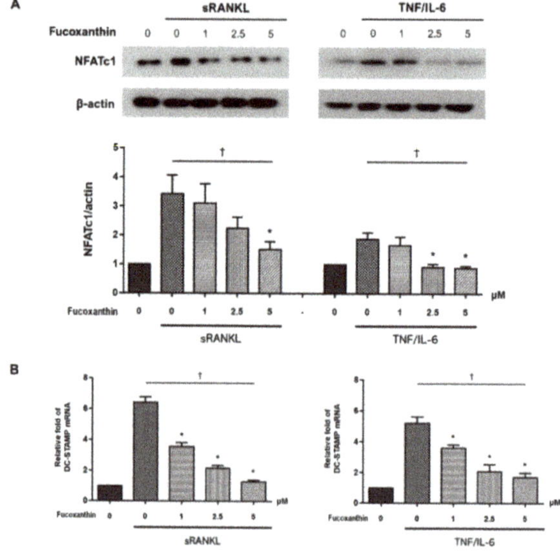

Figure 4. Effect of FX treatment on expression of osteoclast-specific markers. (**A**) Nuclear factor of activated T cell 1 (NFATc1) protein expression and (**B**) dendritic-cell-specific transmembrane protein (*DC-STAMP*) mRNA expression in sRANKL- and TNF/IL-6-stimulated RAW264.7 cells decreased upon treatment with FX in a dose-dependent manner. Data are representative of three independent experiments and expressed as mean ± SEM; * $p < 0.05$ versus FX untreated cells; † $p < 0.05$ by Jonckheere–Terpstra test.

Matrix metallopeptidase (MMP)-9 is a well-established proteolytic effector of osteoclast-mediated bone resorption. To examine whether FX affects MMP-9 production in differentiated osteoclast-like cells, MMP-9 levels in culture supernatant were measured by

ELISA. As shown in Figure 5, cells treated with 5 µM FX showed a significant decrease in MMP-9 levels. Moreover, MMP-9 production decreased following FX treatment in a dose-dependent manner.

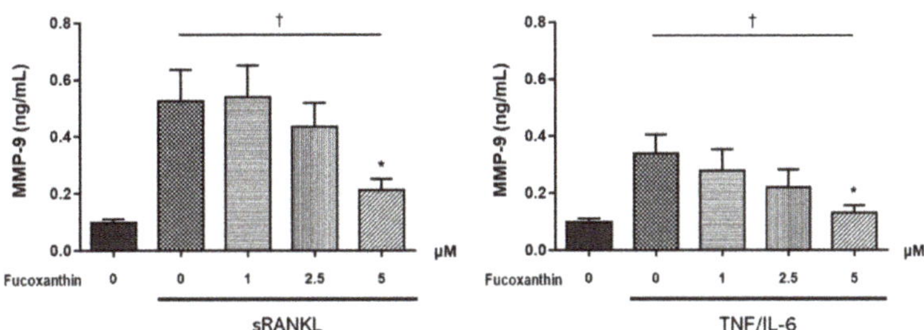

Figure 5. Effect of FX treatment on MMP-9 levels in culture supernatant of osteoclast differentiated RAW264.7 cells. MMP-9 concentration significantly decreased upon treatment with 5 µM FX. Data are representative of three independent experiments and expressed as mean ± SEM; * $p < 0.05$ versus FX untreated cells; † $p < 0.05$ by Jonckheere–Terpstra test.

To better understand the mechanism of the FX-induced inhibition of osteoclast differentiation, we performed immunoblot analysis of RAW264.7 cells to measure the expression of molecules known to be critically involved in the osteoclast signaling pathway, including MAPKs (extracellular signal-regulated kinase (ERK), c-Jun N-terminal kinase (JNK), and p38), NF-κB, and phosphoinositide 3-kinase (PI3K) [31]. The treatment of RAW264.7 cells with FX significantly reduced the phosphorylation of ERK and p38 in a concentration-dependent manner. However, JNK, NF-κB, and PI3K phosphorylation was not altered following FX treatment (Figure 6).

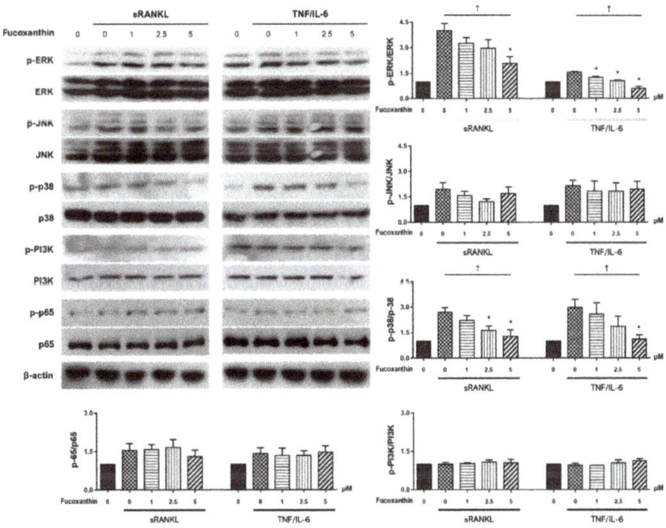

Figure 6. Effect of FX treatment on signaling pathways during osteoclastogenesis. FX inhibited extracellular signal-regulated kinase (ERK) and p38 activation in both RANKL- and TNF-α/IL-6- stimulated conditions. However, c-Jun N-terminal kinase (JNK), phosphoinositide 3-kinase (PI3K), and NF-κB levels were not significantly altered. Data are representative of five independent experiments and expressed as mean ± SEM; * $p < 0.05$ versus FX untreated cells; † $p < 0.05$ by Jonckheere–Terpstra test.

Previous studies demonstrated that nuclear factor erythroid 2-related factor 2 (Nrf2) is a negative regulator of osteoclastogenesis [32–34]. Nrf2 deficiency augments the RANKL-induced activation of ERK and p38 MAP kinases in mouse bone-marrow-derived osteoclast precursor cells [32]. Moreover, FX activates Nrf2 in nonbone and RAW264.7 cells [27,35–37]. We assessed the levels of phosphorylated Nrf2 and Nrf2 proteins in total cell extracts, and the nuclear and cytosolic fractions of cell lysates by Western blotting (Figure 7A). Nrf2 levels in total cell lysates were not affected by FX treatment. Expression of nuclear Nrf2 significantly increased, while cytosolic Nrf2 expression dose-dependently decreased. The proportions of phospho-Nrf2 expression in the nucleus compared to in the cytoplasm were significantly augmented by FX. This effect increased with concentration (Figure 7B). These results suggest that FX induces the dose-dependent phosphorylation and nuclear translocation of Nrf2 in RAW264.7 cells.

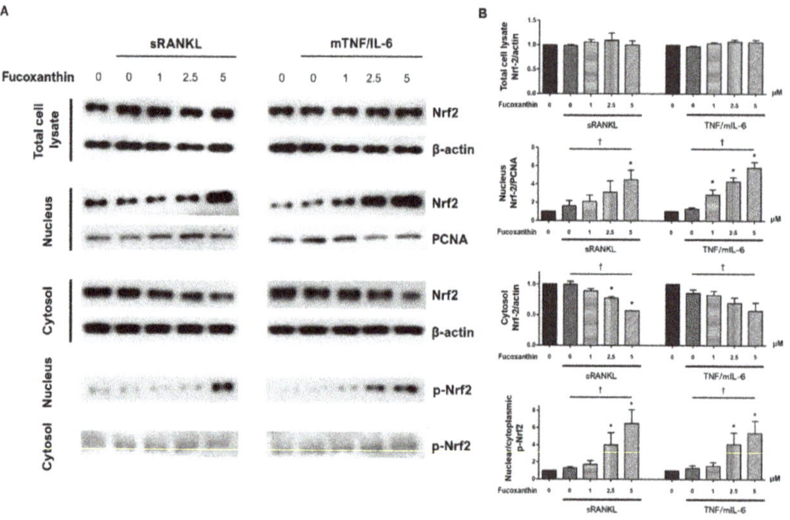

Figure 7. Effect of FX treatment on phosphorylated nuclear factor erythroid 2-related factor 2 (p-Nrf2) expression and nuclear localization of Nrf2 during osteoclast differentiation. Total cellular proteins were extracted from RAW264.7 cells and p-Nrf2 and Nrf2 expression were assessed by Western blotting. Cell lysates were fractionated into nuclear and cytosolic extracts, and identical experiments were performed. (**A**) Representative immunoblots and graphs for Nrf2 in total cell lysate, nucleus, and cytosol, and (**B**) nuclear/cytoplasmic p-Nrf2 from three independent experiments are shown. Data expressed as mean ± SEM; * $p < 0.05$ versus FX untreated cells; † $p < 0.05$ by Jonckheere–Terpstra test.

3. Discussion

Several pharmacotherapeutic drugs such as bisphosphonate, estrogen, and anti-RANKL antibodies are used to treat osteoporosis. However, these drugs are commonly associated with side effects such as medication-related osteonecrosis of the jaw, atrial fibrillation, and esophageal cancer [38–40]. Moreover, osteoporosis is poorly treated globally despite therapeutic advancements [41]. The proportion of patients with a hip fracture who were prescribed bone-protective medication decreased from 40% to 21% between 2001 and 2011 [42]. This undertreatment and poor adherence to drugs may be due to fear of adverse effects [41]. Therefore, therapeutic agents with no or minimal side effects are required. The protective role of carotenoids on bone resorption has been recently gaining attention. Carotenoids present in fruits and vegetables, such as lycopene and β-cryptoxanthin, and marine carotenoid astaxanthin show inhibitory effects on osteoclastogenesis [19,20,43–45].

Prior studies investigated the therapeutic effect of FX—a marine carotenoid—on osteoclastogenesis. Kose et al. [21] investigated the therapeutic effect of FX on alveolar bone resorption in a ligature-induced periodontitis mouse model. FX treatment significantly reduced the number of RANKL-positive osteoclasts located in the resorption lacunae and increased serum bone-specific alkaline phosphatase levels. Das et al. [22] demonstrated that the treatment of osteoclasts differentiated from RAW264.7 cells with 2.5–5 μM FX inhibited RANKL-induced osteoclast differentiation by inducing apoptosis. However, FX did not exert cytotoxic effects in osteoblast-like cells at 2.5 μM. Although not the pure FX compound, extracts containing FX from *Sargassum fusiforme* suppressed osteoclast differentiation, promoted osteoblast formation [46], and exhibited antiresorptive effects in an ovariectomized mice model. These studies suggest that FX functions as a bone-protective agent in osteoclast-mediated skeletal disorders. Here, we showed that FX inhibits both the canonical RANKL-induced osteoclastogenesis and the RANKL-independent TNFα/IL-6-induced differentiation of osteoclasts (Figure 2). Moreover, FX significantly attenuated osteoclastic bone resorption pit in a dose-dependent manner (Figure 3). Unlike the results of Das et al., our results demonstrated that the inhibitory effect of FX at $\leq$5 μM is not mediated by the apoptosis of osteoclast precursor cells (Figure 1B).

To our knowledge, there is no study focusing on the signaling pathway underlying the bone-protective activity of FX. Osteoclast differentiation is a multistep process that involves cell proliferation, commitment, fusion, and activation [2]. During this process, RANKL interacts with RANK to recruit TNF receptor-associated factor (TRAF) adaptor protein and induce downstream targets such as NF-κB, MAPK (JNK, ERK, and p38), PI3K, and Akt [47,48]. Previous studies also reported that TNF-α and IL-6 can independently promote osteoclastogenesis in vitro of RANKL [7,8,49]. Moreover, TNF-α can activate various signaling pathways, including p38 MAPK, ERK, and NF-κB [48,50]. Our study confirmed that FX downregulates ERK and p38 in both RANKL-dependent and -independent pathways, but not JNK or PI3K (Figure 6). The pharmacological action of nitrogen-containing bisphosphonates is also mediated by the inhibition of the MEK/ERK pathway [51].

NFATc1 is characterized as a master molecule of RANKL-induced osteoclast differentiation and can autoamplify its own expression [52]. Previous studies demonstrated that NFATc1 binds to the promoter region of *MMP-9* and *DC-STAMP* in osteoclasts, and increases mRNA expression [53,54]. In line with these findings, our study demonstrated a significant decrease in *DC-STAMP* mRNA expression (Figure 4B) and production of MMP-9 (Figure 5) upon FX treatment, in addition to NFATc1 downregulation.

Nrf-2 is a transcription factor expressed in various cell types, and is known as a regulator of cytoprotective genes against oxidative and chemical injuries [55]. Under normal quiescent conditions, Nrf2 is tethered to cytoplasmic protein Keap1. However, in cells exposed to stressful stimuli, Nrf2 is released from Keap1 and activated via phosphorylation. Nrf2 phosphorylation is important for its stabilization, and phospho-Nrf2 preferentially translocates to the nucleus [55,56]. Nrf-2 can also regulate osteoclast formation and activity. The overexpression of Nrf2 suppressed RANKL-induced osteoclast differentiation by increasing the level of antioxidant enzymes and locally inducing nuclear Nrf2-attenuated osteoclastogenesis [33,34,55]. In the present study, FX treatment promoted the nuclear translocation and phosphorylation of Nrf2 in both RANKL-dependent and -independent pathways (Figure 7). Consistent with this finding, nuclear translocation and phosphorylation of Nrf2 by FX was reported in studies of human keratinocytes and ischemia/reperfusion-induced neuron cells [35,36]. As Park et al. reported, the induction of Nrf2 dramatically suppresses the transcriptional activity of NFATc1 [57], the activation of Nrf2 by FX in our findings suggests the decreased expression of NFATc1 by FX (Figure 4A). Conversely, previous studies using HepG2 cells under oxidative-stress conditions demonstrated that ERK or p38 kinase activation is required for drug-mediated Nrf2 translocation [58,59]. However, this study and a previous study with NRK-52E cells [60] showed that the treatment of cells with FX reduced phosphor-ERK/p38 levels and induced the nuclear translocation of phospho-Nrf2. The interaction between MAPK and Nrf2 pathways may vary depending on cell type, drug, or cell environment.

Limitations exist in this study. Although murine RAW264.7 cells are frequently used as in vitro models of osteoclast differentiation, it is necessary to confirm the beneficial effect of FX in human osteoclast precursors. Furthermore, in vivo studies are required to examine the therapeutic potential of FX in disease models. Nonetheless, to our knowledge, this is the first study to clarify the molecular regulatory mechanisms of FX in osteoclast differentiation.

We demonstrated that FX attenuates both RANKL-dependent and -independent osteoclastogenesis by downregulating ERK and p38 expression, and promoting the nuclear translocation of phospho-Nrf2 in RAW264.7 cells, as summarized in Figure 8. FX is confirmed to have no side effects and can be easily extracted from marine macro/microalgae. Therefore, FX represents a safe and inexpensive candidate drug for the treatment of various diseases accompanying the imbalance between osteoclasts and osteoblasts.

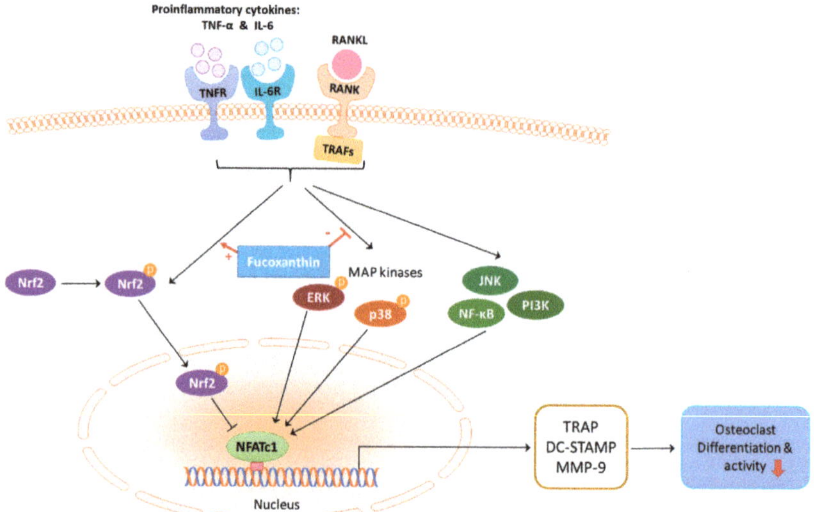

Figure 8. Signaling pathways and effects of FX during osteoclastogenesis. Inhibitory effect of FX is mediated by blocking the activation of ERK and p38, promoting Nrf2 nuclear translocation and phosphorylation, and subsequently downregulating NFATc1. Nrf2 induction was previously reported to suppress NFATc1 transcriptional activity [57].

4. Materials and Methods

4.1. Cell Lines and Reagents

RAW264.7, a murine macrophage cell line, was purchased from the American Type Culture Collection (ATCC; Rockville, MD, USA). Human lung adenocarcinoma cell line NCI-H3122 was a kind gift from Professor Jong-Seok Lee (Seoul National University College of Medicine, Seoul, South Korea). Human peripheral blood mononuclear cells (PMBCs) were obtained from 4 anonymous donors (Koma Biotech, Seoul, Korea). Dulbecco's Modified Eagle's Medium (DMEM) and Minimum Essential Medium Eagle-Alpha Modification (α-MEM) were purchased from Welgene (Daegu, Korea). Fetal bovine serum (FBS) was obtained from Atlas Biologicals (Fort Collins, CO, USA) and penicillin–streptomycin from Gibco (Carlsbad, CA, USA). Recombinant mouse TNF-α, mouse IL-6, and human sRANKL were purchased from PeproTech (Rocky Hill, NJ, USA). CD14 MACS® MicroBeads were purchased from Miltenyi Biotec Inc. (Auburn, CA, USA).

Antibodies against procaspase-3, caspase-3, PARP, cleaved-PARP, ERK, phospho-ERK, p38, phospho-p38, JNK, phospho-JNK, PCNA, PI3K, and phospho-PI3K were purchased from Cell Signaling Technology (Danvers, MA, USA). Anti-β-actin antibody was purchased from Enogene Biotech (New York, NY, USA), and anti-NFATc1, anti-phospho-p65, anti-Nrf2, and phospho-Nrf2 antibodies were from ABcam (Cambridge, UK).

ELISA kits for MMP-9 were obtained from R&D Systems (Minneapolis, MN, USA). The TRAP staining kit was acquired from Takara (Shiga, Japan), and the bone resorption assay kit from Cosmo Bio (Tokyo, Japan). The MTT assay kit was purchased from Sigma–Aldrich (St. Louis, MO, USA).

4.2. Cell Viability Test Using MTT Assay

Cell viability was determined using the MTT assay kit as per the manufacturer's protocol. Cells were seeded into 48 well tissue culture plates at a density of 2×10^3 per well in growth medium. After 24 h, logarithmic phase cells were incubated with different concentrations of FX for 5 days. Thereafter, MTT (5 mg/mL in PBS) was added to each cell. After 4 h, the medium was removed, and dimethylsulfoxide was added to solubilize MTT for an additional 4 h. After extraction with dimethylsulfoxide, optical density (OD) was measured at 495 nm. Percentage viability was calculated as (OD of drug-treated sample/OD of control) $\times$ 100.

4.3. Culture and Differentiation of Cell Lines

RAW264.7 cells were seeded in 48 well plates (2×10^3 per well), and cultured in α-MEM containing 10% FBS and 1% penicillin/streptomycin at 37 °C in 5% CO_2/95% O_2 in a humidified cell incubator. The culture medium was replenished every 3 days. Osteoclast differentiation was induced either by stimulation with sRANKL (50 ng/mL) for 5 days or costimulation with TNF-α (50 ng/mL) and IL-6 (50 ng/mL).

Cryopreserved human PMBCs were thawed and washed, and CD14-positive cells were isolated using anti-CD14 antibody-coated microbeads. CD14+ monocytes were seeded at 1.5×10^5 cells/well in a α-MEM medium containing 10% FBS and M-CSF (50 ng/mL). After confirming that the cells remained attached the following day, they were treated with sRANKL (50 ng/mL) or TNF-α (50 ng/mL)/IL-6 (50 ng/mL) under conditions of 0, 1, 2.5, and 5 μM FX. The culture medium was replaced every 4 days, and multinucleated TRAP-positive cells were counted after 17 days.

4.4. Osteoclast Differentiation from RAW264.7 Cells and Osteoclast Activity Assays

To evaluate the direct effects of FX on osteoclast differentiation, RAW264.7 cells were treated with 0, 1, 2.5, and 5 μM FX under sRANKL or TNF/IL-6 stimulation. Cells were stained with TRAP after 4 days of stimulation, and TRAP-positive multinucleated cells were enumerated under a light microscope.

Osteoclast activity was determined by measuring the area of resorption pits using calcium phosphate-coated 48 well plates according to the manufacturer's recommendation. RAW264.7 cells were washed once with α-MEM containing 10% FBS and seeded onto 48 well plates (2×10^3 per well). The following day, cells were treated with FX under sRANKL or TNF/IL-6 stimulation. On Day 5, the pit area was measured after adding 5% sodium hydrochlorite along the plate wall to remove RAW264.7 cells, and after washing with water. After air drying, microscopic images of all fields were acquired, and the resorbed pit area per well was measured using ImageJ software (NIH, Bethesda, MD, USA).

4.5. Immunoblotting

To determine the expression of intracellular proteins, immunoblotting was performed with the aforementioned antibodies. For NFATc1 and Nrf2 expression, RAW264.7 cells were seeded onto a 6 well plate (2.0×10^4 per well) in α-MEM containing 10% FBS. On the following day, 0, 1, 2.5, and 5 μM FX were added to the culture medium, and cells were grown under sRANKL or TNF-α/IL-6 stimulation for 4 days. For MAPK, p65, and PI3K expression, RAW264.7 cells were cultured in α-MEM containing 10% FBS with 0, 1, 2.5, and 5 µM FX for 4 days. Thereafter, cells were incubated with sRANKL or TNF-α/IL-6 in serum-free media for 30 min.

Total cell lysates were obtained using cold radioimmunoprecipitation assay (RIPA) buffer (25 mM Tris-HCl, pH 7.6; 150 mM NaCl; 1% NP-40; 1% sodium deoxycholate; 0.1% SDS). The crude extract was separated on 10% sodium dodecyl sulfate-polyacrylamide gel electrophoresis (SDS-PAGE) and transferred to polyvinylidene difluoride membranes. Protein bands were detected using an enhanced chemiluminescence system (Amersham Biosciences, Little Chalfont, UK). Nuclear extracts were prepared with the NE-PER Nuclear Cytoplasmic Extraction Reagent kit (Pierce, Rockford, IL, USA) according to the manufacturer's instructions. The relative expression of each protein was determined by densitometric analysis using ImageJ software.

4.6. Reverse Transcription-Polymerase Chain Reaction (RT-PCR) and Real-Time PCR

The expression levels of osteoclast-related genes were measured using RT-PCR with specific primers. cDNA and target-specific primers were added to the power SYBR green PCR master mix (Applied Biosystems, Foster City, CA, USA). PCR cycling parameters were as follows: amplification (1 cycle at 50 °C for 2 min, 1 cycle at 95 °C for 10 min, and 40 cycles at 95 °C for 15 s and 60 °C for 1 min). Fold changes of gene expression were calculated with the ΔΔCt method using ribosomal protein S18 as the reference gene. Specific murine primers are summarized in Table 1.

Table 1. Oligonucleotide primers used for RT-PCR.

Target Gene	GenBank Accession Number		Primer Sequence
18S ribosomal RNA	NR_003278	Forward	5′-GCAATTATTCCCCATGAA CG-3′
		Reverse	5′-GGCCTCACTAAACCATCCAA-3′
DC-STAMP	NM_029422	Forward	5′-TGCCAGGGCTGGAAGTTCAC-3′
		Reverse	5′-AAGGAGCTTCGCATGCAGGT-3′

4.7. ELISA

RAW264.7 cells were seeded onto a 6 well plate (2.0×10^4 per well) in α-MEM containing 10% FBS. On the following day, 0, 1, 2.5, and 5 µM FX were added to the culture medium and incubated with sRANKL or TNF/IL-6 for 4 days. MMP-9 detection was performed using commercial ELISA kits according to the manufacturer's instructions.

4.8. Statistical Analysis

All experiments were performed at least three times, and data are presented as mean ± SEM. Continuous variables were compared using Mann–Whitney U test. For dose-response analyses, the nonparametric Jonckheere–Terpstra trend test was performed. All data were analyzed using STATA® SE, version 15.0 (StataCorp LLC, College Station, TX, USA). A *p* value < 0.05 was considered statistically significant.

5. Conclusions

FX inhibits osteoclast differentiation and bone-resorption activity through downregulating p38 and ERK, and promoting the nuclear translocation of phospho-Nrf2. The results of this study provide useful insight into the molecular mechanisms of FX action. Hence, FX could be used to treat bone diseases caused by excessive osteoclastic activity.

Author Contributions: Conceptualization and experiment design, Y.J.L. and Y.-J.H.; statistical analysis and data interpretation, Y.-J.H., E.H.K., G.K., and Y.-B.P.; cell experiments, Y.S.C. and Y.R.O.; writing—original-draft preparation, Y.-J.H.; writing—review and editing, Y.J.L. and G.K. All authors have read and agreed to the published version of the manuscript.

Funding: This work was supported by grant no. 04-2018-014 from the Seoul National Bundang Hospital Research Fund (to Lee YJ) and the National Research Foundation of Korea (NRF) grant funded by the Korean government (MSIT) (no. 2020R1C1C1010147 to Ha YJ).

Institutional Review Board Statement: Not applicable.

Data Availability Statement: The data presented in this study are available on request from the corresponding author.

Conflicts of Interest: The authors declare no conflict of interest.

References

1. Walsh, M.C.; Kim, N.; Kadono, Y.; Rho, J.; Lee, S.Y.; Lorenzo, J.; Choi, Y. Osteoimmunology: Interplay between the immune system and bone metabolism. *Annu. Rev. Immunol.* **2006**, *24*, 33–63. [CrossRef]
2. Boyle, W.J.; Simonet, W.S.; Lacey, D.L. Osteoclast differentiation and activation. *Nature* **2003**, *423*, 337–342. [CrossRef]
3. Roodman, G.D. Cell biology of the osteoclast. *Exp. Hematol.* **1999**, *27*, 1229–1241. [CrossRef]
4. Teitelbaum, S.L. Bone resorption by osteoclasts. *Science* **2000**, *289*, 1504–1508. [CrossRef] [PubMed]
5. Long, C.L.; Humphrey, M.B. Osteoimmunology: The expanding role of immunoreceptors in osteoclasts and bone remodeling. *BoneKEy Rep.* **2012**, *1*, 59. [CrossRef] [PubMed]
6. Park, J.H.; Lee, N.K.; Lee, S.Y. Current understanding of RANK signaling in osteoclast differentiation and maturation. *Mol. Cells* **2017**, *40*, 706–713. [PubMed]
7. Kobayashi, K.; Takahashi, N.; Jimi, E.; Udagawa, N.; Takami, M.; Kotake, S.; Nakagawa, N.; Kinosaki, M.; Yamaguchi, K.; Shima, N.; et al. Tumor necrosis factor alpha stimulates osteoclast differentiation by a mechanism independent of the ODF/RANKL-RANK interaction. *J. Exp. Med.* **2000**, *191*, 275–286. [CrossRef]
8. Kudo, O.; Sabokbar, A.; Pocock, A.; Itonaga, I.; Fujikawa, Y.; Athanasou, N.A. Interleukin-6 and interleukin-11 support human osteoclast formation by a RANKL-independent mechanism. *Bone* **2003**, *32*, 1–7. [CrossRef]
9. Kim, S.M.; Jung, Y.J.; Kwon, O.N.; Cha, K.H.; Um, B.H.; Chung, D.; Pan, C.H. A potential commercial source of fucoxanthin extracted from the microalga *Phaeodactylum tricornutum*. *Appl. Biochem. Biotechnol.* **2012**, *166*, 1843–1855. [CrossRef]
10. Yang, Y.P.; Tong, Q.Y.; Zheng, S.H.; Zhou, M.D.; Zeng, Y.M.; Zhou, T.T. Anti-inflammatory effect of fucoxanthin on dextran sulfate sodium-induced colitis in mice. *Nat. Prod. Res.* **2020**, *34*, 1791–1795. [CrossRef]
11. Wang, Z.; Li, H.; Dong, M.; Zhu, P.; Cai, Y. The anticancer effects and mechanisms of fucoxanthin combined with other drugs. *J. Cancer Res. Clin. Oncol.* **2019**, *145*, 293–301. [CrossRef] [PubMed]
12. Satomi, Y. Antitumor and cancer-preventative function of fucoxanthin: A marine carotenoid. *Anticancer Res.* **2017**, *37*, 1557–1562. [CrossRef] [PubMed]
13. Das, S.K.; Hashimoto, T.; Kanazawa, K. Growth inhibition of human hepatic carcinoma HepG2 cells by fucoxanthin is associated with down-regulation of cyclin D. *Biochim. Biophys. Acta* **2008**, *1780*, 743–749. [CrossRef]
14. Shiratori, K.; Ohgami, K.; Ilieva, I.; Jin, X.H.; Koyama, Y.; Miyashita, K.; Yoshida, K.; Kase, S.; Ohno, S. Effects of fucoxanthin on lipopolysaccharide induced inflammation In Vitro and In Vivo. *Exp. Eye Res.* **2005**, *81*, 422–428. [CrossRef] [PubMed]
15. Liu, C.L.; Liang, A.L.; Hu, M.L. Protective effects of fucoxanthin against ferric nitrilotriacetate-induced oxidative stress in murine hepatic BNL CL.2 cells. *Toxicology* **2011**, *25*, 1314–1319. [CrossRef]
16. Maeda, H.; Hosokawa, M.; Sashima, T.; Murakami-Funayama, K.; Miyashita, K. Anti-obesity and anti-diabetic effects of fucoxanthin on diet-induced obesity conditions in a murine model. *Mol. Med. Rep.* **2009**, *2*, 897–902. [CrossRef] [PubMed]
17. Das, S.K.; Hashimoto, T.; Shimizu, K.; Yoshida, T.; Sakai, T.; Sowa, Y.; Komoto, A.; Kanazawa, K. Fucoxanthin induces cell cycle arrest at G0/G1 phase in human colon carcinoma cells through up-regulation of p21WAF1/Cipbiochim. *Biophys. Acta* **2005**, *1726*, 328–335. [CrossRef]
18. Ikeda, K.; Kitamura, A.; Machida, H.; Watanabe, M.; Negishi, H.; Hiraoka, J.; Nakano, T. Effect of *Undaria pinnatifida* (Wakame) on the development of cerebrovascular diseases in stroke-prone spontaneously hypertensive rats. *Clin. Exp. Pharm. Physiol.* **2003**, *30*, 44–48. [CrossRef]
19. Costa-Rodrigues, J.; Fernandes, M.H.; Pinho, O.; Monteiro, P.R.R. Modulation of human osteoclastogenesis and osteoblastogenesis by lycopene. *J. Nutr. Biochem.* **2018**, *57*, 26–34. [CrossRef]
20. Hirata, N.; Ichimaru, R.; Tominari, T.; Matsumoto, C.; Watanabe, K.; Taniguchi, K.; Hirata, M.; Ma, S.; Suzuki, K.; Grundler, F.M.W.; et al. Beta-cryptoxanthin inhibits lipopolysaccharide-induced osteoclast differentiation and bone resorption via the suppression of inhibitor of NF-κB kinase activity. *Nutrients* **2019**, *11*, 368. [CrossRef]
21. Kose, O.; Arabaci, T.; Yemenoglu, H.; Kara, A.; Ozkanlar, S.; Kayis, S.; Duymus, Z.Y. Influences of fucoxanthin on alveolar bone resorption in induced periodontitis in rat molars. *Mar. Drugs* **2016**, *14*, 70. [CrossRef]
22. Das, S.K.; Ren, R.; Hashimoto, T.; Kanazawa, K. Fucoxanthin induces apoptosis in osteoclast-like cells differentiated from RAW264.7 cells. *J. Agric. Food Chem.* **2010**, *58*, 6090–6095. [CrossRef]
23. Chiang, Y.F.; Chen, H.Y.; Chang, Y.J.; Shih, Y.H.; Shieh, T.M.; Wang, K.L.; Hsia, S.M. Protective effects of fucoxanthin on high glucose- and 4-hydroxynonenal (4-HNE)-induced injury in human retinal pigment epithelial cells. *Antioxidants* **2020**, *9*, 1176. [CrossRef] [PubMed]

24. Zhang, X.S.; Lu, Y.; Tao, T.; Wang, H.; Liu, G.J.; Liu, X.Z.; Liu, C.; Xia, D.Y.; Hang, C.H.; Li, W.; et al. Fucoxanthin mitigates subarachnoid hemorrhage-induced oxidative damage via sirtuin 1-dependent pathway. *Mol. Neurobiol.* **2020**, *57*, 5286–5298. [CrossRef]

25. Ou, H.C.; Chou, W.C.; Chu, P.M.; Hsieh, P.L.; Hung, C.H.; Tsai, K.L. Fucoxanthin protects against oxLDL-induced endothelial damage via activating the AMPK-Akt-CREB-PGC1α pathway. *Mol. Nutr. Food Res.* **2019**, *63*, e1801353. [CrossRef]

26. Zheng, J.; Piao, M.J.; Keum, Y.S.; Kim, H.S.; Hyun, J.W. Fucoxanthin protects cultured human keratinocytes against oxidative stress by blocking free radicals and inhibiting apoptosis. *Biomol. Ther (Seoul).* **2013**, *21*, 270–276. [CrossRef] [PubMed]

27. Taira, J.; Sonamoto, M.; Uehara, M. Dual biological functions of a cytoprotective effect and apoptosis induction by bioavailable marine carotenoid fucoxanthinol through modulation of the Nrf2 activation in RAW264.7 macrophage cells. *Mar. Drugs* **2017**, *15*, 305. [CrossRef]

28. Hayman, A.R.; Bune, A.J.; Bradley, J.R.; Rashbass, J.; Cox, T.M. Osteoclastic tartrate-resistant acid phosphatase (Acp 5): Its localization to dendritic cells and diverse murine tissues. *J. Histochem. Cytochem.* **2000**, *48*, 219–228. [CrossRef] [PubMed]

29. Zhao, Q.; Wang, X.; Liu, Y.; He, A.; Jia, R. NFATc1: Functions in osteoclasts. *Int. J. Biochem. Cell Biol.* **2010**, *42*, 576–579. [CrossRef]

30. Kukita, T.; Wada, N.; Kukita, A.; Kakimoto, T.; Sandra, F.; Toh, K.; Nagata, K.; Iijima, T.; Horiuchi, M.; Matsusaki, H.; et al. RANKL-induced DC-STAMP is essential for osteoclastogenesis. *J. Exp. Med.* **2004**, *200*, 941–946. [CrossRef]

31. Ono, T.; Nakashima, T. Recent advances in osteoclast biology. *Histochem. Cell Biol.* **2018**, *149*, 325–341. [CrossRef] [PubMed]

32. Park, P.S.U.; Mun, S.H.; Zeng, S.L.; Kim, H.; Bae, S.; Park-Min, K.H. NRF2 Is an upstream regulator of MYC-mediated osteoclastogenesis and pathological bone erosion. *Cells* **2020**, *9*, 2133. [CrossRef] [PubMed]

33. Kanzaki, H.; Shinohara, F.; Kajiya, M.; Fukaya, S.; Miyamoto, Y.; Nakamura, Y. Nuclear Nrf2 induction by protein transduction attenuates osteoclastogenesis. *Free Radic. Biol. Med.* **2014**, *77*, 239–248. [CrossRef]

34. Hyeon, S.; Lee, H.; Yang, Y.; Jeong, W. Nrf2 deficiency induces oxidative stress and promotes RANKL-induced osteoclast differentiation. *Free Radic. Biol. Med.* **2013**, *65*, 789–799. [CrossRef]

35. Zheng, J.; Piao, M.J.; Kim, K.C.; Yao, C.W.; Cha, J.W.; Hyun, J.W. Fucoxanthin enhances the level of reduced glutathione via the Nrf2-mediated pathway in human keratinocytes. *Mar. Drugs* **2014**, *12*, 4214–4230. [CrossRef] [PubMed]

36. Hu, L.; Chen, W.; Tian, F.; Yuan, C.; Wang, H.; Yue, H. Neuroprotective role of fucoxanthin against cerebral ischemic/reperfusion injury through activation of Nrf2/HO-1 signaling. *Biomed. Pharm.* **2018**, *106*, 1484–1489. [CrossRef]

37. Wang, X.; Cui, Y.J.; Qi, J.; Zhu, M.M.; Zhang, T.L.; Cheng, M.; Liu, S.M.; Wang, G.C. Fucoxanthin exerts cytoprotective effects against hydrogen peroxide-induced oxidative damage in L02 cells. *Biomed. Res. Int.* **2018**, *2018*, 1085073. [CrossRef]

38. Khan, A.A.; Morrison, A.; Hanley, D.A.; Felsenberg, D.; McCauley, L.K.; O'Ryan, F.; Reid, I.R.; Ruggiero, S.L.; Taguchi, A.; Tetradis, S.; et al. Diagnosis and management of osteonecrosis of the jaw: A systematic review and international consensus. *J. Bone Min. Res.* **2015**, *30*, 3–23. [CrossRef]

39. Rhee, C.W.; Lee, J.; Oh, S.; Choi, N.K.; Park, B.J. Use of bisphosphonate and risk of atrial fibrillation in older women with osteoporosis. *Osteoporos. Int.* **2012**, *23*, 247–254. [CrossRef]

40. Wysowski, D.K. Reports of esophageal cancer with oral bisphosphonate use. *N. Engl. J. Med.* **2009**, *360*, 89–90. [CrossRef]

41. Compston, J.E.; McClung, M.R.; Leslie, W.D. Osteoporosis. *Lancet* **2019**, *393*, 364–376. [CrossRef]

42. Solomon, D.H.; Johnston, S.S.; Boytsov, N.N.; McMorrow, D.; Lane, J.M.; Krohn, K.D. Osteoporosis medication use after hip fracture in U. S. *patients between 2002 and J. Bone Min. Res.* **2014**, *29*, 1929–1937.

43. Hwang, Y.H.; Kim, K.J.; Kim, S.J.; Mun, S.K.; Hong, S.G.; Son, Y.J.; Yee, S.T. Suppression effect of astaxanthin on osteoclast formation In Vitro and bone loss In Vivo. *Int. J. Mol. Sci.* **2018**, *19*, 912. [CrossRef] [PubMed]

44. Rao, L.G.; Krishnadev, N.; Banasikowska, K.; Rao, A.V. Lycopene I-effect on osteoclasts: Lycopene inhibits basal and parathyroid hormone-stimulated osteoclast formation and mineral resorption mediated by reactive oxygen species in rat bone marrow cultures. *J. Med. Food* **2003**, *6*, 69–78. [CrossRef]

45. Uchiyama, S.; Yamaguchi, M. Inhibitory effect of beta-cryptoxanthin on osteoclast-like cell formation in mouse marrow cultures. *Biochem. Pharm.* **2004**, *67*, 1297–1305. [CrossRef] [PubMed]

46. Koyama, T. Extracts of marine algae show inhibitory activity against osteoclast differentiation. *Adv. Food Nutr. Res.* **2011**, *64*, 443–454.

47. Kim, J.H.; Kim, N. Signaling pathways in osteoclast differentiation. *Chonnam Med. J.* **2016**, *52*, 12–17. [CrossRef]

48. Lee, Z.H.; Kim, H.H. Signal transduction by receptor activator of nuclear factor kappa B in osteoclasts. *Biochem. Biophys. Res. Commun.* **2003**, *305*, 211–214. [CrossRef]

49. Hotokezaka, H.; Sakai, E.; Ohara, N.; Hotokezaka, Y.; Gonzales, C.; Matsuo, K.; Fujimura, Y.; Yoshida, N.; Nakayama, K. Molecular analysis of RANKL-independent cell fusion of osteoclast-like cells induced by TNF-alpha, lipopolysaccharide, or peptidoglycan. *J. Cell. Biochem.* **2007**, *101*, 122–134. [CrossRef]

50. Shinohara, H.; Teramachi, J.; Okamura, H.; Yang, D.; Nagata, T.; Haneji, T. Double stranded RNA-dependent protein kinase is necessary for TNF-α-induced osteoclast formation In Vitro and In Vivo. *J. Cell. Biochem.* **2015**, *116*, 1957–1967. [CrossRef] [PubMed]

51. Tsubaki, M.; Komai, M.; Itoh, T.; Imano, M.; Sakamoto, K.; Shimaoka, H.; Takeda, T.; Ogawa, N.; Mashimo, K.; Fujiwara, D.; et al. Nitrogen-containing bisphosphonates inhibit RANKL- and M-CSF-induced osteoclast formation through the inhibition of ERK1/2 and Akt activation. *J. Biomed. Sci.* **2014**, *21*, 10. [CrossRef]

52. Asagiri, M.; Sato, K.; Usami, T.; Ochi, S.; Nishina, H.; Yoshida, H.; Morita, I.; Wagner, E.F.; Mak, T.W.; Serfling, E.; et al. Autoamplification of NFATc1 expression determines its essential role in bone homeostasis. *J. Exp. Med.* **2005**, *202*, 1261–1269. [CrossRef] [PubMed]
53. Kim, K.; Lee, S.H.; Ha Kim, J.; Choi, Y.; Kim, N. NFATc1 induces osteoclast fusion via up-regulation of Atp6v0d2 and the dendritic cell-specific transmembrane protein (DC-STAMP). *Mol. Endocrinol.* **2008**, *22*, 176–185. [CrossRef]
54. Sundaram, K.; Nishimura, R.; Senn, J.; Youssef, R.F.; London, S.D.; Reddy, S.V. RANK ligand signaling modulates the matrix metalloproteinase-9 gene expression during osteoclast differentiation. *Exp. Cell Res.* **2007**, *313*, 168–178. [CrossRef] [PubMed]
55. Sun, Y.X.; Xu, A.H.; Yang, Y.; Li, J. Role of Nrf2 in bone metabolism. *J. Biomed. Sci.* **2015**, *22*, 101. [CrossRef] [PubMed]
56. Nguyen, T.; Yang, C.S.; Pickett, C.B. The pathways and molecular mechanisms regulating Nrf2 activation in response to chemical stress. *Free Radic. Biol. Med.* **2004**, *37*, 433–441. [CrossRef]
57. Park, C.K.; Lee, Y.; Kim, K.H.; Lee, Z.H.; Joo, M.; Kim, H.H. Nrf2 is a novel regulator of bone acquisition. *Bone* **2014**, *63*, 36–46. [CrossRef]
58. Zipper, L.M.; Mulcahy, R.T. Erk activation is required for Nrf2 nuclear localization during pyrrolidine dithiocarbamate induction of glutamate cysteine ligase modulatory gene expression in HepG2 cells. *Toxicol. Sci.* **2003**, *73*, 124–134. [CrossRef] [PubMed]
59. Shin, J.M.; Lee, K.M.; Lee, H.J.; Yun, J.H.; Nho, C.W. Physalin A regulates the Nrf2 pathway through ERK and p38 for induction of detoxifying enzymes. *BMC Complement. Altern. Med.* **2019**, *19*, 101. [CrossRef] [PubMed]
60. Shopit, A.; Niu, M.; Wang, H.; Tang, Z.; Li, X.; Tesfaldet, T.; Ai, J.; Ahmad, N.; Al-Azab, M.; Tang, Z. Protection of diabetes-induced kidney injury by phosphocreatine via the regulation of ERK/Nrf2/HO-1 signaling pathway. *Life Sci.* **2020**, *242*, 117248. [CrossRef]

 *marine drugs* 

Article

Protective Effects of Fucoxanthin on Hydrogen Peroxide-Induced Calcification of Heart Valve Interstitial Cells

Yi-Fen Chiang [1], Chih-Hung Tsai [2,†], Hsin-Yuan Chen [1,3,†], Kai-Lee Wang [4], Hsin-Yi Chang [5], Yun-Ju Huang [1], Yong-Han Hong [3], Mohamed Ali [6], Tzong-Ming Shieh [7], Tsui-Chin Huang [8], Ching-I Lin [9] and Shih-Min Hsia [1,5,10,11,*]

[1] School of Nutrition and Health Sciences, College of Nutrition, Taipei Medical University, Taipei 11031, Taiwan; yvonne840828@gmail.com (Y.-F.C.); d507104002@tmu.edu.tw (H.-Y.C.); d04641004@ntu.edu.tw (Y.-J.H.)
[2] Yu-Kang Animal Hospital, New Taipei City 220, Taiwan; yukangdvm@yahoo.com.tw
[3] Department of Nutrition, I-Shou University, Kaohsiung 84001, Taiwan; yonghan@isu.edu.tw
[4] Department of Nursing, Ching Kuo Institute of Management and Health, Keelung 20301, Taiwan; kellywang@tmu.edu.tw
[5] Graduate Institute of Metabolism and Obesity Sciences, Taipei Medical University, Taipei 11031, Taiwan; hsinyi.chang@tmu.edu.tw
[6] Clinical Pharmacy Department, Faculty of Pharmacy, Ain Shams University, Cairo 11566, Egypt; mohamed.aboouf@pharma.asu.edu.eg
[7] School of Dentistry, College of Dentistry, China Medical University, Taichung 40402, Taiwan; tmshieh@mail.cmu.edu.tw
[8] Graduate Institute of Cancer Biology and Drug Discovery, College of Medical Science and Technology, Taipei Medical University, Taipei 11031, Taiwan; tsuichin@tmu.edu.tw
[9] Department of Nutrition and Health Sciences, Kainan University, Taoyuan 338, Taiwan; cilin@mail.knu.edu.tw
[10] School of Food and Safety, Taipei Medical University, Taipei 11031, Taiwan
[11] Nutrition Research Center, Taipei Medical University Hospital, Taipei 11031, Taiwan
* Correspondence: bryanhsia@tmu.edu.tw; Tel.: +886-273-61661-6558
† Equal contribution.

Citation: Chiang, Y.-F.; Tsai, C.-H.; Chen, H.-Y.; Wang, K.-L.; Chang, H.-Y.; Huang, Y.-J.; Hong, Y.-H.; Ali, M.; Shieh, T.-M.; Huang, T.-C.; et al. Protective Effects of Fucoxanthin on Hydrogen Peroxide-Induced Calcification of Heart Valve Interstitial Cells. *Mar. Drugs* **2021**, *19*, 307. https://doi.org/10.3390/md19060307

Academic Editors: Masashi Hosokawa and Hayato Maeda

Received: 23 April 2021
Accepted: 22 May 2021
Published: 26 May 2021

Publisher's Note: MDPI stays neutral with regard to jurisdictional claims in published maps and institutional affiliations.

Abstract: Cardiovascular diseases such as atherosclerosis and aortic valve sclerosis involve inflammatory reactions triggered by various stimuli, causing increased oxidative stress. This increased oxidative stress causes damage to the heart cells, with subsequent cell apoptosis or calcification. Currently, heart valve damage or heart valve diseases are treated by drugs or surgery. Natural antioxidant products are being investigated in related research, such as fucoxanthin (Fx), which is a marine carotenoid extracted from seaweed, with strong antioxidant, anti-inflammatory, and anti-tumor properties. This study aimed to explore the protective effect of Fx on heart valves under high oxidative stress, as well as the underlying mechanism of action. Rat heart valve interstitial cells under H_2O_2-induced oxidative stress were treated with Fx. Fx improved cell survival and reduced oxidative stress-induced DNA damage, which was assessed by cell viability analysis and staining with propidium iodide. Alizarin Red-S analysis indicated that Fx has a protective effect against calcification. Furthermore, Western blotting revealed that Fx abrogates oxidative stress-induced apoptosis via reducing the expression of apoptosis-related proteins as well as modulate Akt/ERK-related protein expression. Notably, in vivo experiments using 26 dogs treated with 60 mg/kg of Fx in combination with medical treatment for 0.5 to 2 years showed significant recovery in their echocardiographic parameters. Collectively, these in vitro and in vivo results highlight the potential of Fx to protect heart valve cells from high oxidative stress-induced damage.

Keywords: fucoxanthin; oxidative stress; calcification; heart valve interstitial cell

1. Introduction

The pathogenesis of cardiovascular disease, such as atherosclerosis and aortic valve sclerosis, involves inflammatory reactions in response to a variety of stimuli including high levels of low-density lipoprotein (LDL) and reactive oxygen species (ROS). The latter are triggered by increased oxidative stress, infections, and chemical damage. In cardiovascular diseases, increased load on myocardial cells or insufficient energy supply would cause an imbalance in energy supply and demand, resulting in an increase in energy metabolism and mitochondrial redox. This eventually leads to an increase in reactive oxygen species (ROS) and damage to heart cells [1].

Heart valves, such as the aortic valve, mainly comprise two types of cells: valve endothelial cells (VEC) on the surface and valve interstitial cells (VIC) in the matrix. VECs regulate message transmission and permeability and also prevent thrombosis [2]. VICs maintain the tissue structure of the valves. Normally, healthy valve cells are mainly of the fibroblast type, but cell apoptosis can be induced in cases of increased reactive oxygen species and this leads to valve damage [3]. Apoptosis is a regulatory pathway of programmed cell death in response to signals generated by environmental stimuli. In the process of apoptosis, a DNA repair protein called poly (ADP-ribose) polymerase-1 (PARP-1) is truncated, causing cell apoptosis [4].

Another role of elevated ROS is in affecting extracellular matrix (ECM) remodeling [5]. In heart valves, ECM remodeling occurs in layers enriched with VIC fibroblast cells with subsequent valve fibrosis potential. A previous study showed that valvular osteoblast induction could activate the ECM accumulation and calcification process [6,7]. Moreover, ROS-induced chronic inflammation resulted in an influx of macrophages and increased pro-osteogenic and angiogenic activities [5], which in return increased the production of proteolytic enzymes and triggered ECM remodeling and valve fibrosis [8].

Small-breed dogs weighing less than 20 kg can suffer from heart valve diseases [9]. In Taiwan, Maltese present the highest incidence [10]. The progression of valve disease is correlated with increased age, heart volume enlargement, and the regurgitation of blood flow [10].

Fucoxanthin (Fx) is a carotenoid with high anti-oxidative activity that is abundant in brown seaweeds [11]. Fx has shown a potential ability to lower lipid peroxidation [12] and exert anti-inflammatory [13], anti-tumor [14], and anti-hyperuricemia [15] effects. Additionally, in cardiovascular disease Fx has shown recovery effects related to DNA damage and cardioprotective effects [16]. However, studies of the protective role of Fx in heart valve interstitial cells are lacking. In this study, we investigated the potential mechanism of Fx in protecting the heart valves from fibrosis.

2. Results

2.1. Protective Effect of Fx on VIC Cell Viability from H_2O_2-Induced Oxidative Stress

2.1.1. Effect of H_2O_2 on the VIC Cell Viability

After the cell extraction (Supplementary Figure S1), we treated the cells with H_2O_2 in serial doses for 15 min, 1 h and 4 h to induce high oxidative stress in VIC to measure the cell viability. The results showed that H_2O_2 at doses of 0.5 mM and above significantly reduced cell proliferation at all the time points. Notably, H_2O_2 at 0.5 mM for 15 min (min) decreased the cell viability by 30%, which is considered a moderate effect of oxidative stress to be used in following experiments (Figure 1A).

2.1.2. Effect of Fx on VIC Cell Viability

To evaluate Fx's effect on the VIC cell viability, we treated the cells with different doses of Fx for 24, 48, and 72 h and measured cell viability using and 3-(4,5-Dimethylthiazol-2-yl)-2,5-Diphenyltetrazolium Bromide (MTT) assay. The results showed that Fx did not significantly affect the VIC cell viability, except when the highest dosage (5 mg/mL) was used, probably due to its toxic effect (Figure 1B).

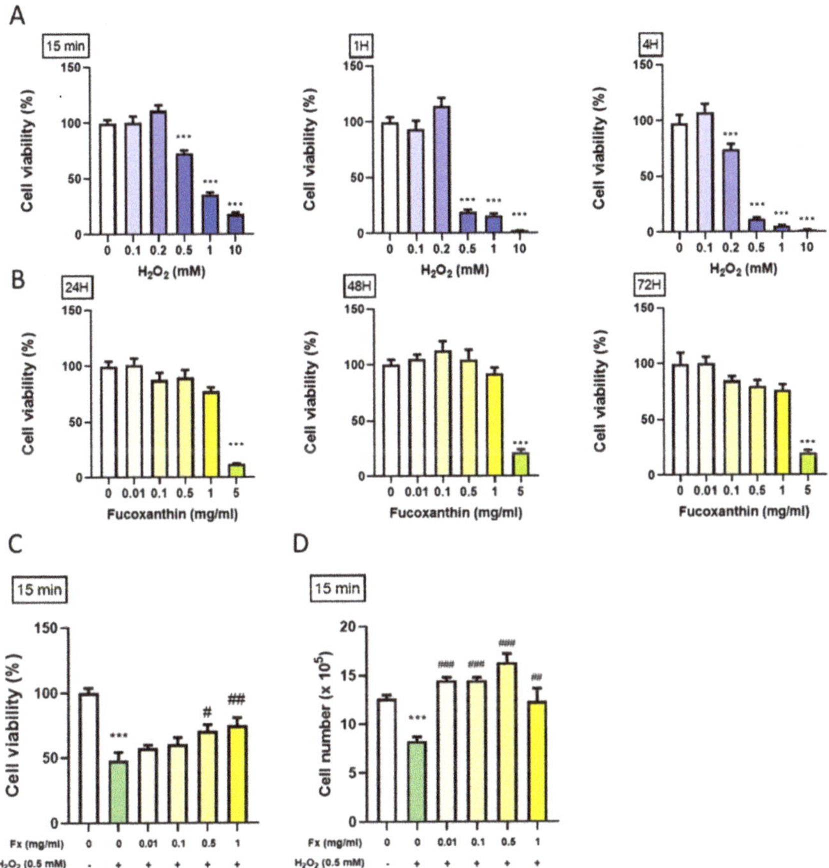

Figure 1. Protective effect of fucoxanthin (Fx) on VIC cell viability following H_2O_2-induced oxidative stress. Rat heart valve interstitial cells (3000 cells/well) were cultured in DMEM/F12 supplemented with 10% FBS for 24 h. (**A**) Treated with H_2O_2 (0.1–10 mM) for 15 min, 1 h and 4 h. (**B**) Treated with Fx (0.01–5 mg/mL) for 24, 48, and 72 h. (**C, D**) Cells were pretreated with Fx (Fx, 0.01–1 mg/mL) for 24 h, then treated with 0.5 mM of H_2O for 15 min. Cell viability was analyzed by MTT assay in (**C**) or through cell counting in (**D**). ***, $p < 0.001$ compared with untreated control cells. #, $p < 0.05$; ##, $p < 0.01$; ###, $p < 0.001$ compared with the H_2O_2-induced group. White bar, control group; Blue bar, different dosage of H_2O_2-induced group; Green bar, H_2O_2-induced group; Yellow bar, Fx-treated group.

2.1.3. Fx-Abrogated H_2O_2-Induced VIC Viability Change

Next, we explored the combined effect of the Fx and H_2O_2 on VIC viability to evaluate the protective effect of Fx. Cells were first pretreated with different doses of Fx for 24 h, then exposed to 0.5 mM of H_2O_2 for 15 min., which earlier was enough to inhibit cell proliferation. MTT assay was used to measure the cell viability. The results showed that Fx could alleviate the inhibitory effect of H2O2-induced oxidative stress on VIC growth (Figure 1C,D).

2.2. Fx-Ameliorated H_2O_2-Induced DNA Damage and Apoptosis-Related Protein Expression

2.2.1. Fx Decreased H_2O_2-Induced VIC Cell Morphology Changes and DNA Damage

To further explore the protective effect of Fx on VIC cells in the context of halting DNA damage, we used PI staining and microscopy to visually observe any morphological changes. Following the pretreatment of the cells with Fx and then H_2O_2, PI staining with fluorescence signal increased, the result showed that Fx was able to prevent oxidative stress-induced DNA damage (Figure 2A,B).

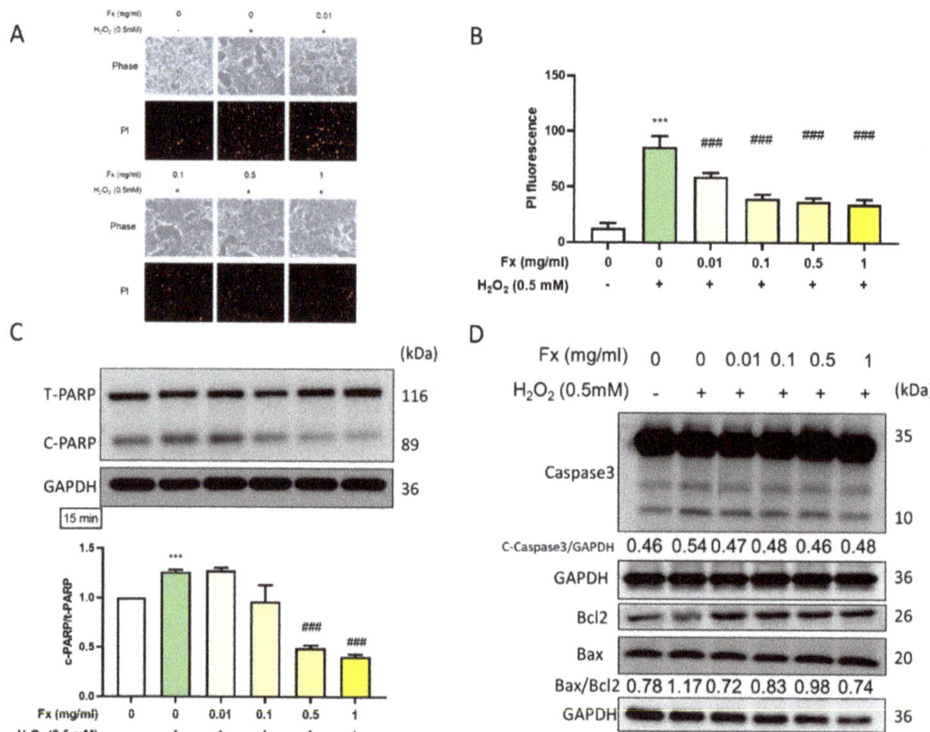

Figure 2. Fucoxanthin (Fx) ameliorated H_2O_2-induced DNA damage and apoptosis-related protein expression. Rat heart valve interstitial cells were cultured in a 6-well plate for 24 h, pretreated with Fx for 24 h, and then treated with H_2O_2 for 15 min. (**A**) Cells were stained with PI solution (1 µg/mL) for 1 h and visualized by microscopy at 200× magnification. (**B**) Quantified by Image J. Western blotting was used to explore the protein expression of (**C**) total and cleaved PARP, and (**D**) apoptosis-related markers cleaved caspase 3, Bcl_2, and Bax. ***, $p < 0.001$ compared with untreated control group. ###, $p < 0.001$ compared with H_2O_2-induced group. White bar, control group; Green bar, H_2O_2-induced group; Yellow bar, Fx-treated group.

2.2.2. Fx-Ameliorated H_2O_2-Induced Apoptosis-Related Protein Expression

Along with the increase in cell damage and decrease in cell viability, we used the Western blotting technique to explore the effect of the administration of both H_2O_2 and Fx on the apoptosis-related protein expression. The H_2O_2-induced group showed a significant increase in the expression of the cleaved form of PARP, while pretreatment with Fx significantly reversed the expression of cleaved PARP (Figure 2C). Furthermore, we explored other apoptosis-related markers, such as cleaved caspase 3 and the Bax/Bcl_2 ratio (Figure 2D). The results confirmed that H_2O_2-induced apoptosis, which was reversed after the Fx treatments. Collectively, these results highlight the anti-apoptotic effect of Fx in VIC cells.

2.3. Effect of Fx on H_2O_2-Induced Reactive Oxygen Species

According to DCFDA's result, treatment with H_2O_2 induced a high level of ROS, which in turn could lead to damage to the valve structure and subsequent calcification [17]. The results show that Fx could alleviate the high oxidative stress-induced ROS level, as shown by a reduction in the fluorescence (Figure 3A) and density (Figure 3B).

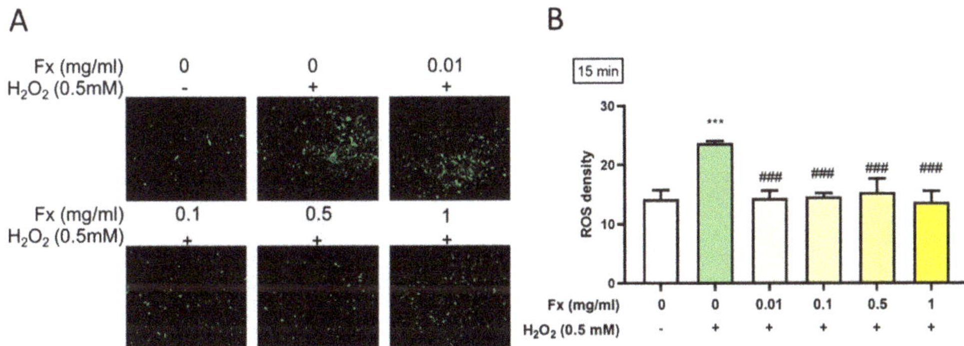

Figure 3. Effect of fucoxanthin (Fx) on the H_2O_2-induced ROS level. Rat heart valve interstitial cells were cultured in DMEM/F12 supplemented with 10% FBS for 24 h, pretreated with 0.01–1 mg/mL of Fx for 24 h, then treated with 0.5 mM of H_2O_2 for 15 min. ROS were visualized using DCFDA. These were assessed by microscopy at 200× magnification (**A**) and we used Image J for the density quantification (**B**). ***, $p < 0.05$ compared with untreated control group; ###, $p < 0.001$ compared with H_2O_2 group. White bar, control group; Green bar, H_2O_2-induced group; Yellow bar, Fx-treated group.

2.4. Effect of Fx on Oxidative Stress-Induced Calcification and the Expression of Its Related Markers in VIC

Oxidative stress was shown to trigger calcification in VIC [18], which involves the activation of the Akt/ERK signaling pathway [19]. Therefore, we evaluated the effect of both H_2O_2 and Fx on the ROS and calcification-related Akt/ERK signaling pathway. The results showed that H_2O_2 treatment significantly increased the phosphorylation of Akt and ERK proteins. Notably, the pre-treatment with Fx was able to decrease such activation (Figure 4A,B). Furthermore, we used Alizarin Red-S for calcification staining and the results showed that calcification was increased in response to H_2O_2 and alleviated by Fx pretreatment (Figure 4C,D). Consistently, the Western blotting results showed that the Fx treatment was able to partially oppose the H_2O_2-induced ECM remodeling of marker matrix metalloproteinase 2 (MMP-2) in VIC cells (Figure 4E).

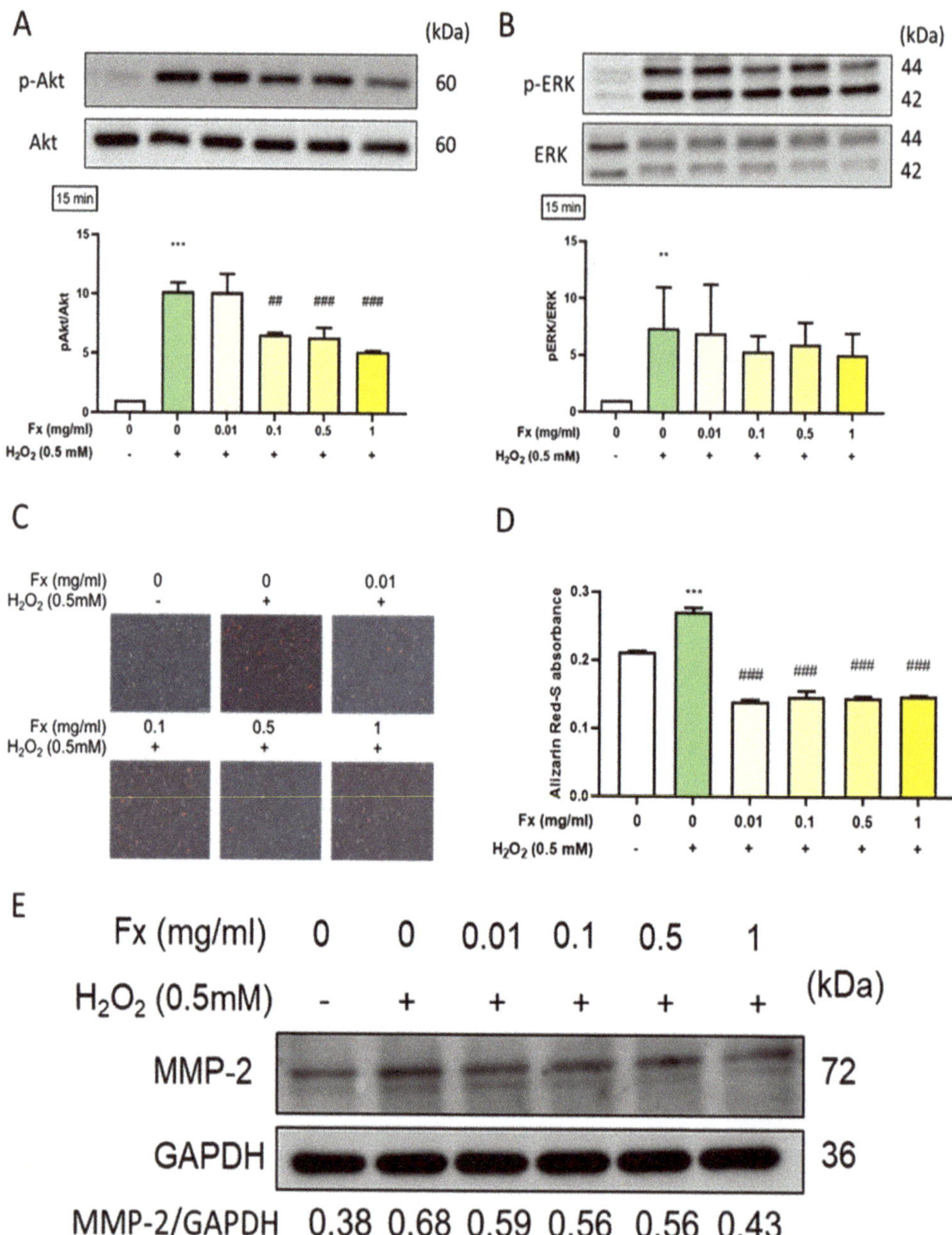

Figure 4. Effect of fucoxanthin (Fx) on oxidative stress-induced calcification and the expression of its related markers in VIC. Rat heart valve interstitial cells were cultured in a 6-well plate for 24 h, pretreated with Fx for 24 h, then induced with H_2O_2 for 15 min. Western blotting was used to analyze the protein expression of (**A**) pAkt/Akt (**B**) pERK/ERK. (**C**) Alizarin Red-S staining was used to visualize calcification, which was assessed by microscopy at a $200\times$ magnification and (**D**) quantified by Image J. (**E**) MMP-2 protein expression was analyzed using Western blotting. **, $p < 0.01$; ***, $p < 0.001$ compared with untreated control group. ##, $p < 0.01$; ###, $p < 0.001$ compared with H_2O_2-induced group. White bar, control group; Green bar, H_2O_2-induced group; Yellow bar, Fx-treated group.

2.5. Long-Term Cardioprotective Effect of Fx Treatment in Dogs

Long-term treatment with Fx in the 26 dogs resulted in a significant decrease in their vertebral heart size (VHS) (Figure 5A). VHS score is a number that normalizes heart size to body size using the mid-thoracic vertebrae as units of measurement, reflecting compensatory cardiac enlargement [20]. In addition, treatment showed the improvement of the left atrium to aortic (LA/AO) dimension ratio (Figure 5B), Tei index (Figure 5C), and E/e value (Figure 5D). The linkage of the mitral valve and tricuspid valve showed a significant decrease in echocardiography (Figure 5E,F). Collectively, Fx supplementation could improve the overall function of ventricular contraction and relaxation, confirming the findings of previous studies [21].

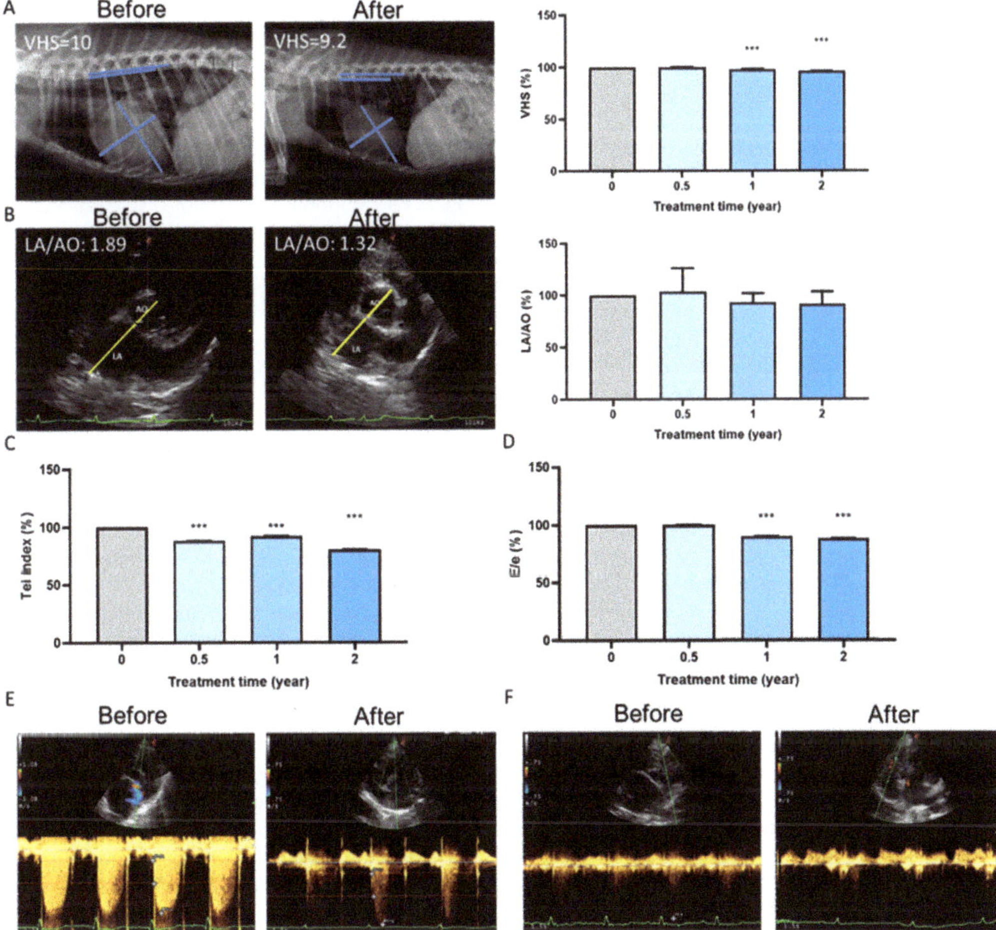

Figure 5. Long-term cardioprotective effect of fucoxanthin (Fx) treatment in dog. After Fx treatments, echocardiography analysis showed (**A**) a significant decrease in the VHS score, (**B**) an improvement in LA/AO, (**C**) a reduction in the Tei index, and (**D**) a decrease in the E/e percentage. The linkage of (**E**) the mitral valve and (**F**) the tricuspid valve decreased. ***, $p < 0.001$ compared with the baseline.

3. Discussion

Oxidative stress and ROS contribute significantly to the pathogenesis of cardiovascular diseases [22], such as stroke and hypertension [23]. Elevated ROS could cause vascular damage via recruiting more leukocytes in blood reperfusion [24]. Moreover, oxidative stress can induce endothelial dysfunction and induce pro-fibrotic and pro-osteoblastic effects with subsequent calcification in the aortic valves, as shown in a mouse calcific aortic valve disease (CAVD) model [25].

Oxidative stress via ROS induces damage to DNA double strands, in addition to the rapid phosphorylation of H2AX by PI3K-related kinases with downstream Akt modulation [26]. Additionally, Akt phosphorylation induced the osteogenic early marker RUNX2 [17], which had the ability to increase osteogenic differentiation [27]. The progression of calcification in VIC cells also involved a significant increase in the osteogenic medium (OM) condition as well as the increased phosphorylation of the NF-κB, PI3K-Akt, TNF, and MAPK signaling pathways [19]. In this study, Fx treatment could alleviate the oxidative stress-induced ROS level (Figure 3) and decrease the progression calcification through Akt- and MAPK-related signaling pathways (Figure 4).

Aortic leaflets are composed of three main layers, named the fibrosa, spongiosa, and ventricularis. This composition builds up the valve function, which is mainly diastole and systole extension [28]. Considering that ECM is also the main component of valves, the structure and the arrangement of ECM play important roles in the valve function.

Increased inflammation and LDL oxidation induce the calcification of the valves [29]. On the other hand, oxidized high-density lipoprotein (ox-HDL) increases in CAVD patients' plasma to protect against the high oxidative stress induced by CAVD [30]. As ROS increase, chronic inflammation, with accompanying immunity components, increases, along with increased growth factors and proteolytic enzymes, such as MMP-9 and MMP-2 for ECM deposition and remodeling effects [31]. Therefore, the up-regulation of MMPs indicates valve remodeling and calcification [32].

Several studies have shown vascular diseases with ROS-related pathology in studies of animals fed with a high-cholesterol or high-fat diet to induce vascular calcification. There are different categories of antioxidant compound. Natural antioxidants, such as gallic acid [33], curcumin [34], and quercetin [35], or synthetic ROS scavengers such as N-acetylcysteine, pyrrolidine dithiocarbamate, and poly(1,8-octamethylene-citrate-co-cysteine) with a reduction in ROS, PI3K/Akt or inflammation-related protein expression could reduce calcification and apoptosis through ROS scavenging [36].

Fx exerts antioxidant ability that has been shown to eliminate pregnancy-related hypertension [37], high glucose-induced diabetes retinopathy [38], and ox-LDL-induced endothelial damage [39] via ROS reduction. In our current in vitro study, Fx showed a potential protective effect against high oxidative stress-induced VIC damage through a reduction in apoptosis (Figure 2) and ROS and modulation of the phosphorylation of Akt and ERK to decrease the calcification and ECM accumulation (Figure 4).

The first-line diagnosis of heart valve disease in dogs includes the use of radiographs, ultrasound, and echocardiography [40]. Indicators of valve disease include an increase in VHS score or LA/AO. In echocardiography assessment, the veterinarian would use the Tei index to evaluate systolic and diastolic function. The E/e' ratio is used to assess the mitral valve inflow and analyze the mitral valve leakage and left ventricular diastolic dysfunction [41]. In our study, fucoxanthin treatments could reverse mitral valve and tricuspid valve leakage. Moreover, using in vivo experiments in dogs, we showed that the long-term supplementation of Fx could improve both compensatory cardiac hypertrophy and valve function (Figure 5).

4. Materials and Methods

4.1. Reagents Preparation

High-stability fucoxanthin (HS Fucoxanthin, HSFUCO, Fx) was obtained from Hi-Q Marine Biotech International Ltd. (Taipei, Taiwan) and dissolved in ddH_2O [38].

4.2. Cell Extraction and Treatment

The extraction of primary rat valve interstitial cells (VIC) was carried out as shown in a previous study [42]. Briefly, after harvesting all the leaflets, pellet was centrifuged and incubated with collagenase II (Thermo) for 2 h to obtain VIC debris. Cells were then cultured in Dulbecco's Modified Eagle's Medium (DMEM/F12) (CASSION, Taichung City, Taiwan) combined with 100 units/mL of penicillin, 100 µg/mL of streptomycin (CORNING, Manassas, VA, USA), sodium bicarbonate (2.438 g/L; Bio-Shop, Burlington, ON, Canada), and 10% fetal bovine serum (FBS; CORNING, Manassas, VA, USA). To explore the protective effect of Fx against oxidative stress, VIC were pretreated with Fx for 24 h and then treated with H_2O_2 for 15 min.

4.3. 3-(4,5-Dimethylthiazol-2-yl)-2,5-Diphenyltetrazolium Bromide (MTT) Assay

VIC cells were seeded in 96-well plates (3000 cells/well). The cells were pretreated with different doses of Fx (mg/mL) for 24 h and then with H_2O_2 for 15 min. After the treatment, we used the MTT assay (Abcam, Cambridge, MA, USA) for the analysis of cell viability. We added 1 mg/mL of MTT for 3 h until the crystal precipitation formed. Then added 100 µL/well of dimethyl sulfoxide (DMSO; ECHO Chemical Co. Ltd., Taipei, Taiwan) to dissolve the crystal formation. We used VERSA Max microplate reader (Molecular Devices, San Jose, CA, USA) to measure the optical density at 570 and 630 nm.

4.4. Cell Counting

After the treatments, cell pellets were mixed with 0.4% trypan blue solution (Gibco, Grand Island, NY, USA). We used a hemocytometer (Hausser scientific company, Horsham, PA, USA) to calculate the cell number at 200× magnification.

4.5. PI Staining

Propidium iodide (PI) solution (500 µg/mL) (Sigma-Aldrich, St. Louis, MO, USA) was dissolved with sterile ddH_2O and stained with PI (1 µg/mL) solution for 1 h. We used microscopy (Olympus, Tokyo, Japan) to carry out fluorescence imaging at 200× magnification.

4.6. ROS Density Measurement

VIC cells were cultured in 6-well plates. We used 25 µM $2',7'$–dichlorofluorescin di-acetate (DCFDA, Cayman, Ann Arbor, MI, USA) staining for 30 min. Then, we used microscopy to capture the fluorescence image. We used the Image J software (Version 1.52t, NIH, Bethesda, MD, USA) to carry out ROS density quantification in single cells.

4.7. Alizarin Red-S Staining Assay

The calcification progression was assessed by Alizarin Red-S staining (Sciencell, Carlsbad, CA, USA). After treatment, the cells were fixed with 4% paraformaldehyde (Sigma-Aldrich) for 10 min and then stained with Alizarin Red-S for 30 min. We used a fluorescence microscope to image the stained area. We used the VERSA Max microplate reader to measure the absorbance at 405 nm.

4.8. Protein Extraction and Western Blot

Cells were lysed in radioimmunoprecipitation assay (RIPA) lysis buffer with added protease and phosphatase inhibitors (Roche, Mannheim, Baden-Württemberg, Germany). Additionally, we quantified the cells with a bicinchoninic acid (BCA) assay, used sodium dodecyl sulfate polyacrylamide gel electrophoresis (SDS-PAGE), and transferred to a polyvinylidene fluoride (PVDF) membrane. Blocking with 5% bovine serum albumin (BSA) solution for 1 H was carried out. We used primary antibodies—poly (ADP-ribose) polymerase (PARP) (1:1000; Cell Signaling, Boston, MA, USA), glyceraldehyde 3-phosphate dehydrogenase (GAPDH) (1:10,000; Proteintech, Rosemont, IL, USA), p-Akt (1:1000, Cell signaling), Akt (1:1000, Cell signaling), p44/42 MAPK (Erk1/2) (1:1000, Cell signaling), phospho-p44/42 MAPK (Erk1/2) (1:1000, Cell signaling), MMP-2 (1:1000, Abcam), BCL2

Associated X (Bax) (1:1000, Cell signaling), and B-cell lymphoma 2 (Bcl-2) (1:500, Santa Cruz, Santa Cruz, CA, USA)—cultured the mixture at 4 °C overnight, washed it three times, and then stained it with horseradish peroxidase (HRP)-conjugated secondary antibody (1:5000–10,000) for 2 h. The signal was captured by the eBlot Touch Imager[tm] (eBlot Photoelectric Technology, Shanghai, China). The band densities were determined using the Image J software program version 1.52 t (NIH, Bethesda, MD, USA). The expression level of these target proteins was analyzed in three individual experiments.

4.9. In Vivo Animal Experiments

With the help of a veterinarian, we recruited 26 heart disease-diagnosed dogs for the in vivo experiment. The dogs were treated with Fuco Pets HeartFight® (contained 60 mg/kg Fx) twice daily from Hi-Q Marine Biotech International Ltd. (Taipei, Taiwan) combined with medical treatments for 0.5 to 2 years. We used conventional echocardiography and standard Doppler examination to follow up the valve function. Esaote's MyLab™ClassC® (Italy) equipped with a PA-122 probe cardio phased array (frequency range of 3–8 MHz) was used to obtain all the echocardiographic data.

The left atrium to aorta (LA/AO) ratio was measured using B-mode images acquired from a short axis five-chamber view of the right sternum wall.

4.10. Statistical Analysis

The data are expressed as the mean ± standard deviation (SD). We used GraphPad Prism 8.0 for the analysis. Student's t-test was used for the comparisons between the two groups. One-way ANOVA tests were used to compare multiple groups, followed by Tukey's post hoc test. A p-value of less than 0.05 was considered significant. The p values are presented as *, $p < 0.05$; **, $p < 0.01$; ***, $p < 0.001$; or [#], $p < 0.05$; [##], $p < 0.01$; and [###], $p < 0.001$.

5. Conclusions

Treatment with Fx was demonstrated to effectively protect against the harmful effects of high H_2O_2-induced oxidative stress in heart valve interstitial cells through the antioxidant potential of Fx as follows: (1) Fx can recover H_2O_2-induced cell viability impairment, (2) Fx can oppose H_2O_2-induced apoptosis, (3) Fx can inhibit the Akt/ERK-related signaling pathway to reduce heart valve calcification, (4) long-term treatment with Fx could recover the heart valve function and leakage in dogs. These data show that Fx has the potential to protect heart valve cells from damage caused by high oxidative stress (Figure 6).

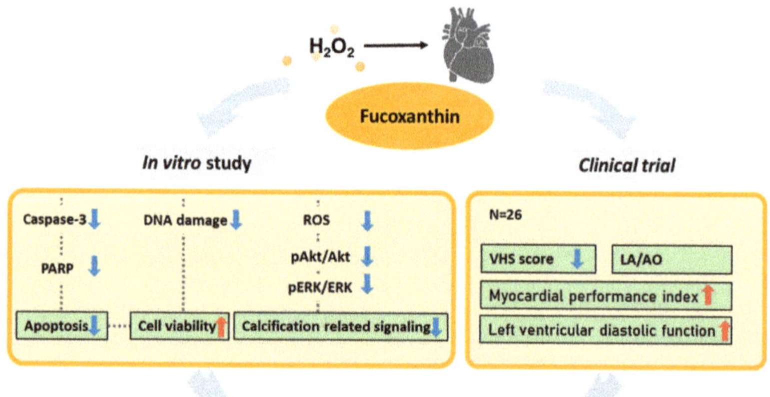

Figure 6. Schematic representation of the potential effects of Fx on protection against high oxidative stress-related cell apoptosis and the ROS-related calcification signaling pathway. Furthermore, in vivo experiment shows Fx's ability to protect against valve-related disease in dogs.

Supplementary Materials: The following are available online at https://www.mdpi.com/article/. 10.3390/md19060307/s1: Figure S1. Protein marker expression in rat valve interstitial cells (A) Used western blot Rat valve interstitial cells (VICs) were negative for the endothelial cell marker, CD31, positive with vimentin and α-SMA. (B) Used microscopy in 200× magnification to capture the morphology of VICs.

Author Contributions: Conceptualization, Y.-F.C., H.-Y.C. (Hsin-Yuan Chen) and S.-M.H.; experimentation, Y.-F.C. and C.-H.T.; data analysis and figure preparation, Y.-F.C. and H.-Y.C. (Hsin-Yuan Chen); methodology and resources, Y.-F.C., C.-H.T., T.-M.S., Y.-J.H., Y.-H.H., K.-L.W., H.-Y.C. (Hsin-Yi Chang), C.-I.L. and T.-C.H.; writing—original draft preparation, Y.-F.C.; writing—review and editing, Y.-F.C., M.A. and S.-M.H.; editing and approval of the final version of the manuscript, S.-M.H. All authors have read and agreed to the published version of the manuscript.

Funding: This study was supported by grants (MOST109-2314-B-038-059, MOST 109-2628-B-038-015, MOST 109-2320-B-254-001, MOST 109-2811-B-038-523 and MOST 109-2320-B-424-001) from the Ministry of Science and Technology, Taiwan, and grants (MOE-RSC-108RSN0005) from the Ministry of Education, Taiwan.

Institutional Review Board Statement: The animal studies were conducted according to the protocols approved by the Institutional Animal Care and Use Committee (IACUC) of Taipei Medical University (NO. LAC-2020-0149).

Informed Consent Statement: Not applicable.

Data Availability Statement: The data presented in this study are available on request from the corresponding author.

Conflicts of Interest: The authors declare no conflict of interest.

Ethical Approval for Experiments on Animals: The study was approved by the Experimental Animal Care and Use Committee of Taipei Medical University (NO. LAC-2020-0149). All animals received humane care in compliance with the Principles of Laboratory Animal Care and the Guide for the Care and Use of Laboratory Animals, published by the National Science Council, Taiwan.

References

1. Wasmus, C.; Dudek, J. Metabolic Alterations Caused by Defective Cardiolipin Remodeling in Inherited Cardiomyopathies. *Life* **2020**, *10*, 277. [CrossRef] [PubMed]
2. Gould, S.T.; Srigunapalan, S.; Simmons, C.A.; Anseth, K.S. Hemodynamic and Cellular Response Feedback in Calcific Aortic Valve Disease. *Circ. Res.* **2013**, *113*, 186–197. [CrossRef]
3. Towler, D.A. Oxidation, inflammation, and Aortic Valve Calcification Peroxide Paves an Osteogenic Path. *J. Am. Coll. Cardiol.* **2008**, *52*, 851–854. [CrossRef] [PubMed]
4. Chaitanya, G.V.; Steven, A.J.; Babu, P.P. PARP-1 Cleavage Fragments: Signatures of Cell-death Proteases in Neurodegeneration. *CCS* **2010**, *8*, 1–11. [CrossRef] [PubMed]
5. Di Vito, A.; Donato, A.; Presta, I.; Mancuso, T.; Brunetti, F.S.; Mastroroberto, P.; Amorosi, A.; Malara, N.; Donato, G. Extracellular Matrix in Calcific Aortic Valve Disease: Architecture, Dynamic and Perspectives. *Int. J. Mol. Sci.* **2021**, *22*, 913. [CrossRef] [PubMed]
6. Ma, X.; Zhao, D.; Yuan, P.; Li, J.; Yun, Y.; Cui, Y.; Zhang, T.; Ma, J.; Sun, L.; Ma, H.; et al. Endothelial-to-Mesenchymal Transition in Calcific Aortic Valve Disease. *Acta Cardiol. Sin.* **2020**, *36*, 183–194. [PubMed]
7. Nagy, E.; Eriksson, P.; Yousry, M.; Caidahl, K.; Ingelsson, E.; Hansson, G.K.; Franco-Cereceda, A.; Bäck, M. Valvular Osteoclasts in Calcification and Aortic Valve Stenosis Severity. *Int. J. Cardiol.* **2013**, *168*, 2264–2271. [CrossRef]
8. Nsaibia, M.J.; Boulanger, M.C.; Bouchareb, R.; Mkannez, G.; Le Quang, K.; Hadji, F.; Argaud, D.; Dahou, A.; Bossé, Y.; Koschinsky, M.L.; et al. OxLDL-derived Lysophosphatidic Acid Promotes the Progression of Aortic Valve Stenosis Through A LPAR1-Rhoa-NF-KB Pathway. *Cardiovasc. Res.* **2017**, *113*, 1351–1363. [CrossRef]
9. Keene, B.W.; Atkins, C.E.; Bonagura, J.D.; Fox, P.R.; Häggström, J.; Fuentes, V.L.; Oyama, M.A.; Rush, J.E.; Stepien, R.; Uechi, M. ACVIM Consensus Guidelines for the Diagnosis and Treatment of Myxomatous Mitral Valve Disease in Dogs. *J. Vet. Intern Med.* **2019**, *33*, 1127–1140. [CrossRef]
10. Tsai, C.-H.; Chen, H.-Y.; Hsia, S.-M. Four case reports: Effects of Fucoidan and Fucoxanthin on the Treatment of Degenerative Heart Valve Disease in Dogs. *J. Anim. Sci. Vet. Med.* **2020**, *5*, 114–122.
11. Liu, Y.; Liu, M.; Zhang, X.; Chen, Q.; Chen, H.; Sun, L.; Liu, G. Protective Effect of Fucoxanthin Isolated from Laminaria japonica against Visible Light-Induced Retinal Damage Both in Vitro and in Vivo. *J. Agric. Food Chem.* **2016**, *64*, 416–424. [CrossRef]
12. Iwasaki, S.; Widjaja-Adhi, M.A.K.; Koide, A.; Kaga, T.; Nakano, S.; Beppu, F.; Hosokawa, M.; Miyashita, K. In Vivo Antioxidant Activity of Fucoxanthin on Obese/Diabetes KK-A^y Mice. *Food Nutr. Sci.* **2012**, *3*, 1491–1499.

13. Prasedya, E.S.; Martyasari, N.W.R.; Abidin, A.S.; Pebriani, S.A.; Ilhami, B.T.K.; Frediansyah, A.; Sunarwidhi, A.L.; Widyastuti, S.; Sunarpi, H. Macroalgae Sargassum cristaefolium Extract Inhibits Proinflammatory Cytokine Expression in BALB/C Mice. *Scientifica (Cairo)* **2020**, *2020*, 9769454. [PubMed]

14. Le Goff, M.; Le Ferrec, E.; Mayer, C.; Mimouni, V.; Lagadic-Gossmann, D.; Schoefs, B.; Ulmann, L. Microalgal Carotenoids And Phytosterols Regulate Biochemical Mechanisms Involved in Human Health and Disease Prevention. *Biochimie* **2019**, *167*, 106–118. [CrossRef] [PubMed]

15. Chau, Y.-T.; Chen, H.-Y.; Lin, P.-H.; Hsia, S.-M. Preventive Effects of Fucoidan and Fucoxanthin on Hyperuricemic Rats Induced by Potassium Oxonate. *Mar. Drugs* **2019**, *17*, 343. [CrossRef]

16. Hentati, F.; Tounsi, L.; Djomdi, D.; Pierre, G.; Delattre, C.; Ursu, A.V.; Fendri, I.; Abdelkafi, S.; Michaud, P. Bioactive Polysaccharides from Seaweeds. *Molecules* **2020**, *25*, 3152. [CrossRef] [PubMed]

17. Branchetti, E.; Sainger, R.; Poggio, P.; Grau, J.B.; Patterson-Fortin, J.; Bavaria, J.E.; Chorny, M.; Lai, E.; Gorman, R.C.; Levy, R.J.; et al. Antioxidant Enzymes Reduce DNA Damage and Early Activation of Valvular Interstitial Cells in Aortic Valve Sclerosis. *Arterioscler. Thromb. Vasc. Biol.* **2013**, *33*, e66–e74. [CrossRef]

18. Singh, S.; Torzewski, M. Fibroblasts and Their Pathological Functions in the Fibrosis of Aortic Valve Sclerosis and Atherosclerosis. *Biomolecules* **2019**, *9*, 472. [CrossRef]

19. Zhou, T.; Wang, Y.; Liu, M.; Huang, Y.; Shi, J.; Dong, N.; Xu, K. Curcumin Inhibits Calcification of Human Aortic Valve Interstitial Cells by Interfering NF-κB, AKT, and ERK Pathways. *Phytother. Res.* **2020**, *34*, 2074–2081. [CrossRef]

20. Nakayama, H.; Nakayama, T.; Hamlin, R.L. Correlation of Cardiac Enlargement as Assessed by Vertebral Heart Size and Echocardiographic and Electrocardiographic Findings in Dogs with Evolving Cardiomegaly due to Rapid Ventricular Pacing. *J. Vet. Intern. Med.* **2001**, *15*, 217–221. [CrossRef] [PubMed]

21. Galizzi, A.; Bagardi, M.; Stranieri, A.; Zanaboni, A.M.; Malchiodi, D.; Borromeo, V.; Brambilla, P.G.; Locatelli, C. Factors Affecting the Urinary Aldosterone-to-Creatinine Ratio in Healthy Dogs and Dogs with Naturally Occurring Myxomatous Mitral Valve Disease. *BMC Vet. Res.* **2021**, *17*, 15. [CrossRef]

22. Heistad, D.D. Oxidative stress and Vascular Disease: 2005 Duff lecture. *Arterioscler. Thromb. Vasc. Biol.* **2006**, *26*, 689–695. [CrossRef]

23. Griendling, K.K.; FitzGerald, G.A. Oxidative Stress and Cardiovascular Injury: Part II: Animal and Human studies. *Circulation* **2003**, *108*, 2034–2040. [CrossRef]

24. Wei, E.P.; Kontos, H.A.; Christman, C.W.; DeWitt, D.S.; Povlishock, J.T. Superoxide Generation and Reversal of Acetylcholine-induced Cerebral Arteriolar Dilation after Acute Hypertension. *Circ. Res.* **1985**, *57*, 781–787. [CrossRef]

25. Weiss, R.M.; Ohashi, M.; Miller, J.D.; Young, S.G.; Heistad, D.D. Calcific Aortic Valve Stenosis in Old Hypercholesterolemic Mice. *Circulation* **2006**, *114*, 2065–2069. [CrossRef]

26. Misteli, T.; Soutoglou, E. The Emerging Role of Nuclear Architecture in DNA Repair and Genome Maintenance. *Nat. Rev. Mol. Cell Biol.* **2009**, *10*, 243–254. [CrossRef]

27. Byon, C.H.; Javed, A.; Dai, Q.; Kappes, J.C.; Clemens, T.L.; Darley-Usmar, V.M.; McDonald, J.M.; Chen, Y. Oxidative Stress Induces Vascular Calcification through Modulationo of the Osteogenic Transcription Factor Runx2 by AKT Signaling. *J. Biol. Chem.* **2008**, *283*, 15319–15327. [CrossRef]

28. Scott, M.; Vesely, I. Aortic Valve Cusp Microstructure: The Role of Elastin. *Ann. Thorac. Surg.* **1995**, *60*, S391–S394. [CrossRef]

29. O'Brien, K.D.; Reichenbach, D.D.; Marcovina, S.M.; Kuusisto, J.; Alpers, C.E.; Otto, C.M. Apolipoproteins B, (a), and E accumulate in the Morphologically Early Lesion of 'Degenerative' Valvular Aortic Stenosis. *Arterioscler. Thromb. Vasc. Biol.* **1996**, *16*, 523–532. [CrossRef]

30. Farrar, E.J.; Huntley, G.D.; Butcher, J. Endothelial-derived Oxidative Stress drives Myofibroblastic Activation and Calcification of the Aortic Valve. *PLoS ONE* **2015**, *10*, e0123257. [CrossRef]

31. Pawade, T.A.; Newby, D.E.; Dweck, M.R. Calcification in Aortic Stenosis: The Skeleton Key. *J. Am. Coll. Cardiol.* **2015**, *66*, 561–577. [CrossRef]

32. Helske, S.; Syväranta, S.; Lindstedt, K.A.; Lappalainen, J.; Oörni, K.; Mäyränpää, M.I.; Lommi, J.; Turto, H.; Werkkala, K.; Kupari, M.; et al. Increased Expression of Elastolytic Cathepsins S, K, and V and their Inhibitor Cystatin C in Stenotic Aortic Valves. *Arterioscler. Thromb. Vasc. Biol.* **2006**, *26*, 1791–1798. [CrossRef]

33. Kee, H.J.; Cho, S.N.; Kim, G.R.; Choi, S.Y.; Ryu, Y.; Kim, I.K.; Hong, Y.J.; Park, H.W.; Ahn, Y.; Cho, J.G.; et al. Gallic Acid Inhibits Vascular Calcification Through the Blockade of BMP2-Smad1/5/8 Signaling Pathway. *Vascul. Pharmacol.* **2014**, *63*, 71–78. [CrossRef]

34. Roman-Garcia, P.; Barrio-Vazquez, S.; Fernandez-Martin, J.L.; Ruiz-Torres, M.P.; Cannata-Andia, J.B. Natural Antioxidants and Vascular Calcification: A Possible Benefit. *J. Nephrol.* **2011**, *24*, 669–672. [CrossRef]

35. Lu, T.S.; Lim, K.; Molostvov, G.; Yang, Y.C.; Yiao, S.Y.; Zehnder, D.; Hsiao, L.L. Induction of Intracellular Heat-shock protein 72 prevents the Development of Vascular Smooth Muscle Cell Calcification. *Cardiovasc. Res.* **2012**, *96*, 524–532. [CrossRef]

36. Chao, C.-T.; Yeh, H.-Y.; Tsai, Y.-T.; Chuang, P.-H.; Yuan, T.-H.; Huang, J.-W.; Chen, H.-W. Natural and Non-Natural Antioxidative Compounds: Potential Candidates for Treatment of Vascular Calcification. *Cell Death Discov.* **2019**, *5*, 145. [CrossRef]

37. Menichini, D.; Alrais, M.; Liu, C.; Xia, Y.; Blackwell, S.C.; Facchinetti, F.; Sibai, B.M.; Longo, M. Maternal Supplementation of Inositols, Fucoxanthin, and Hydroxytyrosol in Pregnant Murine Models of Hypertension. *Am. J. Hypertens.* **2020**, *33*, 652–659. [CrossRef]

38. Chiang, Y.F.; Chen, H.Y.; Chang, Y.J.; Shih, Y.H.; Shieh, T.M.; Wang, K.L.; Hsia, S.M. Protective Effects of Fucoxanthin on High Glucose- and 4-Hydroxynonenal (4-HNE)-Induced Injury in Human Retinal Pigment Epithelial Cells. *Antioxidants (Basel)* **2020**, *9*, 1176. [CrossRef]
39. Ou, H.C.; Chou, W.C.; Chu, P.M.; Hsieh, P.L.; Hung, C.H.; Tsai, K.L. Fucoxanthin Protects Against oxLDL-Induced Endothelial Damage via Activating the AMPK-Akt-CREB-PGC1α Pathway. *Mol. Nutr. Food Res.* **2019**, *63*, e1801353. [CrossRef]
40. Casamián-Sorrosal, D.; Barrera-Chacón, R.; Fonfara, S.; Belinchón-Lorenzo, S.; Gómez-Gordo, L.; Galapero-Arroyo, J.; Fernández-Cotrina, J.; Cristobal-Verdejo, J.I.; Duque, F.J. Association of Myocardial Parasitic Load with Cardiac Biomarkers and Other Selected Variables in 10 Dogs with Advanced Canine Leishmaniasis. *Vet. Rec.* **2021**, e198. [CrossRef]
41. Lee, S.W.; Park, M.C.; Park, Y.B.; Lee, S.K. E/E′ Ratio is more Sensitive Than E/A Ratio for Detection of Left Ventricular Diastolic Dysfunction in Systemic Lupus Erythematosus. *LUPUS* **2008**, *17*, 195–201. [CrossRef] [PubMed]
42. Lin, C.; Zhu, D.; Markby, G.; Corcoran, B.M.; Farquharson, C.; Macrae, V.E. Isolation and Characterization of Primary Rat Valve Interstitial Cells: A New Model to Study Aortic Valve Calcification. *J. Vis. Exp.* **2017**, 56126. [CrossRef] [PubMed]

marine drugs

Article

NaCl Promotes the Efficient Formation of *Haematococcus pluvialis* Nonmotile Cells under Phosphorus Deficiency

Feng Li [1,2,3,4], Ning Zhang [1], Yulei Zhang [1], Qingsheng Lian [1], Caiying Qin [1], Zuyuan Qian [1], Yanqi Wu [1], Zhiyuan Yang [1], Changling Li [1], Xianghu Huang [1] and Minggang Cai [1,2,3,4,*]

[1] College of Fisheries, Guangdong Ocean University, Zhanjiang 524088, China; lifeng2318@gdou.edu.cn (F.L.); zhangn@gdou.edu.cn (N.Z.); zhangyl@gdou.edu.cn (Y.Z.); lianqingsheng@stu.gdou.edu.cn (Q.L.); tancaiying@stu.gdou.edu.cn (C.Q.); qianzuyuan@stu.gdou.edu.cn (Z.Q.); wuyanqi@stu.gdou.edu.cn (Y.W.); yangzhiyuan@stu.gdou.edu.cn (Z.Y.); licl@gdou.edu.cn (C.L.); huangxh@gdou.edu.cn (X.H.)
[2] Fujian Provincial Key Laboratory for Coastal Ecology and Environmental Studies, Xiamen University, Xiamen 361102, China
[3] Key Laboratory of Marine Chemistry and Applied Technology, Xiamen University, Xiamen 361102, China
[4] College of Ocean and Earth Science, Xiamen University, Xiamen 361102, China
* Correspondence: mgcai@xmu.edu.cn; Tel.: +86-0592-288-6188

Abstract: Natural astaxanthin helps reduce the negative effects caused by oxidative stress and other related factors, thereby minimizing oxidative damage. Therefore, it has considerable potential and broad application prospects in human health and animal nutrition. *Haematococcus pluvialis* is considered to be the most promising cell factory for the production of natural astaxanthin. Previous studies have confirmed that nonmotile cells of *H. pluvialis* are more tolerant to high intensity of light than motile cells. Cultivating nonmotile cells as the dominant cell type in the red stage can significantly increase the overall astaxanthin productivity. However, we know very little about how to induce nonmotile cell formation. In this work, we first investigated the effect of phosphorus deficiency on the formation of nonmotile cells of *H. pluvialis*, and then investigated the effect of NaCl on the formation of nonmotile cells under the conditions of phosphorus deficiency. The results showed that, after three days of treatment with 0.1% NaCl under phosphorus deficiency, more than 80% of motile cells had been transformed into nonmotile cells. The work provides the most efficient method for the cultivation of *H. pluvialis* nonmotile cells so far, and it significantly improves the production of *H. pluvialis* astaxanthin.

Keywords: *Haematococcus pluvialis*; astaxanthin; nonmotile cells; phosphorus deficiency

Citation: Li, F.; Zhang, N.; Zhang, Y.; Lian, Q.; Qin, C.; Qian, Z.; Wu, Y.; Yang, Z.; Li, C.; Huang, X.; et al. NaCl Promotes the Efficient Formation of *Haematococcus pluvialis* Nonmotile Cells under Phosphorus Deficiency. *Mar. Drugs* **2021**, *19*, 337. https://doi.org/10.3390/md19060337

Academic Editors: Masashi Hosokawa and Hayato Maeda

Received: 25 May 2021
Accepted: 10 June 2021
Published: 13 June 2021

1. Introduction

Astaxanthin is a keto-carotenoid that has a wide range of applications in aquaculture, food, cosmetics, and human health due to its strong antioxidant and coloring functions [1,2]. Currently, there are mainly two types of astaxanthin on the market: artificial synthetic astaxanthin and natural astaxanthin [3]. Artificial synthetic astaxanthin accounts for about 95% of the market and is mainly used in aquaculture [2]. Although synthetic astaxanthin has the advantages of lower cost and price, it has not been approved for human consumption due to possible safety issues [3]. In contrast, natural astaxanthin is significantly better than artificial synthetic astaxanthin in terms of stability, antioxidant activity, absorption effect and biosafety, and has been approved by China, the United States, Japan, and some EU countries for aquaculture, dietary supplements, cosmetic ingredients, and other uses [2,4]. However, due to technical limitations in the production of raw materials, the global production of natural astaxanthin is low, resulting in high market prices [5,6], and the retail prices of nutraceutical grade astaxanthin have even reached US $100,000 per kilogram [7].

Haematococcus pluvialis is the most competitive natural source for commercial astaxanthin production and the global annual production capacity is about 800 tons [8], which is still far behind the 10,000-ton for spirulina and 1000-ton for chlorella. The high cell death rate during the production process is the main reason for the low productivity of *H. pluvialis* astaxanthin [9,10]. The accumulation of biomass and the synthesis of astaxanthin are two important factors that must be considered in the production of astaxanthin in *H. pluvialis*. However, culture conditions suitable for rapid cell growth and culture conditions for astaxanthin accumulation are mutually exclusive [4]. A two-stage culture strategy is widely adopted by *H. pluvialis* industry [9], because it is the most effective strategy for solving the contradiction between cell fast growth and astaxanthin accumulation at present [11]. In actual production, however, when the vegetative cells are transferred from the first stage (green stage) to the second stage (red stage), a large number of cell deaths occur due to high intensity of light in combination with nutrient depletion stress. As a result, the overall astaxanthin productivity is very low [9]. Therefore, the reduction of cell mortality in the red stage has become the key to increasing the overall astaxanthin production of *H. pluvialis*.

The cells of *H. pluvialis* usually go through vegetative green stage, intermediate palmella stage and cyst stage, in which several types of cells are distinguished: motile cells, nonmotile palmella cells, and haematocysts (aplanospores) [4]. Previous studies found that the nonmotile cells of *H. pluvialis* were more tolerant to photooxidative stress than motile cells [12]. Using nonmotile cells as the main cell type for the astaxanthin production can significantly reduce the cell mortality and increase astaxanthin productivity in the red stage [13]. However, there are few reports on the induction of *H. pluvialis* nonmotile cells formation. In this study, we first investigated the effect of phosphorus deficiency on the formation of nonmotile cells of *H. pluvialis*, and then the effect of NaCl on the formation of nonmotile cells under the conditions of phosphorus deficiency. Our results showed that the addition of NaCl effectively promoted the formation of nonmotile cells under the condition of phosphorus deficiency. The work provides the most cost-efficient method for the preparation of *H. pluvialis* nonmotile cells so far, and this is of great significance for improving the production of *H. pluvialis* astaxanthin by using cell regulation technology.

2. Results

2.1. The Effect of Phosphorus Deficiency on the Formation of Nonmotile Cells of H. pluvialis

As shown in Figure 1, the total number of cells and the number of nonmotile cells in the two experimental groups both showed a trend of increasing with time. After 9 days of cultivation, the total number of cells and nonmotile cells in the two groups both reached the maximum value. The total number of cells in the control group reached 147.5×10^4 cells mL^{-1}, which was about 59% higher than that in the P-deficiency group. The maximum number of nonmotile cells in the P-deficiency group was 37.5×10^4 cells mL^{-1}, which was 4.17 times that of the control group. We also calculated the daily percentage growth rate of nonmotile cells, and the results showed that the percentage growth rate of nonmotile cells under phosphorus deficiency condition increased by an average of about 4.5% per day, which was 6.7 times that under normal conditions, indicating that phosphorus deficiency can significantly promote the formation of *H. pluvialis* nonmotile cells.

As shown in Figure 2 and Table 1, after 9 days of phosphorus deficiency treatment, more than 40% of the motile cells have been transformed into nonmotile cells. During this process, red pigmentation due to astaxanthin accumulation appears towards the center of nonmotile cells. By comparison, most of the cells in the control group are still motile cells in green color, with only about 6% nonmotile cells. In addition, some dead cells were observed in P-deficiency group and the cell mortality reached 9.4%, which was about 5.5% higher than that of the control group, indicating that phosphorus deficiency can increase the cell death rate.

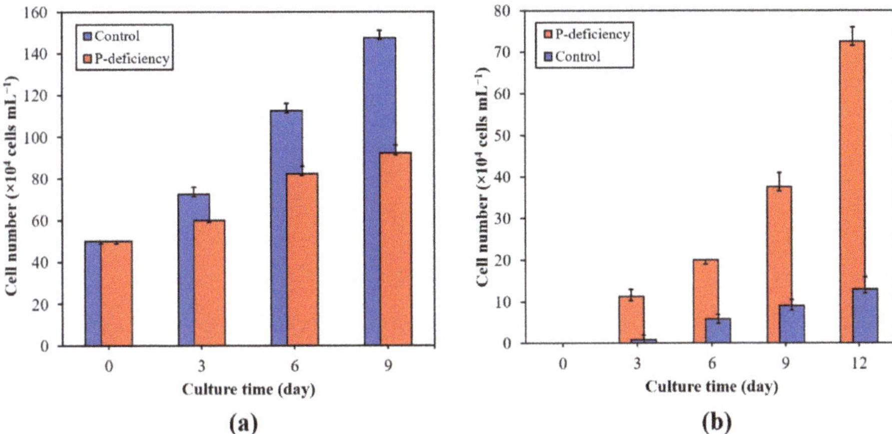

Figure 1. Changes on the total cell number (**a**) and nonmotile cells number (**b**) of *H. pluvialis* in control- and P-deficiency treatment group. The results were presented as mean + SD.

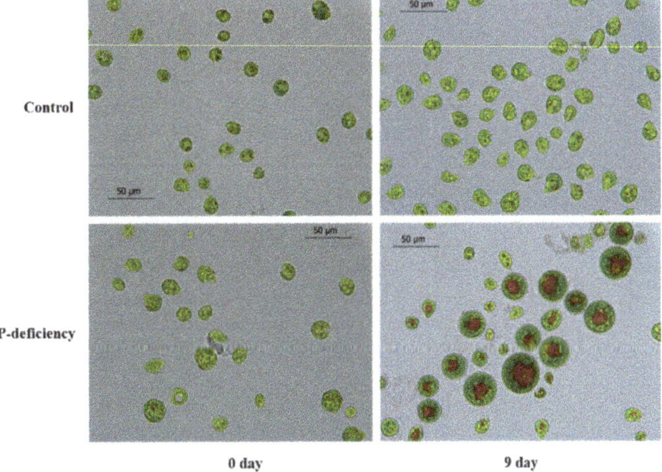

Figure 2. The cell morphology of *H. pluvialis* on day 0 and day 9 in control- and P-deficiency treatment groups.

Table 1. The percentage of nonmotile cells, daily percentage growth rate of nonmotile cells, and cell mortality of control- and P-deficiency treatment groups.

Parameters	Control Group	P-Deficiency Treatment
The percentage of nonmotile cells (%)	6.1	40.5
Daily percentage growth rate of nonmotile cells (% day^{-1})	0.67	4.50
Cell mortality (%)	3.9	9.4

2.2. The Effect of Adding NaCl on the Formation of Nonmotile Cells under Phosphorus Deficiency Condition

Under phosphorus deficiency conditions, adding different concentrations of NaCl has significant effects on the growth, cell morphology, nonmotile cells formation rate, and cell mortality of *H. pluvialis*. As shown in Figure 3a, the total number of cells in both 0.1% NaCl and 0.4% NaCl treatment groups increased first, followed by a decrease, and then increased again. The total number of cells in the 0.2% NaCl treatment group also showed this trend. After 72 h of treatment, the total number of cells in the 0.1% NaCl and 0.2% NaCl treatment groups reached the maximum value, which were 75.0×10^4 cells mL^{-1} and 70.0×10^4 cells mL^{-1}, respectively. The maximum total cell number in the 0.4% NaCl treatment group appeared at the 12th h, and the value was 69.25×10^4 cells mL^{-1}.

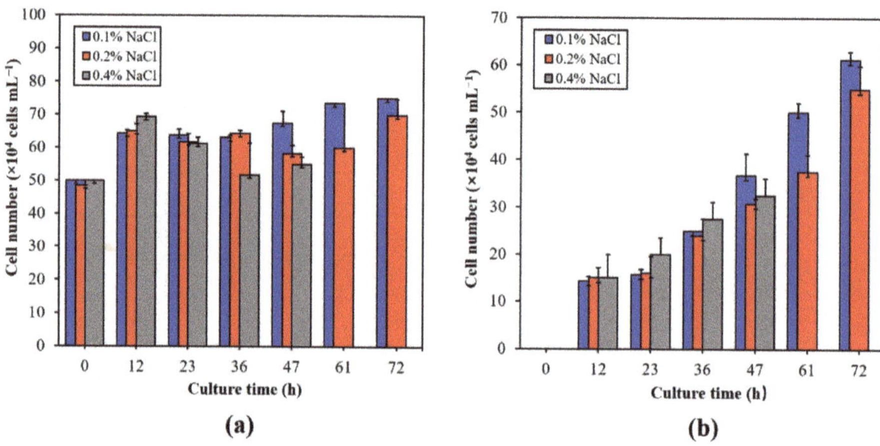

Figure 3. Changes on the total cell number (**a**) and nonmotile cells number (**b**) of *H. pluvialis* in 0.1%, 2%, and 0.4% NaCl treatment groups. The data at 61 h and 72 h in the 0.4% NaCl treatment group are not shown due to cell adhesion has affected the accurate determination of cell number. The results were presented as mean + SD.

It can be seen from Figure 3b that the number of nonmotile cells in the 0.1% NaCl and 0.2% NaCl treatment groups showed a rapid increase after 23 h of treatment. After 72 h of treatment, the number of nonmotile cells in the 0.1% NaCl treatment group reached 61.25×10^4 cells mL^{-1}, which was about 11.4% higher than that in the 0.2% NaCl treatment group. The increase of nonmotile cells number in the 0.4% NaCl treatment group was the fastest among the three groups within 36 h of treatment, and then the increase rate slowed down. After 47 h of treatment, cell adhesion occurred in the 0.4% NaCl treatment group and accompanied by a large number of cell deaths, which made it impossible to accurately determine the cell number. Therefore, the data of the 0.4% NaCl treatment group at 61 h and 72 h were not shown in Figure 3.

As shown in Figure 4 and Table 2, after 72 h of treatment, 81.7% of the motile cells in the 0.1% NaCl treatment group had been transformed into green nonmotile cells, the daily percentage growth rate of nonmotile cells reached 27.2% day^{-1}, and the cells was in good shape. In the 0.2% NaCl treatment group, after 36 h of treatment, the cell color gradually changed from green to orange-green due to the accumulation of carotenoids. After 72 h of treatment, 78.6% of the motile cells transformed into nonmotile cells. The daily percentage growth rate of nonmotile cells was 26.2% day^{-1}, and 3.4% of the cells died due to stress.

When the concentration of NaCl added to the phosphorus deficiency medium reached 0.4% (w/v), it was observed that some cells were damaged after 23 h of treatment. After 36 h of treatment, carotenoids began to accumulate inside the cells. After treatment for 47 h, cell adhesion occurred and over 38% of the cells died (Table 2). By comparison, the cell mortality in 0.1% NaCl and 0.2% NaCl treatment groups were only 1.8% and 3.4%, indicating that high concentrations of NaCl can significantly increase cell mortality rate.

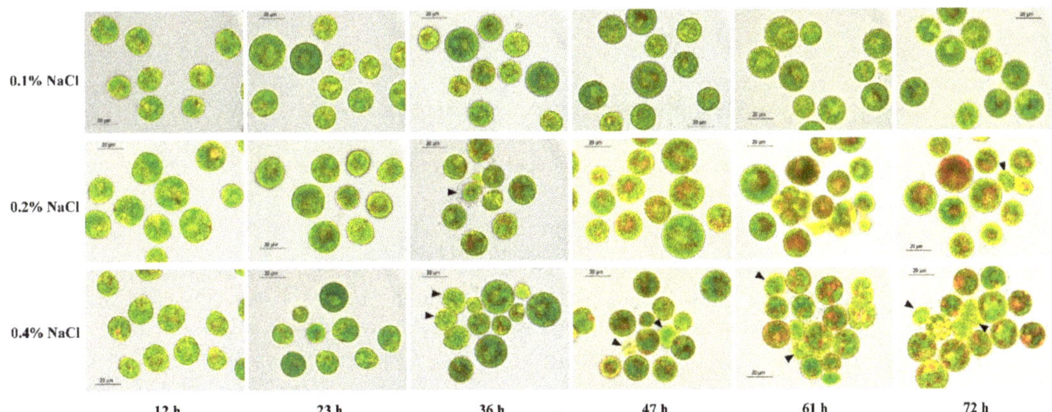

Figure 4. Morphological changes of *H. pluvialis* cells in 0.1%, 0.2%, and 0.4% NaCl treatment groups. The damaged or dead cells are indicated by arrows.

Table 2. The percentage of nonmotile cells, daily percentage growth rate of nonmotile cells, and cell mortality in 0.1, 0.2, and 0.4 NaCl treatment groups.

Parameters	0.1% NaCl	0.2% NaCl	0.4% NaCl
The percentage of nonmotile cells (%)	81.7 [1]	78.6 [1]	59.1 [2]
Daily percentage growth rate of nonmotile cells (% day^{-1})	27.2	26.2	29.5
Cell mortality (%)	1.8 [2]	3.4 [3]	38.2 [2]

[1] Obtained after 72 h of treatment. [2] Obtained after 47 h of treatment. [3] Obtained after 61 h of treatment.

3. Materials and Methods

3.1. Algal Strain and Culture Conditions

H. pluvialis CCMA-451 was obtained from the Center for Collections of Marine Algae in Xiamen University, China. The motile cells were grown photoautotrophically at 20 μmol photons m^{-2} s^{-1} in liquid Bold Basal Medium (BBM) with 3 times NaNO$_3$.

The 5-day-old green motile cells were collected by centrifugation (2000 rpm, 2 min), and transferred into a fresh phosphorus-free BBM medium at an initial optical density of 0.5 (OD$_{680}$). Then NaCl was added to the cultures to adjust the concentrations to 1%, 2%, and 4% (*m/v*). All experiments were performed in triplicate in a 1-L glass columns (inner diameter 5 cm) at 25 ± 1 °C under continuous illumination (30 μmol photons m^{-2} s^{-1}) for 3 days. Culture mixing was provided continuously by bubbling of filtered air enriched with 1.5% (*v/v*) CO$_2$ at a flow rate of 100 mL min^{-1}.

3.2. Morphological Observation

The morphological changes of the algal cells were observed using a Leica DM750 light microscope (Leica Microsystems, Wetzlar, Germany) and photos were taken with a Leica ICC50 W camera (Leica Microsystems, Wetzlar, Germany).

3.3. Determination of Cell Number

The samples were fixed with Lugol's iodine solution first. Then, cell numbers were counted using a Neubauer improved cell counting chamber under Leica DM750 light microscope and measured as cells mL^{-1}.

The daily percentage growth rate of nonmotile cells (% day^{-1}) was calculated as follows:

$$\text{Daily percentage growth rate of nonmotile cells} \left(\% \text{ day}^{-1}\right) = \frac{\frac{C_{Nt}}{C_{Tt}} \times 100\%}{t} \tag{1}$$

where C_{Nt} and C_{Tt} were the nonmotile cells number and total cells number on day t, respectively.

3.4. Statistical Analysis

Statistical analysis was performed with the SPSS for Windows statistical software package (IBM SPSS v22.0, Inc., 2010; Chicago, IL, USA). The difference was considered significant when $p < 0.05$ and the results were presented as mean + SD.

4. Discussion

The goal of improving the cultivation technology of *H. pluvialis* is to maximize the production of biomass and astaxanthin. Achieving cell growth and astaxanthin accumulation simultaneously under the same cultural conditions is difficult. The major reason for this is that the rapid growth of *H. pluvialis* requires favorable environmental conditions [14–16] and the accumulation of astaxanthin requires unfavorable cultural conditions [11,17–19]. The two-stage cultivation strategy has effectively solved the contradiction between cell growth and astaxanthin accumulation. However, it fails to solve the high cell mortality during the red stage, so this technology still has a lot of room for improvement.

Previous studies have confirmed that the motile cells are susceptible to photooxidative stress and tend to die under high intensity of light conditions [9,12,20]. Compared with motile cells, nonmotile cells with thick cell walls are not only more tolerant to stress [12], but also more effectively use chemical energy accumulated during cell transformation to rapidly synthesize and accumulate astaxanthin under stress conditions [12,21,22]. Li et al. [20] suggested that using nonmotile cells as the dominant cell type in the red stage can reduce cell mortality more than 70%, while biomass and astaxanthin content can increase up to 2.12 times and 3.5 times, respectively. Therefore, by inducing motile cells to transform into nonmotile cells with more tolerance and astaxanthin accumulation ability before entering the red stage, the problem of high cell mortality is expected to be effectively solved.

The formation of nonmotile cells of *H. pluvialis* is accompanied by encystment, which is considered to be a self defense mechanism [22–24]. Previous studies have shown that the encystment of *H. pluvialis* may be related to carbon assimilation [23,25] and phosphorus plays a key role in photosynthetic carbon assimilation [26]. The phosphate translocator is an important structure in the chloroplast membrane, which can transport triose phosphate (the first sugar produced in photosynthesis) produced during the Calvin cycle to the cytoplasm for the synthesis of sucrose, and at the same time transport the released inorganic phosphate (Pi) to chloroplast matrix. When phosphorus is deficient, the Pi exchanged with triose phosphate decreases, resulting in a reduction of Pi in the chloroplast and a decrease in the ATP/ADP ratio, which affects the transport in the C3 pathway and the progress of related reactions, thereby reducing the photosynthetic rate. At the same time, triose phosphate is converted into starch due to phosphorus deficiency and stored in the chloroplast. The reduction of triose phosphate transported from the chloroplast to the cytosol further affects the synthesis of sucrose, thereby inhibiting the photosynthesis of algae and promoting the formation of encystment [14] to maintain the cell integrity, structure, and function. Although phosphorus deficiency can promote the formation of nonmotile cells, the efficiency is low. According to our results, the addition of NaCl under phosphorus deficiency conditions significantly promotes the formation of nonmotile cells. It made the daily percentage growth rate of nonmotile cells increase from 4.5% to more than 26%. We speculate that NaCl promotes the formation of nonmotile cells under the condition of phosphorus deficiency owing to the following reasons. (1) The increase of NaCl concentration affects the oxygen metabolism in algal cells, accelerates the production of reactive oxygen species (ROS), and reduces the function of the scavenging system, which leads to the accumulation of ROS in the cells. (2) Higher salinity increases the osmotic pressure and promotes loss of water from cells, thereby affecting the normal metabolic activities of the cells. (3) The increase of NaCl concentration affects the ion homeostasis in the cells, thereby causing nutritional stress. Thus, it is necessary to carry out further physiological and biochemical studies.

We have to point out that we did not use algae cultured without NaCl for 72 h as a control. However, it doesn't seem to affect our conclusion that adding NaCl to phosphorus-deficiency medium can significantly promote the formation efficiency of *H. pluvialis* non-motile cells. We have also determined the most efficient induction method (synergistic induction by phosphorus deficiency and 0.1% NaCl) of nonmotile cells. After three days of cultured under the optimal conditions developed in this study, more than 80% of motile cells transformed into nonmotile cells. The astaxanthin production efficiency of these non-motile cells should be evaluated by extraction and HPLC in future. This will be profitable for the *Haematococcus* industry.

Author Contributions: Data curation, N.Z.; formal analysis, N.Z., Y.Z. and F.L.; funding acquisition, M.C.; investigation, F.L., Q.L., C.Q., Z.Q., Y.W. and Z.Y.; methodology, N.Z. and Y.Z.; project administration, M.C.; resources, X.H. and C.L.; writing—original draft, F.L.; writing—review & editing, F.L. and M.C. All authors have read and agreed to the published version of the manuscript.

Funding: This research was mainly supported by Xiamen Southern Ocean Technology Center of China under Grant (number 14CZP035HJ09) and partly funded by Xiamen Scientific and Technologic Projects under Grant (number 3052Z20031086 and 3052Z20123004); Marine Science Base Scientific Research Training and Scientific Research Ability Enhancement Project of Xiamen University under Grant (number J1210050); Xiamen University Training Program of Innovation and Entrepreneurship for Undergraduates under Grant (number 2016X0619); and the Special Funds for Scientific Research of Marine Public Welfare Industry under Grant (number 201305016).

Conflicts of Interest: The authors declare no conflict of interest.

References

1. Dave, D.; Liu, Y.; Pohling, J.; Trenholm, S.; Murphy, W. Astaxanthin recovery from Atlantic shrimp (*Pandalus borealis*) processing materials. *Bioresour. Technol. Rep.* **2020**, *11*, 100535. [CrossRef]
2. Li, X.; Wang, X.Q.; Duan, C.L.; Yi, S.S.; Gao, Z.Q.; Xiao, C.W.; Agathos, S.N.; Wang, G.C.; Li, J. Biotechnological production of astaxanthin from the microalga *Haematococcus pluvialis*. *Biotechnol. Adv.* **2020**, *43*, 107602. [CrossRef] [PubMed]
3. Capelli, B.; Bagchi, D.; Cysewski, G.R. Synthetic astaxanthin is significantly inferior to algal-based astaxanthin as an antioxidant and may not be suitable as a human nutraceutical supplement. *Nutrafoods* **2013**, *12*, 145–152. [CrossRef]
4. Shah, M.M.R.; Liang, Y.M.; Chen, J.J.; Daroch, M. Astaxanthin-producing green microalga *Haematococcus pluvialis*: From single cell to high value commercial products. *Front. Plant Sci.* **2016**, *7*, 531. [CrossRef] [PubMed]
5. Molino, A.; Mehariya, S.; Iovine, A.; Larocca, V.; Sanzo, G.D.; Martino, M.; Casella, P.; Chianese, S.; Musmarra, D. Extraction of astaxanthin and lutein from microalga *Haematococcus pluvialis* in the red phase using CO_2 supercritical fluid extraction technology with ethanol as co-solvent. *Mar. Drugs* **2018**, *16*, 432. [CrossRef]
6. Pereira, S.; Otero, A. *Haematococcus pluvialis* bioprocess optimization: Effect of light quality, temperature and irradiance on growth, pigment content and photosynthetic response. *Algal Res.* **2020**, *51*, 102027. [CrossRef]
7. Oslan, S.N.H.; Shoparwe, N.F.; Yusoff, A.H.; Rahim, A.A.; Chang, C.S.; Tan, J.S.; Oslan, S.N.; Arumugam, K.; Ariff, A.B.; Sulaiman, A.Z.; et al. A review on *Haematococcus pluvialis* bioprocess optimization of green and red stage culture conditions for the production of natural astaxanthin. *Biomolecules* **2021**, *11*, 256. [CrossRef]
8. Gao, F.Z.; Ge, B.S.; Xiang, W.Z.; Qin, S. Development of microalgal industries in the past 60 years due to biotechnological research in China: A review. *Sci. Sin. Vitae* **2021**, *51*, 26–39. (In Chinese)
9. Wang, J.F.; Han, D.X.; Sommerfeld, M.R.; Lu, C.M.; Hu, Q. Effect of initial biomass density on growth and astaxanthin production of *Haematococcus pluvialis* in an outdoor photobioreactor. *J. Appl. Phycol.* **2013**, *25*, 253–260. [CrossRef]
10. Li, F.; Cai, M.G.; Lin, M.W.; Huang, X.H.; Wang, J.; Ke, H.W.; Wang, C.H.; Zheng, X.H.; Chen, D.; Yang, S.H. Enhanced biomass and astaxanthin production of *Haematococcus pluvialis* by a cell transformation strategy with optimized initial biomass density. *Mar. Drugs* **2020**, *18*, 341. [CrossRef]
11. Fábregas, J.; Otero, A.; Maseda, A.; Domínguez, A. Two-stage cultures for the production of astaxanthin from *Haematococcus pluvialis*. *J. Biotechnol.* **2001**, *89*, 65–71. [CrossRef]
12. Han, D.X.; Wang, J.F.; Sommerfeld, M.; Hu, Q. Susceptibility and protective mechanisms of motile and non-motile cells of *Haematococcus pluvialis* (Chlorophyceae) to photooxidative stress. *J. Phycol.* **2012**, *48*, 693–705. [CrossRef]

13. Li, F.; Cai, M.G.; Lin, M.W.; Huang, X.H.; Wang, J.; Zheng, X.H.; Wu, S.T.; An, Y. Accumulation of astaxanthin was improved by the nonmotile cells of *Haematococcus pluvialis*. *BioMed Res. Int.* **2019**, *2019*, 8191762. [CrossRef] [PubMed]

14. Boussiba, S. Carotenogenesis in the green alga *Haematococcus pluvialis*: Cellular physiology and stress response. *Physiol. Plant.* **2000**, *108*, 111–117. [CrossRef]

15. Aflalo, C.; Meshulam, Y.; Zarka, A.; Boussiba, S. On the relative efficiency of two-vs. one-stage production of astaxanthin by the green alga *Haematococcus pluvialis*. *Biotechnol. Bioeng.* **2007**, *98*, 300–305. [CrossRef]

16. Río, E.D.; Acién, F.G.; García-Malea, M.C.; Rivas, J.; Molina-Grima, E.; Guerrero, M.G. Efficiency assessment of the one-step production of astaxanthin by the microalga *Haematococcus pluvialis*. *Biotechnol. Bioeng.* **2008**, *100*, 397–402.

17. Orosa, M.; Franqueira, D.; Cid, A.; Abalde, J. Analysis and enhancement of astaxanthin accumulation in *Haematococcus pluvialis*. *Bioresour. Technol.* **2005**, *96*, 373–378. [CrossRef]

18. He, P.; Duncan, J.; Barber, J. Astaxanthin accumulation in the green alga *Haematococcus pluvialis*: Effects of cultivation parameters. *J. Integr. Plant Biol.* **2007**, *49*, 447–451. [CrossRef]

19. Choi, Y.E.; Yun, Y.S.; Park, J.M.; Yang, J.W. Determination of the time transferring cells for astaxanthin production considering two-stage process of *Haematococcus pluvialis* cultivation. *Bioresour. Technol.* **2011**, *102*, 11249–11253. [CrossRef]

20. Li, F.; Cai, M.G.; Lin, M.W.; Huang, X.H.; Wang, J.; Ke, H.W.; Zheng, X.H.; Chen, D.; Wang, C.H.; Wu, S.T.; et al. Differences between motile and nonmotile cells of *Haematococcus pluvialis* in the production of astaxanthin at different light intensities. *Mar. Drugs* **2019**, *17*, 39. [CrossRef] [PubMed]

21. Boussiba, S.; Vonshak, A. Astaxanthin accumulation in the green alga *Haematococcus pluvialis*. *Plant Cell Physiol.* **1991**, *32*, 1077–1082. [CrossRef]

22. Kobayashi, M.; Kurimura, Y.; Tsuji, Y. Light-independent, astaxanthin production by the green microalga *Haematococcus pluvialis* under salt stress. *Biotechnol. Lett.* **1997**, *19*, 507–509. [CrossRef]

23. Kakizono, T.; Kobayashi, M.; Nagai, S. Effect of carbon/nitrogen ratio on encystment accompanied with astaxanthin formation in a green alga, *Haematococcus pluvialis*. *J. Ferment. Bioeng.* **1992**, *74*, 403–405. [CrossRef]

24. Kobayashi, M.; Kakizono, T.; Nagai, S. Enhanced carotenoid biosynthesis by oxidative stress in acetate-induced cyst cells of a green unicellular alga, *Haematococcus pluvialis*. *Appl. Environ. Microbiol.* **1993**, *59*, 867–873. [CrossRef] [PubMed]

25. Droop, M.R. Some factors governing encystment in *Haematococcus pluvialis*. *Arch. Für Mikrobiol.* **1955**, *21*, 267–272. [CrossRef] [PubMed]

26. Rychter, A.M.; Rao, I.M. Role of phosphorus in photosynthetic carbon metabolism. *Handb. Photosynth.* **2005**, *2*, 123–148.

Article

Distribution of the Water-Soluble Astaxanthin Binding Carotenoprotein (AstaP) in Scenedesmaceae

Hiroki Toyoshima [1], **Ami Miyata** [1], **Risako Yoshida** [1], **Taichiro Ishige** [2], **Shinichi Takaichi** [3] **and Shinji Kawasaki** [1,3,*]

[1] Department of Bioscience, Tokyo University of Agriculture, 1-1-1 Sakuragaoka, Setagaya-ku, Tokyo 156-8502, Japan; hiroki.toyoshima91@gmail.com (H.T.); e-perfect7-x.hbk@i.softbank.jp (A.M.); lisakoyosida@icloud.com (R.Y.)

[2] NODAI Genome Research Centre, Tokyo University of Agriculture, 1-1-1 Sakuragaoka, Setagaya-ku, Tokyo 156-8502, Japan; t3ishige@nodai.ac.jp

[3] Department of Molecular Microbiology, Tokyo University of Agriculture, 1-1-1 Sakuragaoka, Setagaya-ku, Tokyo 156-8502, Japan; st206165@nodai.ac.jp

* Correspondence: kawashin@nodai.ac.jp; Tel.: +81-3-5477-2764

Abstract: Photooxidative stress-inducible water-soluble astaxanthin-binding proteins, designated as AstaP, were identified in two Scenedesmaceae strains, *Coelastrella astaxanthina* Ki-4 and *Scenedesmus obtusus* Oki-4N; both strains were isolated under high light conditions. These AstaPs are classified as a novel family of carotenoprotein and are useful for providing valuable astaxanthin in water-soluble form; however, the distribution of AstaP orthologs in other microalgae remains unknown. Here, we examined the distribution of AstaP orthologs in the family Scenedesmaceae with two model microalgae, *Chlamydomonas reinhardtii* and *Chlorella variabilis*. The expression of AstaP orthologs under photooxidative stress conditions was detected in cell extracts of Scenedesmaceae strains, but not in model algal strains. Aqueous orange proteins produced by Scenedesmaceae strains were shown to bind astaxanthin. The protein from *Scenedesmus costatus* SAG 46.88 was purified. It was named ScosAstaP and found to bind astaxanthin. The deduced amino acid sequence from a gene encoding ScosAstaP showed 62% identity to Ki-4 AstaP. The expression of the genes encoding AstaP orthologs was shown to be inducible under photooxidative stress conditions; however, the production amounts of AstaP orthologs were estimated to be approximately 5 to 10 times lower than that of Ki-4 and Oki-4N.

Keywords: astaxanthin; *Scenedesmus*; *Coelastrella*; carotenoprotein; astaxanthin-binding protein

Citation: Toyoshima, H.; Miyata, A.; Yoshida, R.; Ishige, T.; Takaichi, S.; Kawasaki, S. Distribution of the Water-Soluble Astaxanthin Binding Carotenoprotein (AstaP) in Scenedesmaceae. *Mar. Drugs* 2021, 19, 349. https://doi.org/10.3390/md19060349

Academic Editors: Masashi Hosokawa and Hayato Maeda

Received: 19 May 2021
Accepted: 18 June 2021
Published: 20 June 2021

Publisher's Note: MDPI stays neutral with regard to jurisdictional claims in published maps and institutional affiliations.

1. Introduction

Under the unfavorable conditions combined with high light irradiation, photosynthetic organisms feel photooxidative stress [1,2]. Plants are known to use carotenoids to dissipate excess light energy. Carotenoids are distributed in many organisms, such as chloroplasts of plants and algae, animal skins, and fish eggs, and more than 800 natural carotenoids have been identified [3]. They are hydrophobic; however, only a few are known to be water-soluble by binding proteins. Water-soluble carotenoproteins have been identified, such as astaxanthin-binding crustacyanin from crustaceans [4,5], zeaxanthin-binding GSTP1 (glutathione S-transferase like protein) from the human eye [6], lutein-binding protein from silkworm [7], cyanobacterial orange carotenoid protein (OCP) [8], and eukaryotic microalgal AstaP (astaxanthin binding protein) [9,10]. These water-soluble carotenoproteins did not show structural relationships with each other [9].

In photosynthetic organisms, water-soluble carotenoproteins (WSCPs) have been well characterized in cyanobacteria named orange carotenoid protein (OCP) [8,11,12]. In our previous study, two types of novel WSCPs were identified from the eukaryotic microalgae, *Coelastrella astaxanthina* Ki-4 and *Scenedesmus obtusus* Oki-4N [9,10]. These two strains were shown to produce novel astaxanthin-binding water-soluble proteins. Both AstaPs belonged to the fasciclin protein family. This family is characterized as secreted and cell surface proteins; however, to our knowledge, none of the proteins have been reported to bind lipids, including carotenoids. Based on the photooxidative stress inducible profiles and the potent activities toward $^1O_2$ quenching with astaxanthin binding, these novel types of astaxanthin-binding proteins are proposed to be involved in a unique function of photooxidative stress protection in plants. In this study, we investigated the distribution and characteristics of WSCPs in taxonomically related Scenedesmaceae strains.

2. Results

2.1. Effect of Photooxidative Stresses on the Tested Strains

Scenedesmaceae strains were obtained from culture collections, as described previously [11]. *S. obtusus* and *C. striolata* were selected because they are the type species of the genera *Scenedesmus* and *Coelastrella*, respectively. Other strains were chosen as the relatively well-used research strains in the genus *Coelastrella*. According to the locality information in the literature, these strains were isolated from non-stressed environmental conditions, such as lakeside and peat bog [13]. The test strains showed good growth in the medium used for the Ki-4 and Oki-4N strains [10]; therefore, these strains were subjected to photooxidative stress after growing under non-stressed conditions.

Strains Ki-4 and Oki-4N showed the optimum expression level of AstaP when they were grown under 0.7 M and 0.5 M NaCl stress with high light exposure conditions (800 µmol photons m^{-2} s^{-1}), respectively, where the color of cells changed from green to orange [9,10]. The tested Scenedesmaceae strains' cell color turned from green to white under the same stress conditions as Ki-4; therefore, we examined the upper limits of salt concentrations for each strain under the same high light intensity. Figure 1 shows the change in cell color of the strains after one to two weeks from the start of stress treatments using the upper limit of salt concentrations for each strain under high light exposure conditions. *S. obtusus* and *S. obliquus* turned green to white under 0.3 M NaCl and 0.25 M NaCl with high light exposure conditions; therefore, these strains were stressed by 0.25 M and 0.2 M NaCl, respectively, under the high light exposure conditions. Other strains were stressed with 0.5 M NaCl (for *C. vacuolata*) or 0.4 M NaCl (for *S. costatus* and *C. striolata*) with high light exposure.

A 1.0 g aliquot of wet cells stressed by salt with a high light exposure was suspended in 9.0 mL of 50 mM Tris-HCl buffer, pH 7.5. Cells were broken by a bead beater instrument, and the aqueous supernatants were obtained after ultracentrifugation. Strain Ki-4 and Oki-4N produced orange supernatants with broad absorption peak at 484 nm (Table 1) [9,10]. Except for *S. obliquus*, the test strains produced pale orangish or yellowish supernatants. The color of aqueous supernatants from the Scenedesmaceae strains were difficult to detect; therefore, the supernatants were concentrated by using a protein concentrator, and the colors and the absorption spectrum for each strain were shown in Figure 2. The amount of pigment produced per aliquot of cells from each test strain was estimated to be less than 5–10 times lower than that of Ki-4 and Oki-4N (Table 1).

These orange supernatants were passed through gel-filtration column chromatography, and two major peaks were detected (Figure 3). The first peaks from each test strain were detected around the HPLC retention time of 18 min, which were estimated to have a molecular weight greater than 670 kDa, and the absorption peaks were observed at approximately 454, 484, and 674 nm. These data indicated that the first peaks contained chlorophyll, which resembled that of the first peak of the strain Oki-4N described in our previous study [10]. The second major peak in each strain was detected at the HPLC retention time of 28 min (*S. obtusus*) or 30–35 min, which showed a broad absorption peak at 484 nm. In our previous study, AstaP orthologs were classified into two groups: glycosylated large-sized AstaPs and non-glycosylated small-sized AstaPs. The difference in the second peak retention time indicated variations in molecular weight among the test strains.

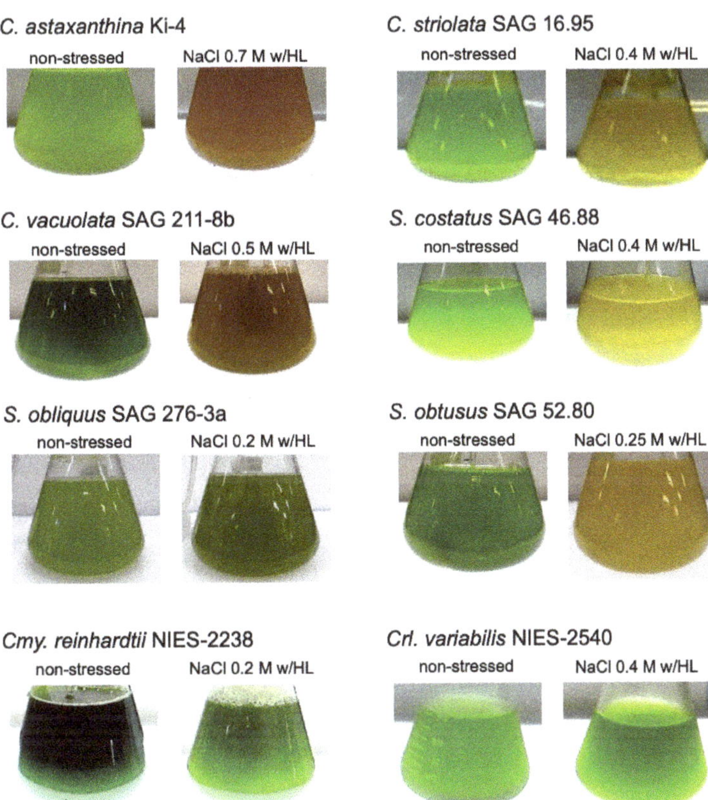

Figure 1. Survival of test strains under salt stress conditions with high light exposure. Salt concentrations for each test strains were decided based on the upper limit of salt concentrations with high light (w/HL, 800 µmol photons m^{-2} s^{-1}). Each test strain was cultivated under non-stressed conditions and stressed by adding salt w/HL for one to two weeks as described in Materials and Methods.

Table 1. Measurements of the amount of aqueous pigments after ultracentrifugation of cell extracts.

Species and Strain	Stress Conditions (with/ HL)	OD$_{480}$ (1 g Wet Cell/10 mL)	Reference
Coelastrella astaxanthina Ki-4	0.7 M NaCl	1.3	[9]
Scenedesmus obtusus Oki-4N	0.5 M NaCl	1.0	[10]
Coelastrella striolata SAG 16.95	0.4 M NaCl	0.14	This study
Coelastrella vacuolata SAG 211-8b	0.5M NaCl	0.12	This study
Scenedesmus costatus SAG 46.88	0.4 M NaCl	0.16	This study
Scenedesmus obliquus SAG 276-3a	0.2 M NaCl	0.02	This study
Scenedesmus obtusus SAG 52.80	0.25 M NaCl	0.15	This study
Chlamydomonas reinhardtii NIES-2238	0.2 M NaCl	0.04	This study
Chlorella variabilis NIES-2540	0.4 M NaCl	0.33	This study

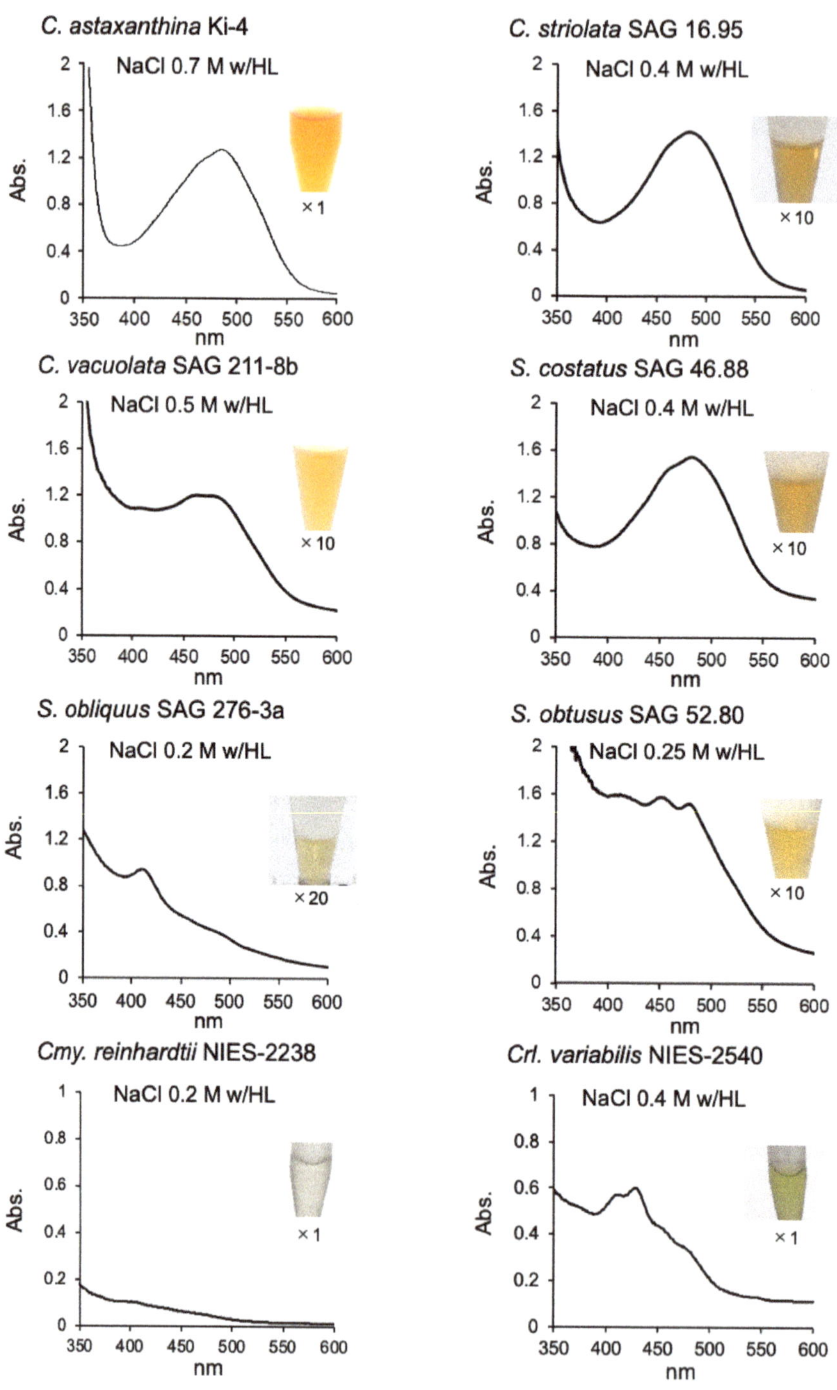

Figure 2. Absorption spectrum and the color of aqueous cell extracts. Aqueous cell extracts from photooxidative stressed cells were obtained after ultracentrifugation and concentrated by using a protein concentrator. ×1, ×10, ×20 means the rate of concentration by using a protein concentrator.

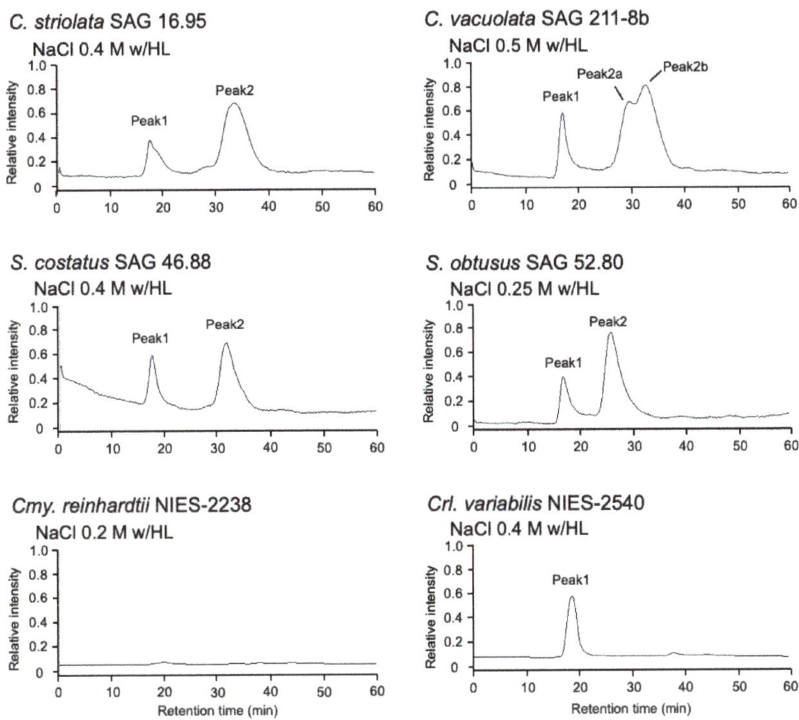

Figure 3. Separation of carotenoid pigments by gel-filtration column chromatography. Aqueous cell extracts obtained after ultracentrifugation were loaded into the gel-filtration column chromatography.

The protein fractions from the second peak in each test strain were collected, and the binding pigments were extracted as described previously [9,10]. The elution profiles were compatible with those from the strains Ki-4 and Oki-4N (Figure 4), and the pigments from each strain were identified to be astaxanthin and adonixanthin based on C_{18}-HPLC retention time and absorption spectrum. These results indicated that the test strains also produced AstaP-like water-soluble proteins under the photooxidative stress conditions.

2.2. Effect of Photooxidative Stress Response on Model Microalgae

In our previous study, AstaP orthologs are found in the genome annotation data of *Chlamydomonas reinhardtii* and *Chlorella variabilis* [8]. In this study, we analyzed the production of aqueous pigments from two strains under photooxidative stress conditions. *Cmy. reinhardtii* and *Crl. variabilis* turned green to white under 0.25 M and 0.5 M NaCl conditions with high light intensity, respectively. Therefore, photooxidative stresses were archived under 0.2 M and 0.4 M NaCl, respectively, with high light for six days. The cell-free extracts from photooxidative-stressed cells were obtained by a bead beater, ultracentrifuged, and their color was pale yellow. A yellow fraction was detected using gel filtration HPLC of the aqueous supernatant at 18 min in *Crl. variabilis* but not in *Cmy. reinhardtii*. The yellow fraction of *Crl. variabilis* showed absorption peaks at 454, 484, and 674 nm, indicating the presence of chlorophyll-binding protein similar to that of other Scenedesmaceae strains. These data suggested that the production of aqueous carotenoid-binding protein was not detectable in these model microalgae.

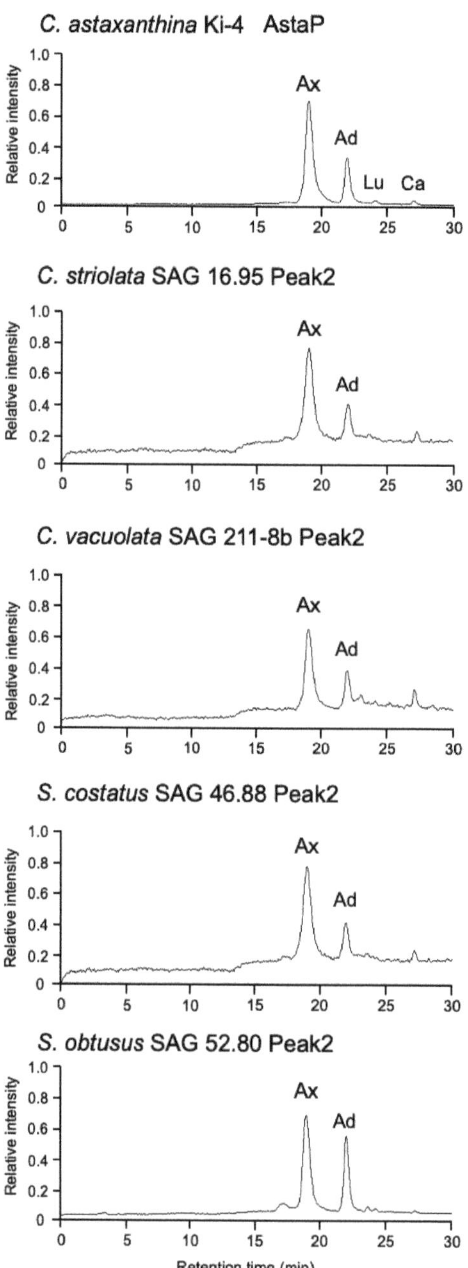

Figure 4. C_{18}-HPLC elution profiles of the binding carotenoids from the collected Peak2 fractions after gel filtration column chromatography. Ax, astaxanthin; Ad, adonixanthin; Lt, lutein; and Ca, canthaxanthin.

2.3. Purification and Characterization of Water-Soluble Astaxanthin-Binding Protein from S. costatus SAG 46.88

In order to characterize the existence of AstaP orthologs in the test strains, we chose *S. costatus* for further purification of astaxanthin binding protein because of its relatively high expression level (Table 1). The aqueous orange fraction obtained by gel-filtration column chromatography from the stressed *S. costatus* CFE was further purified by isoelectric focusing. An orange band that migrated to the position corresponding to pI 9 was detected (Figure 5A), and the purity of the excised orange protein band was determined by SDS-PAGE (Figure 5B). A single band appeared as an apparent molecular mass of 27 kDa, and the N-terminal amino acid sequence was determined to be AVPEAKTT, which showed partial homology to the N-terminal sequence of Ki-4-AstaP (ATPKANAT). The binding pigments of the purified protein by isoelectric focusing were analyzed by C_{18}-HPLC and were astaxanthin and adonixanthin, which were identical to those of Ki-4 AstaP [9]; therefore, we named this protein ScosAstaP.

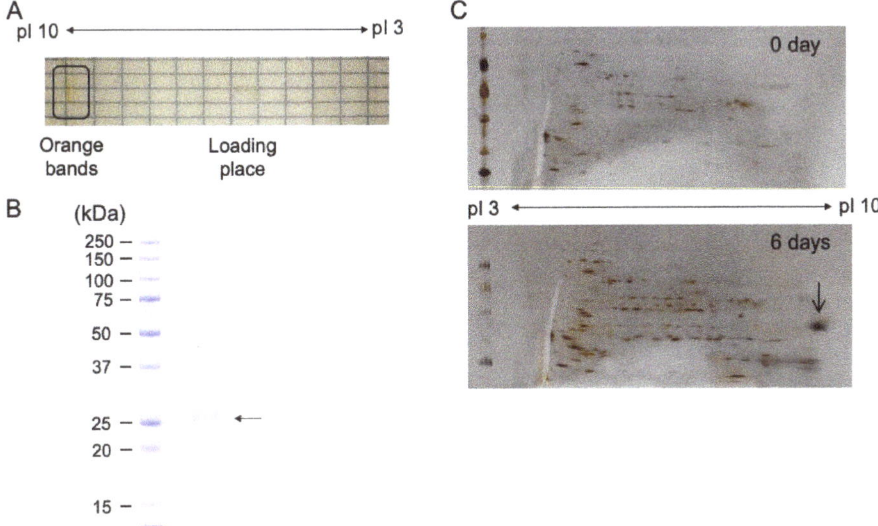

Figure 5. Purification and characterization of AstaP from *S. costatus*. (**A**) isoelectric focusing of aqueous pigments. The circle indicates the migrated orange bands after the electrophoresis; (**B**) the migrated orange band in A excised and electrophoresed by SDS-PAGE. The purified protein is shown by an arrow. Molecular mass standards are shown (kDa); (**C**) two-dimensional electrophoresis of the total proteins of *S. costatus*. *S. costatus* was cultivated under high light conditions and subjected to 0.4 M NaCl for six days. An arrow indicates the excised spot for N-terminal sequencing.

Two-dimensional polyacrylamide gel electrophoresis (2D PAGE) was performed to detect the photooxidative stress-inducible spots in *S. costatus*, and a large spot was detected in the area of alkaline pI after exposure to salt stress with high light (Figure 5). The N-terminal amino acid sequence (AVPEAKTT) from the large spot coincided with that of the purified protein. These results indicated that the orange protein was an AstaP ortholog inducible under the photooxidative stress conditions.

2.4. AstaP Ortholog in S. costatus Is a Fasciclin-Like Glycosylated Protein

To obtain the gene encoding ScosAstaP, a cDNA library was constructed from the stressed *S. costatus* mRNA. The gene encoding the N-terminal amino acid sequence was found in the de novo sequencing data of the cDNA library, and the full-length cDNA was obtained by PCR using the *S. costatus* cDNA library as a template. The deduced amino acid sequence from a gene encoding ScosAstaP showed 62% identity with Ki-4 AstaP

(Figure 6A). Bioinformatics analyses revealed that ScosAstaP conserved 20 amino acid residues of the N-terminal hydrophobic signal sequence and two fasciclin-like H1 and H2 domains. These protein primary structures are similar to Ki-4-AstaP; however, only one putative N-glycosylation Asn-x-Thr site was detected in ScosAstaP compared to the five sites in Ki-4 AstaP (Figure 6A). The calculated molecular mass of the mature protein deduced from ScosAstaP cDNA, excluding 20 amino acid residues of the N-terminus signal peptide, was 21.4 kDa. The reason for the difference of calculated molecular weight and the measured molecular weight (apparent molecular weight from SDS-PAGE was 27 kDa) is supposedly due to the glycosylation of ScosAstaP like as Ki-4 AstaP (Ki-4 AstaP was 21.2 kDa of the calculated molecular mass and the apparent molecular weight from SDS-PAGE was 33 kDa). Northern blot analysis indicated that the gene encoding ScosAstaP was inducible by photooxidative stress (Figure 7).

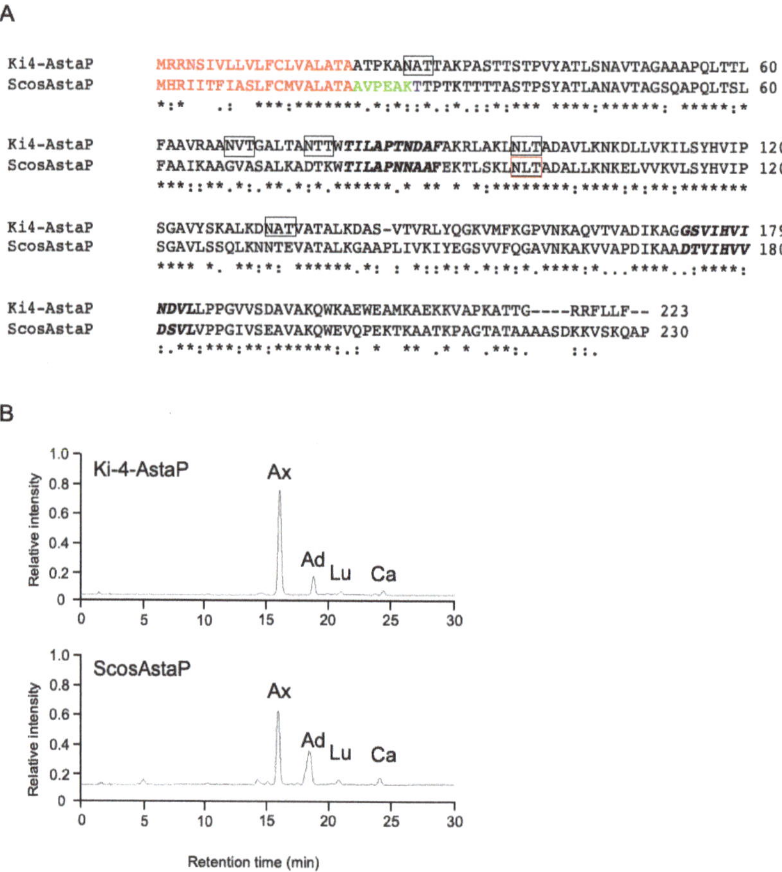

Figure 6. Structural comparison of Ki-4 AstaP and *S. costatus* AstaP orthologs. (**A**) comparison of the deduced amino acid sequences of Ki-4 AstaP and ScosAstaP. N-terminal signal peptides are shown in red font. The experimentally detected N-terminal sequence is highlighted in green. Potential sites for N-linked glycosylation are boxed (black: Ki-4 AstaP, red: ScosAstaP); (**B**) C$_{18}$-HPLC elution profile of the binding carotenoids from purified Ki-4 AstaP [9] and the purified ScosAstaP. Ax, astaxanthin; Ad, adonixanthin; Lt, lutein; and Ca, canthaxanthin.

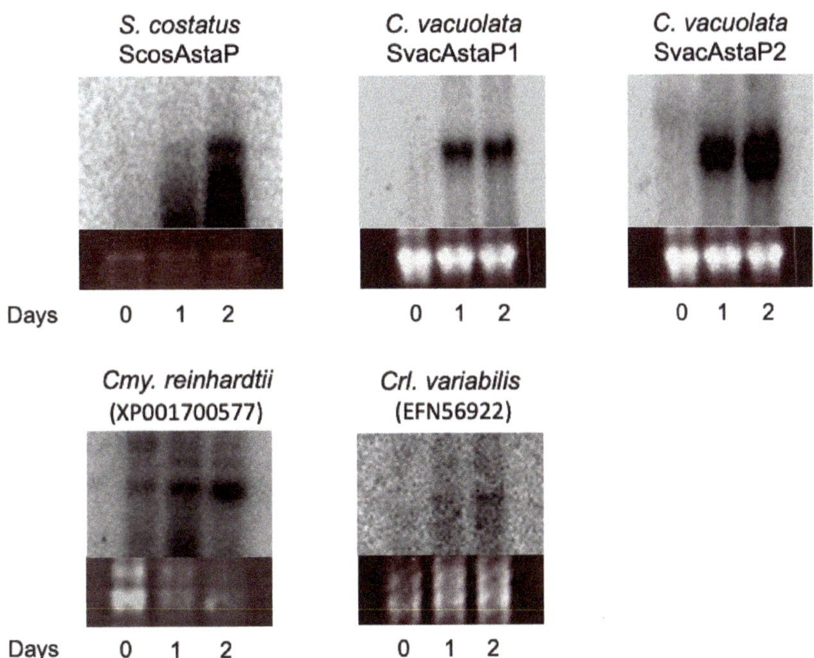

Figure 7. Northern blots of total RNA probed with the gene encoding AstaP orthologs from *S. costatus*, *C. vacuolata*, *Cmy. reinhardtii*, and *Crl. variabilis*. Strains were cultivated under photooxidative stress conditions. 0, just before the start of stress, 1–2, after one and two days of stress. Gene accession numbers are shown in parentheses. Ethidium-bromide staining of the ribosomal RNA (rRNA) to confirm the equal loading is shown below the autoradiogram.

2.5. Identification of AstaP Orthologs in C. vacuolata and Two Model Microalgae

C. vacuolata showed two peaks of astaxanthin-binding protein by gel-filtration chromatography (Figure 3); therefore, the presence of two types of AstaP orthologs was predicted. Two cDNA sequences that encode AstaP orthologs were found in the cDNA sequence data from the photooxidative-stressed *C. vacuolata* cDNA library. Both full-length cDNA clones were amplified by PCR, and nucleotide sequences were confirmed. The deduced amino acid sequence indicated that *C. vacuolata* possesses two types of AstaP orthologs. These two proteins showed different pI (calculated pI were 6.7 and 9.7), and we named them CvacAstaP1 and CvacAstaP2, respectively. Both orthologs possess N-glycosylation sites and N-terminal hydrophobic signal peptides for cell surface secretion, and they showed 44% identity with each other (Supplementary Figures S1 and S2). ScosAstaP1 had ten putative N-linked glycosylation sites, whereas ScosAstaP2 had only two putative N-glycosylation sites. The number of N-glycosylation sites was expected to affect the difference in molecular weight detected by gel-filtration column chromatography, as shown in Figure 3.

AstaP orthologs were found in the model algae *Cmy. reinhardtii* and *Crl. variabilis*, and phylogenetic analysis was performed in our previous study. Both cDNAs were amplified by PCR and confirmed the coincidence of the deduced amino acid sequence in the database (Supplementary Figure S1). To determine the gene expression profiles of AstaP orthologs in model algae and *C. vacuolata*, the genes encoding *Cmy. reinhardtii*, *Crl. variabilis*, and two AstaP orthologs from *C. vacuolata* were amplified and analyzed these expression profiles by Northern blot analysis. Both AstaP orthologs were induced by photooxidative stress treatments.

The *Scenedesmus actus* AstaP ortholog was identified as a translation product of a gene induced by Cr stress [14], but its binding of astaxanthin has not been determined.

Phylogenetic analysis revealed that the AstaP orthologs were classified into three groups: glycosylated basic pI, glycosylated acidic pI, and non-glycosylated acidic pI (Figure 8).

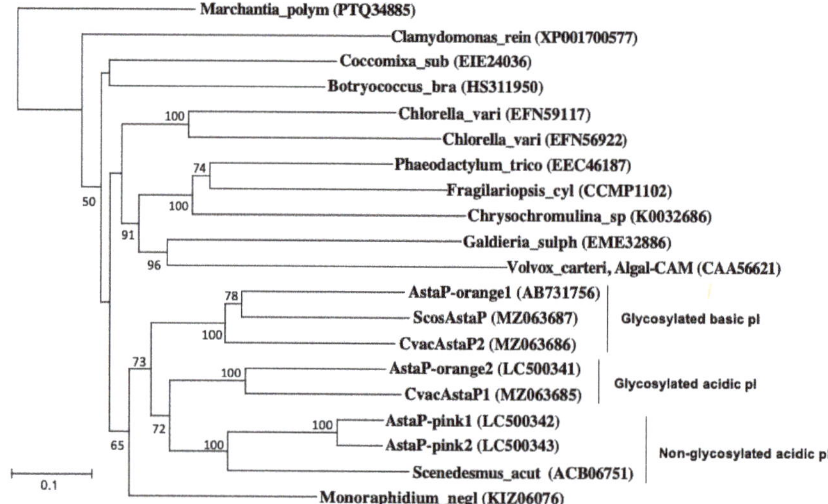

Figure 8. Neighbor-joining phylogenetic tree of the deduced sequence of the AstaP orthologs. A livewort homologue (Marchantia_polym: *Marchantia polymorpha*) was used as an outgroup. Chlamydomonas rein: *Chlamydomonas reinhardtii*, Chlolella_vari: *Chlorella variabilis*, Monoraphidium_negl: *Monoraphidium neglectum*, Scenedesmus_acut: *Scenedesmus acutus*, Botryococcus_bra: *Botryococcus braunii*, Coccomyxa_sub: *Coccomyxa subellipsoidea*, Fragilariopsis_cyl: *Fragilariopsis cylindrus*, Chrysochromulina sp: *Chrysochromulina tobinii*, Phaeodactylum_trico: *Phaeodactylum tricornutum*, Volvox_carteri: *Volvox carteri*, and Galdieria_sulph: *Galdieria sulphuraria*. Accession numbers for each protein are shown in parentheses. The bootstrap values >50 are indicated at the branch points.

3. Discussion

In this study, we investigated the distribution of AstaP orthologs from Scenedesmaceae strains. Although AstaP proteins have not been identified except strains that were isolated under high light conditions in our previous study, new AstaP orthologs from Scenedesmaceae strains were identified in the strains from culture collections. All the AstaP orthologs commonly conserved H1 and H2 fasciclin domains and N-terminal hydrophobic signal peptide for secretion; however, the number of glycosylation sites, pI, and molecular weights were found to vary depending on the strain (Supplementary Figure S1). *Cmy. reinhardtii* and *Crl. variabilis* were shown to possess photooxidative stress inducible AstaP orthologs; however, we could not detect the expression of water-soluble carotenoproteins in both strains. Further analysis will be needed to make the function of these gene products clear including their possibilities for carotenoid binding.

Scenedesmaceae strains from the culture collection were isolated under non-stressful conditions, such as in lake (*S. costatus*) [15], bark (*C. vacuolata*) [16], and peat bog (*C. striolata*) [13,17,18]. All the test strains, except *S. obliquus*, were found to express astaxanthin-binding water-soluble protein under the experimental conditions in this study; however, the amounts of production were estimated to be low compared to those of the Ki-4 and Oki-4N. We conclude that AstaP is widely distributed in Scenedesmaceae, and the expression levels of AstaP may correlate with the ability of photooxidative stress tolerance of each strain.

In conclusion, although the functional differences among these AstaP orthologs remain unclear, the presence or absence of glycosylation, the difference in pI, and the presence or absence of signal peptide for secretion may reflect the subcellular localization of AstaPs. In our previous study, Ki-4 AstaP, which is classified as a glycosylated basic pI group, was predicted to be localized at the cell surface [9,19]. On the other hand, Oki-4N AstaP-pinks, which were classified as a non-glycosylated acidic pI group, were suggested to be localized at the endoplasmic reticulum or vacuole but not at the cell surface based on the fluorescent microscopic analysis [10]. Further characterization of AstaP orthologs will clarify the distribution, localization, and specific functions under the photooxidative stress conditions.

4. Materials and Methods

4.1. Microalgal Strains and Growth Conditions

Scenedesmaceae strains were obtained from culture collection center as previously described [13]. Strain Ki-4 and Oki-4N were isolated in our laboratory as previously described [9,10]. *Scenedesmus* sp. Oki-4N was named to *Scenedesmus obtusus* Oki-4N based on a high similarity of homology score of 18S rDNA and ITS2 sequence with the morphological identity with *Scenedesmus obtusus* as previously described [10]. The strain name *Coelastrella vacuolata* (synonym: *Chlorella fusca* var *vacuolata* = *Scenedesmus vacuolatus* = *Graesiella vacuolata*) [13,16,18,20] was used in this study. The composition of the modified A3 medium was the same as previously described [9,10]. Algal strains were cultivated in A3 medium with a 16h/8h light/dark regime at 26 °C under low light conditions (60 μmol photons m^{-2} s^{-1}) as previously described [18]. Photooxidative stress was induced by the addition of sterilized NaCl when the growth reached an OD_{750} value of 1.0 cm^{-1} under conditions of high light exposure (~800 μmol photons m^{-2} s^{-1}) as previously described [9,10]. Cells did not change color to orange under high light conditions (800 μmol photons m^{-2} s^{-1}) without salt, or low light conditions with salt.

4.2. Gel Filtration Column Chromatography

The water-soluble astaxanthin binding protein fractions were obtained by gel filtration column chromatography as previously described. Briefly, stressed cells were harvested, and 1.0 g of stressed wet cells were suspended in 9.0 mL of 50 mM Tris-HCl buffer at pH 7.5. Cells were broken by a multi-beads shocker (Yasui Kikai, Osaka, Japan), dissolved in Tris-buffer pH 7.5. Cell extracts (CFEs) were ultracentrifuged at 100,000× *g* for 2 h to remove cell debris and lipids. The CFEs were passed through a Sephacryl S-200 HR gel filtration column at a flow rate of 1.0 mL/min (1.6 × 60 cm, GE Healthcare, Chicago, IL, USA). The elution profiles were monitored using a photodiode array detector LaChrome Elite software (Hitachi Ltd., Tokyo, Japan), and fractions of orange eluates were collected. Proteins were concentrated by using Amicon Ultra (Merck, Darmstadt, Germany).

4.3. Pigment Extraction and Identification

The binding pigments of water-soluble carotenoprotein were extracted using the Bligh–Dyer method as previously described [9,10,19,21]. Pigments were extracted with methanol: chloroform: H_2O = 12:5:3 by gentle mixing in a tube. After the addition of chloroform and water (2:3), the organic phase was obtained and evaporated to dryness under N_2 gas. The extracted pigments were completely dissolved in acetone and analyzed by C_{18}-HPLC (CAPCEL PAK C18 reversed-phase column, 150 × 4.6 mm, flow rate 1.0 mL/min). The pigment was identified based on the absorption spectra obtained using an HPLC photodiode array detector, HPLC retention times, and molecular masses from high-resolution LC/MS analysis in comparison with those of standard compounds.

4.4. Purification of the Water Soluble Astaxanthin Binding Protein

The water-soluble astaxanthin binding protein of *S. costatus* was purified by isoelectric focusing. Isoelectric focusing was performed horizontally with Maltiphor II (GE Healthcare, Chicago, IL, USA). Migrated orange band was excised and electrophoresed by SDS-PAGE. As standard markers, Precision Plus Protein^{TM} standard kit (Bio-Rad, Berkeley, CA, USA) was used.

4.5. Determination of Peptide Sequence

The N-terminal amino acid sequence was determined by the Edman degradation method using a PPSQ30 peptide sequencer (Shimadzu, Tokyo, Japan) as previously described [9]. Target protein band was obtained using a transferred protein on a PVDF membrane stained with Coomassie Brilliant Blue and digested with trypsin (Promega, Madison, WI, USA).

4.6. Isolation of RNA and Construction of a cDNA Library

Isolation of RNA and construction of a cDNA library were performed as previously described [9,10]. To prepare a cDNA library, *S. costatus* and *C. vacuolata* cells that were subjected to NaCl stress under high light stress and total RNA were extracted with Trizol reagent (Invitrogen, Waltham, MA, USA). Poly(A)$^+$ mRNA was isolated from total RNA and used to generate a full-length cDNA library with a Smart-Infusion PCR cloning system (Clontech, Palo Alto, CA, USA). The cDNA libraries were sequenced using the Illumina HiSeq 2500 system (Illumina, San Diego, CA, USA) and assembled de novo sequence by CLC Genomics Workbench (Qiagen, Hilden, Germany).

4.7. Bioinformatics

Bioinformatic analyses were performed as previously described [9,10]. Briefly, database searches for sequence homology were performed using the programs BLASTp (http://www.ncbi.nlm.nih.gov/BLAST/, accessed on 19 June 2021) and FASTA (http://www.genome.jp/tools/fasta/, accessed on 19 June 2021) set to standard parameters. The putative cell localization of AstaP was predicted using PSORT (http://psort.hgc.jp/form.html, accessed on 19 June 2021), SignalP (http://www.cbs.dtu.dk/services/SignalP/, accessed on 19 June 2021) and TargetP (http://www.cbs.dtu.dk/services/TargetP/, accessed on 19 June 2021). The *O*-linked glycosylation of Ser and Thr residues was predicted by the program NetOGlyc 3.1 (http://www.cbs.dtu.dk/services/NetOGlyc/, accessed on 19 June 2021). The *N*-linked glycosylation site was predicted by the program NetNGlyc 1.0 (http://www.cbs.dtu.dk/services/NetNGlyc/, accessed on 19 June 2021). The isoelectric point (pI) and molecular mass were predicted by GENETYX-MAC software (GENETYX Corporation, Tokyo, Japan).

4.8. Northern Hybridization

Northern hybridization was performed as previously described [9,10]. Briefly, total RNA was isolated using TRIzol (Invitrogen, Waltham, MA, USA) from the cells that were harvested at several time points after applying salt stress under high light conditions. The total RNA (10 µg) was subjected to electrophoresis in 1.0% agarose gels, blotted onto nylon Hybond N$^+$ membranes (Amersham, Chicago, IL, USA) were probed with the PCR-amplified DNA fragment encoding the target region. The identity of the amplified DNA fragment was confirmed by size and nucleotide sequence. The RNA was pre-hybridized at 60 °C for 30 min, and the ^{32}P-labeled DNA probe was hybridized to RNA on the membrane at 60 °C for 12 h.

4.9. Two-Dimensional Electrophoresis

Two-dimensional electrophoresis was performed as previously described [9,10]. Untreated *S. costatus* cells and exposed to 0.4 M NaCl stress under high light conditions were harvested and promptly frozen in liquid nitrogen. Proteins of several samples were sedimented by acetone, and the protein pellet was dried in vacuo. Dried pellet was dissolved in a two-dimensional (2D) electrophoresis sample buffer containing 8 M urea, 30 mM DTT, 2% (*v*/*v*) Pharmalyte 3–10 (Pharmacia), and 0.5% Triton X-100. Both isoelectric focusing (IEF) and SDS-PAGE were performed horizontally with Maltiphor II (Pharmacia, Peapark, NJ, USA). For electrophoresis, 40 µg of total protein was loaded onto gels. For detection, the gels were silver-stained with a Silver Staining Kit (Invitrogen).

Supplementary Materials: The following are available online at https://www.mdpi.com/article/10.3390/md19060349/s1, Figure S1: The deduced amino acid sequences of the AstaP orthologs. Figure S2: Comparison of amino acid sequences among AstaP-proteins in *C. vacuolata*.

Author Contributions: Conceptualization, H.T., A.M., and S.K.; methodology, H.T., A.M., R.Y., S.T., and S.K.; investigation, H.T., A.M., R.Y., T.I., and S.K.; data curation, H.T., and S.K.; writing—original draft preparation, H.T., and S.K.; writing—review and editing, H.T., S.T., and S.K.; visualization, H.T. and S.K.; funding acquisition, S.K. All authors have read and agreed to the published version of the manuscript.

Funding: This study was supported in part by a grant-in-aid from the Japan Society for the Promotion of Science (No. 26440155), the Institute for Fermentation, Osaka (IFO, G-2014-2-072), and the Cosmetology Research Foundation (J-15-2) (to SK), and MEXT-Supported Program for the Strategic Research Foundation at Private Universities (S1311017).

Institutional Review Board Statement: Not applicable.

Informed Consent Statement: Not applicable.

Data Availability Statement: The cDNA sequence data of CvacAstaP1, CvacAstaP2, and ScosAstaP have been deposited in the NCBI database under the accession numbers MZ063685–MZ063687, respectively.

Acknowledgments: The authors thank many colleagues for advice, discussion, and technical assistance, especially Keita Yamazaki, Natsumi Abe, Ayae Furukawa, Takumi Satoh, and Youichi Niimura. The authors thank Eri Kubota for performing RNA sequencing.

Conflicts of Interest: The authors declare no conflict of interests.

Abbreviations

Cmy	Chlamydomonas
Crl	Chlorella
WSCP	Water-Soluble Carotenoprotein
w/HL	with High Light

References

1. Asada, K. The water-water cycle in chloroplasts: Scavenging of active oxygens and dissipation of excess photons. *Annu. Rev. Plant Physiol. Plant Mol. Biol.* **1999**, *50*, 601–639. [CrossRef] [PubMed]
2. Foyer, C.H.; Shigeoka, S. Understanding oxidative stress and antioxidant functions to enhance photosynthesis. *Plant Physiol.* **2011**, *155*, 93–100. [CrossRef]
3. Britton, G.; Liaaen-Jensen, S.; Pfander, H. (Eds.) *Carotenoids Handbook*; Birkhauseer: Basel, Switzerland, 2004.
4. Keen, J.N.; Caceres, I.; Eliopoulos, E.E.; Zagalsky, P.F.; Findlay, J.B.C. Complete sequence and model for the A2 subunit of the carotenoid pigment complex, crustacyanin. *Eur. J. Biochem.* **1991**, *197*, 407–417. [CrossRef]
5. Keen, J.N.; Caceres, I.; Eliopoulos, E.E.; Zagalsky, P.F.; Findlay, J.B.C. Complete sequence and model for the C1 subunit of the carotenoprotein crustacyanin, and model for the dimer, β-crustacyanin, formed from the C1 and A2 subunits with astaxanthin. *Eur. J. Biochem.* **1991**, *202*, 31–40. [CrossRef] [PubMed]
6. Bhosale, P.; Larson, A.J.; Frederick, J.M.; Southwick, K.; Thulin, C.D.; Bernstein, P.S. Identification and characterization of a Pi isoform of glutathione S-transferase (GSTP1) as a zeaxanthin-binding protein in the macula of the human eye. *J. Biol. Chem.* **2004**, *279*, 49447–49454. [CrossRef] [PubMed]
7. Tabunoki, H.; Sugiyama, H.; Tanaka, Y.; Fujii, H.; Banno, Y.; Jouni, Z.E.; Kobayashi, M.; Sato, R.; Maekawa, H.; Tsuchida, K. Isolation, characterization, and cDNA sequence of a carotenoid binding protein from the silk gland of Bombyx mori larvae. *J. Biol. Chem.* **2002**, *277*, 32133–32140. [CrossRef] [PubMed]
8. Kay Holt, T.; Krogmann, D.W. A carotenoid-protein from cyanobacteria. *BBA Bioenerg.* **1981**, *637*, 408–414. [CrossRef]
9. Kawasaki, S.; Mizuguchi, K.; Sato, M.; Kono, T.; Shimizu, H. A novel astaxanthin-binding photooxidative stress-inducible aqueous carotenoprotein from a eukaryotic microalga isolated from asphalt in midsummer. *Plant Cell Physiol.* **2013**, *54*, 1027–1040. [CrossRef] [PubMed]
10. Kawasaki, S.; Yamazaki, K.; Nishikata, T.; Ishige, T.; Toyoshima, H.; Miyata, A. Photooxidative stress-inducible orange and pink water-soluble astaxanthin-binding proteins in eukaryotic microalga. *Commun. Biol.* **2020**, *3*, 490. [CrossRef] [PubMed]
11. Wilson, A.; Kinney, J.N.; Zwart, P.H.; Punginelli, C.; D'Haene, S.; Perreau, F.; Klein, M.G.; Kirilovsky, D.; Kerfeld, C.A. Structural determinants underlying photoprotection in the photoactive orange carotenoid protein of cyanobacteria. *J. Biol. Chem.* **2010**, *285*, 18364–18375. [CrossRef] [PubMed]

12. Kerfeld, C.A.; Sawaya, M.R.; Brahmandam, V.; Cascio, D.; Ho, K.K.; Trevithick-Sutton, C.C.; Krogmann, D.W.; Yeates, T.O. The crystal structure of a cyanobacterial water-soluble carotenoid binding protein. *Structure* **2003**, *11*, 55–65. [CrossRef]
13. Kawasaki, S.; Yoshida, R.; Kiichi, O.A.; Toyoshima, H. *Coelastrella astaxanthina* sp. nov. (Sphaeropleales, Chlorophyceae), a novel microalga isolated from an asphalt surface in midsummer in Japan. *Phycol. Res.* **2019**, *68*, 107–114. [CrossRef]
14. Torelli, A.; Marieschi, M.; Castagnoli, B.; Zanni, C.; Gorbi, G.; Corradi, M.G. Identification of S2-T A63: A cDNA fragment corresponding to a gene differentially expressed in a Cr-tolerant strain of the unicellular green alga *Scenedesmus acutus*. *Aquat. Toxicol.* **2008**, *86*, 495–507. [CrossRef] [PubMed]
15. Schmidle, W. Beiträge zur alpinen Algenflora. *Österreichische Bot. Z.* **1895**, *45*, 305–311. [CrossRef]
16. Hegewald, E.; Hanagata, N. Phylogenetic studies on Scenedesmaceae (Chlorophyta). *Algol. Stud.* **2000**, *100*, 29–49. [CrossRef]
17. Chodat, R. Matériaux pour l'histoire des algues de la Suisse. *Bull. Soc. Bot Geneve Ser.* **1921**, *2*, 66–114.
18. Guiry, M.D.; Guiry, G.M. *AlgaeBase*; World-Wide Electronic Publication, National University of Ireland: Galway, Ireland, 2021.
19. Toyoshima, H.; Takaichi, S.; Kawasaki, S. Water-soluble astaxanthin-binding protein (AstaP) from *Coelastrella astaxanthina* Ki-4 (Scenedesmaceae) involving in photo-oxidative stress tolerance. *Algal Res.* **2020**, *50*, 101988. [CrossRef]
20. Kaufnerová, V.; Eliáš, M. The demise of the genus *Scotiellopsis* Vinatzer (Chlorophyta). *Nov. Hedwigia* **2013**, *97*, 415–428. [CrossRef]
21. Bligh, E.G.; Dyer, W.J. A rapid method of total lipid extraction and purification. *Can. J. Biochem. Physiol.* **1959**, *37*, 911–917. [CrossRef] [PubMed]

marine drugs

MDPI

Article

Transcriptomics and Metabolomics Analyses Provide Novel Insights into Glucose-Induced Trophic Transition of the Marine Diatom *Nitzschia laevis*

Xuemei Mao [1,2,3], Mengdie Ge [1,3], Xia Wang [1,3], Jianfeng Yu [1,3], Xiaojie Li [1,3], Bin Liu [1,3,*] and Feng Chen [1,3,*]

[1] Shenzhen Key Laboratory of Marine Microbiome Engineering, Institute for Advanced Study, Shenzhen University, Shenzhen 518060, China; maoxm@szu.edu.cn (X.M.); gemengdie2020@email.szu.edu.cn (M.G.); xiawang6632@szu.edu.cn (X.W.); j.yu@szu.edu.cn (J.Y.); lixiaojie@szu.edu.cn (X.L.)
[2] College of Physics and Optoelectronic Engineering, Shenzhen University, Shenzhen 518060, China
[3] Institute for Innovative Development of Food Industry, Shenzhen University, Shenzhen 518060, China
* Correspondence: liubin@szu.edu.cn (B.L.); sfchen@szu.edu.cn (F.C.); Tel.: +86-0755-26539262 (B.L.); +86-0755-26539265 (F.C.)

Abstract: Diatoms have important ecological roles and are natural sources of bioactive compounds. *Nitzschia laevis* is a member of marine diatoms that accumulates high-value products including fucoxanthin and eicosapentaenoic acid (EPA). In this study, physiological data showed that comparing to autotrophic growth, mixotrophic cultivation with glucose supplementation led to a decrease of chlorophyll and fucoxanthin content in *N. laevis*, and an increase of biomass density and EPA yield. To further examine the metabolic barriers for fucoxanthin and EPA biosynthesis, comparative transcriptomic and metabolome analyses were conducted, with a focus on the genes related to carotenoids biosynthesis and fatty acid metabolism. The results indicated that phytoene desaturase (PDS) and zeta-carotene isomerase (ZISO) could be the rate-limiting enzymes in carotenoid biosynthesis. The transcription regulation of 3-ketoacyl-CoA synthase (KCS) and elongation of very long chain fatty acids protein (EVOVL) are important contributors associated with polyunsaturated fatty acids (PUFAs) accumulation. Furthermore, we also investigated the glucose-associated regulatory genes using weighted gene co-expression network analysis, and identified potential hub genes linked with cell cycle, carbohydrate metabolism, purine biosynthesis, and lipid metabolism. This study offers a high-quality transcriptome resource for *N. laevis* and provides a molecular framework for further metabolic engineering studies on fucoxanthin and EPA production.

Keywords: diatom; fucoxanthin; eicosapentaenoic acid (EPA); transcriptomics; trophic transition

Citation: Mao, X.; Ge, M.; Wang, X.; Yu, J.; Li, X.; Liu, B.; Chen, F. Transcriptomics and Metabolomics Analyses Provide Novel Insights into Glucose-Induced Trophic Transition of the Marine Diatom *Nitzschia laevis*. *Mar. Drugs* **2021**, *19*, 426. https://doi.org/10.3390/md19080426

Academic Editors: Masashi Hosokawa and Hayato Maeda

Received: 28 June 2021
Accepted: 24 July 2021
Published: 27 July 2021

Publisher's Note: MDPI stays neutral with regard to jurisdictional claims in published maps and institutional affiliations.

1. Introduction

Diatoms are important producers in the aquatic systems, as they are estimated to undertake 20–25% of the global carbon fixation and contribute ~40% of net primary productions [1,2]. They also play important roles in taking part in the biogeochemical cycling of silicon and carbon in the ecosystem [3]. Diatoms could grow under phototrophic conditions using CO_2 as the carbon source and light as energy. Some diatoms have developed strategies in the form of mixotrophic or heterotrophic growth to utilize organic carbon sources such as glucose, glycerol in certain niches. Diatoms are important sources of bioactive compounds like fucoxanthin, polyunsaturated fatty acids (PUFAs), e.g., eicosapentaenoic acid (EPA), flavonoids, phenolic compounds, as well as lipid production for biofuels [4,5]. Fucoxanthin, one of the most abundant carotenoids in diatoms, has promising applications in human health due to its antioxidant, anti-inflammatory, anticancer, anti-obesity, antidiabetic, antiangiogenic, and antimalarial activities [6,7]. EPA, a bioactive PUFA, plays a cardioprotective role in preventing the occurrence of cardiovascular diseases [8]. Chrysolaminarin (β-1,3-glucan), the principal storage polysaccharide in the diatom stored in vacuoles, has anti-tumor bioactivity [9]. Chrysolaminarin and lipids are the two major storage substances for carbon skeletons in diatoms; lipids (primarily triacylglycerols (TAGs)) are also considered as potential sources for biofuels [10].

The morphology and physiology of diatoms are distinctive to other photosynthetic organisms, especially on their intracellular compartmentation and carbon partitioning pathways due to its complex evolutionary history [11]. Several diatom species have been used as research models, including a centric diatom *Thalassiosira pseudonana* for silica biomineralization, and a pennate diatom *Phaeodactylum tricornutum* for carbohydrate metabolism, xanthophyll cycle, and lipid metabolism [12,13]. With the advances in genomic sequencing and multi-omics analyses, the metabolic pathways in the model diatoms are being elucidated, and genetic engineering approaches have been developed [13–16]. One of the challenges is that the metabolism of diatoms is complex and substantially different among diatom species [11], therefore some questions could not be demonstrated in existing model systems. For example, *P. tricornutum* is unable to assimilate glucose due to the lack of a transporter gene [17]. *Nitzschia laevis* can grow under autotrophic, mixotrophic, and heterotrophic conditions, and reach high cell density when glucose is supplemented as the sole carbon resource [18,19]. Thus, *N. laevis* is likely to have unique metabolic characteristics during the glucose-induced trophic transition. In addition, *N. laevis* has also been demonstrated to be a robust industrial strain for fucoxanthin and EPA production [20], thus it is an important organism to be investigated and further optimized for industrial.

Glucose is an important source of carbon to many organisms, it has been reported that the assimilation of glucose has a profound impact to the gene expression pattern, metabolic networks and cell growth in higher plants and green algae [21,22]. The physiological impact of glucose assimilation to diatoms is yet clear. Previously, we found that supplementation of glucose could induce a shift of physiological parameters, e.g., growth rate and biomolecular composition; to elucidate the underlying molecular mechanisms that associated with growth trophic transitions in response to glucose, we conducted metabolomics and transcriptomics analyses, the data are presented in this work.

2. Results and Discussion

2.1. Physiological Changes in N. laevis under Glucose Addition

With the addition of glucose (+G), the cultivation shifted from autotrophy to mixotrophy (or pure heterotrophy), resulting in a significant increase in biomass yield in comparison to the control group (−G) (Figure 1A) ($p < 0.01$). Moreover, the cellular abundance of fucoxanthin was decreased in mixotrophic cells after 4-day culture with glucose (Figure 1B). In diatom, fucoxanthin and chlorophyll a/c are primary pigments in the light-harvesting complex associated with photosystem I and II [23,24]. The decrease of both chlorophyll (Figure 1B) and fucoxanthin content indicates a general downregulation of the photosynthetic apparatus (including PSI and PSII) under glucose addition. TFA content displayed a minor decrease on Day 4 (Figure 1C). As both lipid and carbohydrates production were dependent on the carbon precursors and used as stored substances, the synthesis of these biomolecules are potential competitors for carbon backbones. In microalgae, the most abundant carbohydrate molecules are species-dependent, as that in cyanobacteria is glycogen, whereas that in green algae Chlamydomonas is starch [25]. The main carbohydrate in diatom like *Phaeodactylum*, is chrysolaminaran [26]. In this study, the intracellular soluble polysaccharides, likely in the form of chrysolaminaran, were analyzed and showed a significant increase upon glucose addition (Figure 1C) ($p < 0.05$). The percentage of EPA in TFA was increased on Day 4 with glucose addition (Figure 1D), and the EPA yield reached 38.18 mg/L in +G group while only 16.09 mg/L in −G group ($p < 0.01$). A previous study also has concluded that mixotrophic culture is preferred for the production of EPA from *N. laevis* [27].

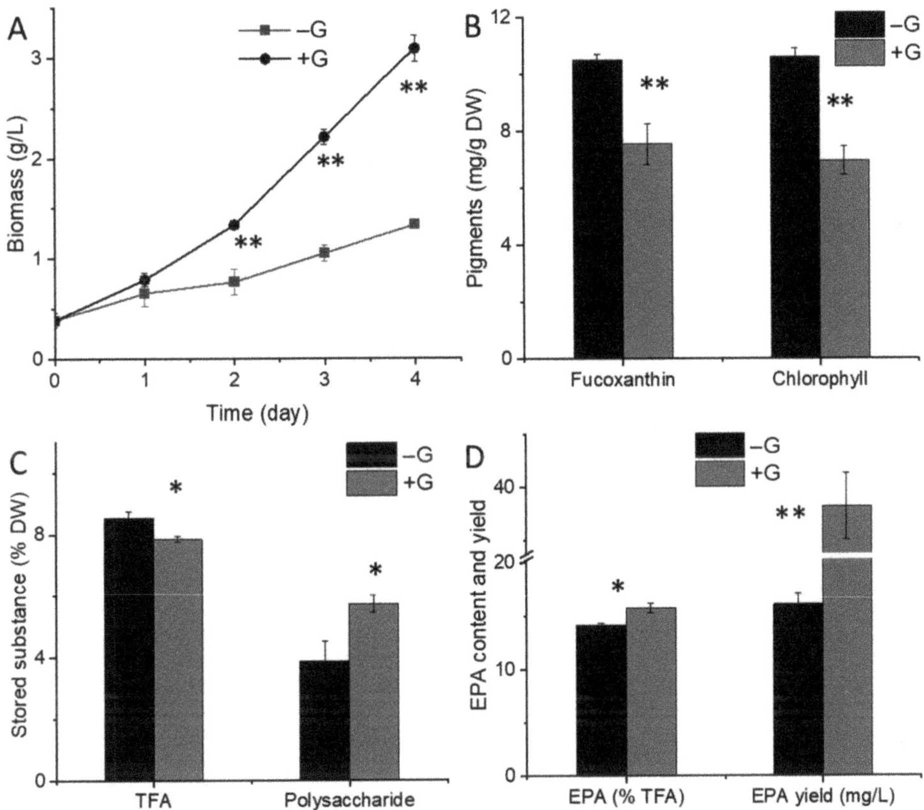

Figure 1. Physiological changes in *N. laevis* after glucose addition. (**A**) Biomass growth, (**B**) fuco-xanthin and total chlorophyll content, (**C**) total fatty acids (TFA) and intracellular soluble polysaccharides content, and (**D**) eicosapentaenoic acid (EPA) content and yield after 4-days culture. * indicates $p < 0.05$, and ** indicates $p < 0.01$. +G, with glucose addition; −G, without glucose addition.

2.2. The Changes of Transcriptome during the Transition from Autotrophy to Mixotrophy

RNA-seq is a powerful tool to provide insights to gene expression and function, the association of genes to metabolic pathways, and their regulations. Data generated from RNA-seq can be used to complement genomic studies for functional annotations and evolution of gene clusters and metabolic pathways. To elucidate the underlying mechanisms that associated with physiological changes in *N. laevis*, we investigated transcriptome-wide differences in gene expression during the transition from autotrophy to mixotrophy. Samples were collected for RNA-seq at 3 h, 6 h, and 12 h from both the glucose group (with glucose addition, +G) and control group (without glucose addition, +G). De novo sequences were prepared from 18 samples (referred as +G3h, +G6h, +G12h, −G3h, −G6h, −G12h). The de novo assembly generated 16,622 unigenes in total (Table S1). To reflect the gene expressional correlation between samples, the Pearson correlation coefficients for all gene expression levels between two samples were shown in Table S2. Moreover, the principal component analysis (PCA) results also revealed the relationship between the samples (Figure S1A). The transcription profiles among +G and −G groups showed significant divergence, and time points also attributed to the divergence among groups, while the three biological replicates were comparable. The box plot showed an overview of the gene expression level in each sample (Figure S1B). Through the sequence similarity search in NR (Non-Redundant Protein Sequence) database, functions information of genes was predicted and the similarity between the transcript sequences of *N. laevis* and the similar species was obtained. As shown in Figure S1C, 3163 genes were matched with homologous genes from *Pseudo-nitzschia multistriata*, and 2665 homologous genes were identified in *Fragilariopsis cylindrus*. Moreover, these genes accounted for ~52.8% of total genes, and genes conserved across all three species of *Phaeodactylum tricornutum*, *Fistulifera solaris*, and *Thalassiosira oceanica* accounted for ~37.0% of total genes. These results revealed the close phylogenetic links between *N. laevis* and other diatoms. The abundance of transcripts from +G group were compared with respective counterparts from −G group, and interpreted as $\log_2$FoldChange (FC), and genes with $\log_2$FC $\geq$ 1 or $\leq$ −1 and padj $\leq$ 0.05 were defined as differentially expressed genes (DEGs). The volcano plots shown the distribution of DEGs (Figure 2A–C): There were 1068 upregulated DEGs and 1058 downregulated DEGs in the +G3h group, 699 upregulated DEGs and 728 downregulated DEGs in the +G6h group, and 822 upregulated DEGs and 781 downregulated DEGs in the +G12h group. As the +G3h group yielded the largest number of DEGs, the KEGG pathways enrichment for the DEGs of +G3h group was performed and pathways significantly enriched ($p < 0.05$) were shown in Figure 2D. The top 10 enriched pathways of DEGs were listed with relative DEG IDs, including several amino acid metabolism pathways ('alanine, aspartate and glutamate metabolism', 'arginine biosynthesis', 'valine, leucine and isoleucine biosynthesis', 'valine, leucine and isoleucine degradation'); two carbohydrate metabolism pathways ('pyruvate metabolism', 'propanoate metabolism'); two lipid metabolism pathways ('fatty acid degradation', 'fatty acid elongation'); one pathway of translation ('ribosome biogenesis in eukaryotes'); and 'vitamin B6 metabolism'.

To identify gene families that quickly respond to the transition from autotrophy to mixotrophy, the 100 most up- and downregulated DEGs at 3 h were listed for hierarchical cluster analysis to group these DEGs with their expressional patterns during the first 12 h (Figure S2A,B). Even though most of these DEGs were unknown in function according to sequence alignment results in NR and KEGG database, some genes were annotated and classified. The results showed that most of these DEGs were associated with carbohydrate metabolism, amino acid metabolism, and lipid metabolism (Figure S2C,D). For carbohydrate metabolism, most of upregulated DEGs were involved in pyruvate metabolism and glycolysis/gluconeogenesis, while most of downregulated DEGs were involved in fructose and mannose metabolism, glyoxylate and dicarboxylate metabolism, and propanoate metabolism (Table S3). For lipid metabolism, one gene in fatty acid biosynthesis, one in glycerophospholipid metabolism and one in glycerolipid metabolism were downregulated, while three genes in fatty acid degradation, two genes in biosynthesis of unsaturated fatty acids, and one in fatty acid biosynthesis were upregulated (Table S3). Besides, note that there were 2 genes (hexaprenyl-diphosphate synthase (hexPS) and diphosphomevalonate decarboxylase (MVD)) related in terpenoid backbone biosynthesis pathway in the list of top 100 upregulated DEGs (Table S3), and terpenoid backbone biosynthesis is directly associated with carotenoids biosynthesis, thus influence the production of fucoxanthin.

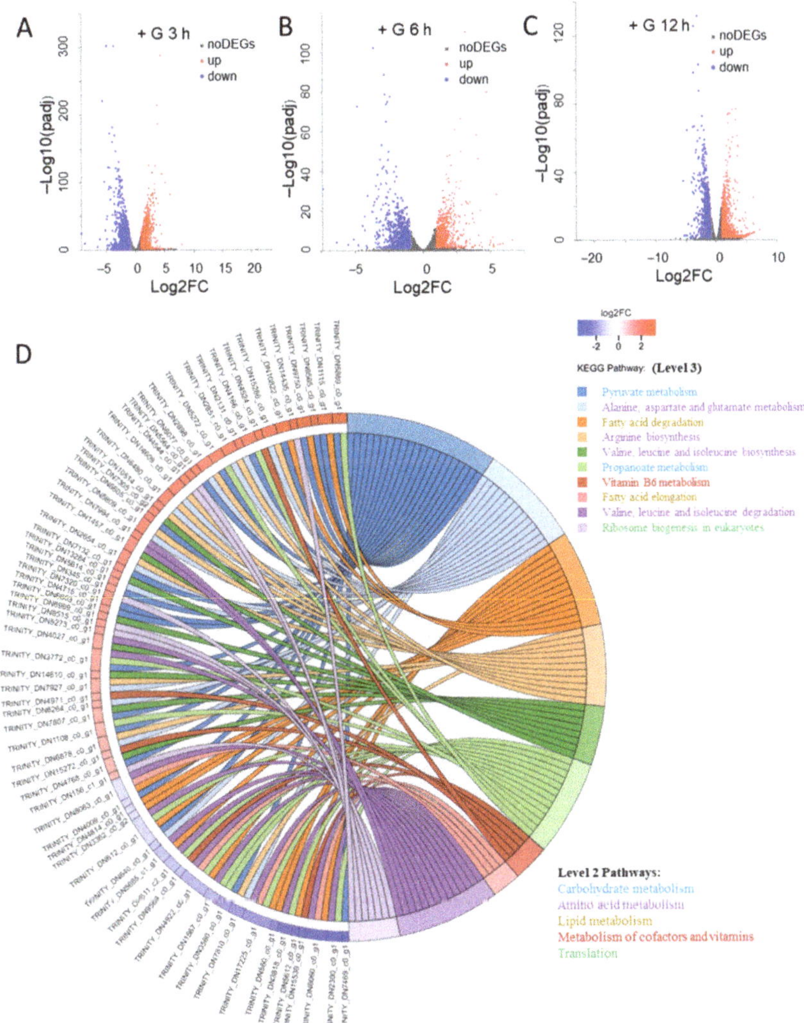

Figure 2. Differentially expressed genes (DEGs) overview during the transition from autotrophy to mixotrophy. (**A**) The volcano plots of DEGs at 3 h. (**B**) The volcano plots of DEGs at 6 h. (**C**) The volcano plots of DEGs at 12 h. (**D**) KEGG pathways enrichment for the DEGs in the +G3h group. DEGs, differential expressed genes; FC, fold change.

2.3. The Changes of Metabolites Profile during the Transition from Autotrophy to Mixotrophy

Metabolic pathways in microalgae are responsive to environmental changes. Even with transcriptomic data, biochemical regulation is subject to many other regulatory controls (e.g., regulated degradation, post translational modification, and allostery), so the conclusions remain tentative and hypothesis-generating rather than conclusive. However, the metabolomic data greatly strengthen the hypotheses produced by the transcriptomic data. Mass spectrometry were used to collect data on metabolites analysis, by applying a filtering coefficient of 30% variation of peak size, there were 13,139 and 1316 peaks identified under Positive ion mode (ESI+) and negative ion mode (ESI−), respectively (Table S4). To illustrate the differences of metabolomics between −G and +G groups, metabolites with fold change ≥ 1.2 or ≤ 0.83, and *p*-value <0.05 were defined as having significant changes. In total, there were 847 decreased metabolites and 856 increased metabolites under ESI+ in the +G microalgal cells (Figure S3A); while the signals of 136 metabolites under ESI- were decreased and that of 122 metabolites were increased (Figure S3B). With the exception of unknown compounds, most of these differential metabolites belong to the following chemical groups: lipids and lipid-like molecules, benzenoids, organoheterocyclic compounds, and organic acids (Table S4). As designated in the KEGG databases, most of identified metabolites belong to amino acid metabolism, metabolism of cofactors and vitamins, carbohydrate metabolism and lipid metabolism pathway (Figure S3C). In contrast to that of −G group, the valine content was increased by 57% (Figure 3). Valine is present in many proteins, mostly in the interior of globular proteins helping to determine the three-dimensional structure [28]. α-Ketoglutarate was significantly increased by 147% in the +G group, which is an intermediate compound of the tricarboxylic acid cycle (TCA cycle) and amino acid metabolism, nitrogen metabolism. Correlated with glucose addition, pyruvate, the product of glycolysis and precursor of carotenoids, fatty acids and many amino acids, was increased by 86% (Figure 3). Due to the crosstalk between amino acids and central carbon metabolism, the increase of valine, α-ketoglutarate, and pyruvate was consistent with the upregulation of several amino acid metabolic pathways revealed by transcriptomic analysis, including 'valine, leucine and isoleucine biosynthesis', 'arginine biosynthesis' and 'alanine, aspartate and glutamate metabolism'. The upregulation of amino acid biosynthesis under mixotrophic culture was expected, as after glucose addition, amino acids and proteins synthesis were required for cell proliferation. The contents of C16:0, EPA, and DHA were greatly increased, while C12:2 was significantly decreased (Figure 3), indicating that the addition of glucose enhanced the biosynthesis of long-chain fatty acids. This result was also consistent with the transcriptomic result that fatty acid elongation pathway was upregulated by glucose addition.

2.4. Identification of the Candidate Genes Involved in Carotenoids Biosynthesis and Their Expression Profiles

From this section, we focused on the transcriptome results of selected functional pathways involved in fucoxanthin and fatty acid biosynthesis. Carotenoids biosynthesis includes two metabolic steps: terpenoid backbone biosynthesis and carotenoid biosynthesis. Carotenoids are derived from terpene backbone precursors, isopentenyl pyrophosphate (IPP) and dimethylallyl pyrophosphate (DMAPP). Previous studies have revealed that green microalgae produce IPP and DMAPP via the chloroplastic 2-C-methylerythritol 4-phosphate (MEP) pathway [29,30], while diatoms and higher plants commonly have both cytosolic mevalonate (MVA) pathway and chloroplastic MEP pathway for IPP and DMAPP synthesis [31,32]. These two pathways are likely to generate precursors that are allocated to different isoprenoid end-products. It is currently known that mono- and diterpenes are produced in the plastid compartment via the MEP pathway; whereas the cytosolic MVA pathway is used for the biosynthesis of triterpenes, including the sterol precursor, cycloartenol; however, the exchange of terpene precursors between the two compartments increased the complexity of studying their metabolism [33]. In this study, all genes encoding the enzymes of the MEP pathway and most genes of the MVA pathway were identified in *N. laevis* (Table S5). This study is the first report on the genes and expression profiles of the MEP and the MVA pathways in *N. laevis*.

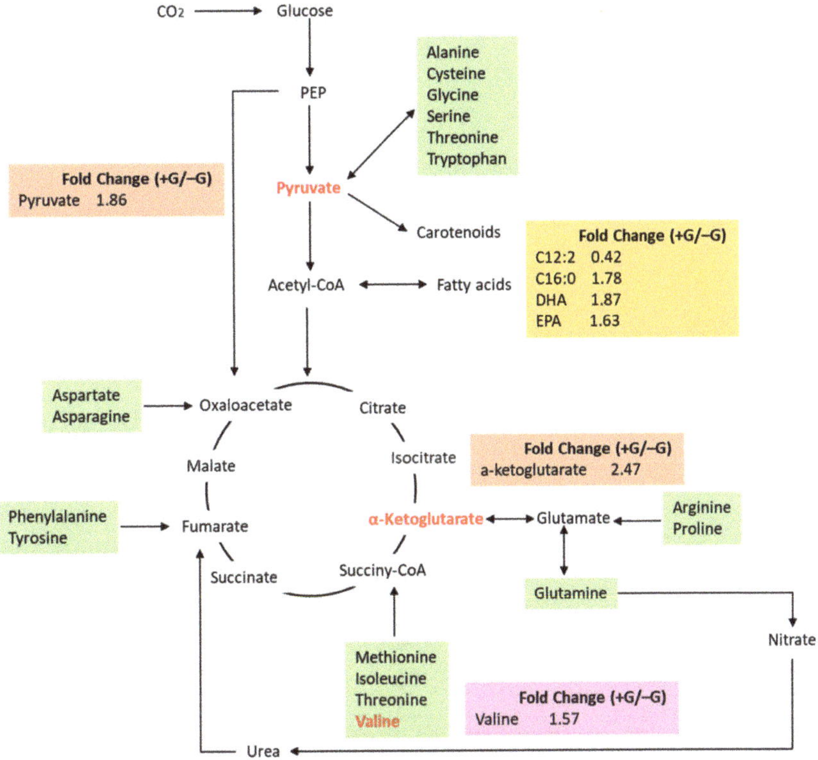

Figure 3. Brief overview of amino acid and central carbon metabolism based on metabolomic analysis. The red color of words indicates the content of the metabolite was increased under glucose addition.

The transcription level of genes relate to terpene precursors and carotenoid biosynthesis were further illustrated with the respective time points (Figure 4). All genes in MEP pathway showed no significant change through the 12 h period (Figure 4A). In the MVA pathway, acetyl-CoA C-acetyltransferase (ACAT) was slightly downregulated only at 3 h, while diphosphomevalonate decarboxylase (MVD) was apparently upregulated at 3 h and 6 h. Besides, isopentenyl-diphosphate delta-isomerase (idi) catalyzed the reversible conversion of IPP and DMAPP, and one of the idi isoforms in *N. laevis* was significantly upregulated at 3 h and 6 h. IPP and DMAPP are converted to form geranylgeranyl diphosphate (GGPP) catalyzed by the geranyl diphosphate synthase (GPPS), farnesyl diphosphate synthase (FDPS), and geranylgeranyl diphosphate synthase (GGPPS). Among these three enzymes, the gene encoding GGPPS had a relatively higher transcription level than the other two genes, and the gene encoding FDPS showed elevated expression under glucose addition. Therefore, the gene transcription levels of terpenoid backbone biosynthesis were overall upregulated by glucose addition.

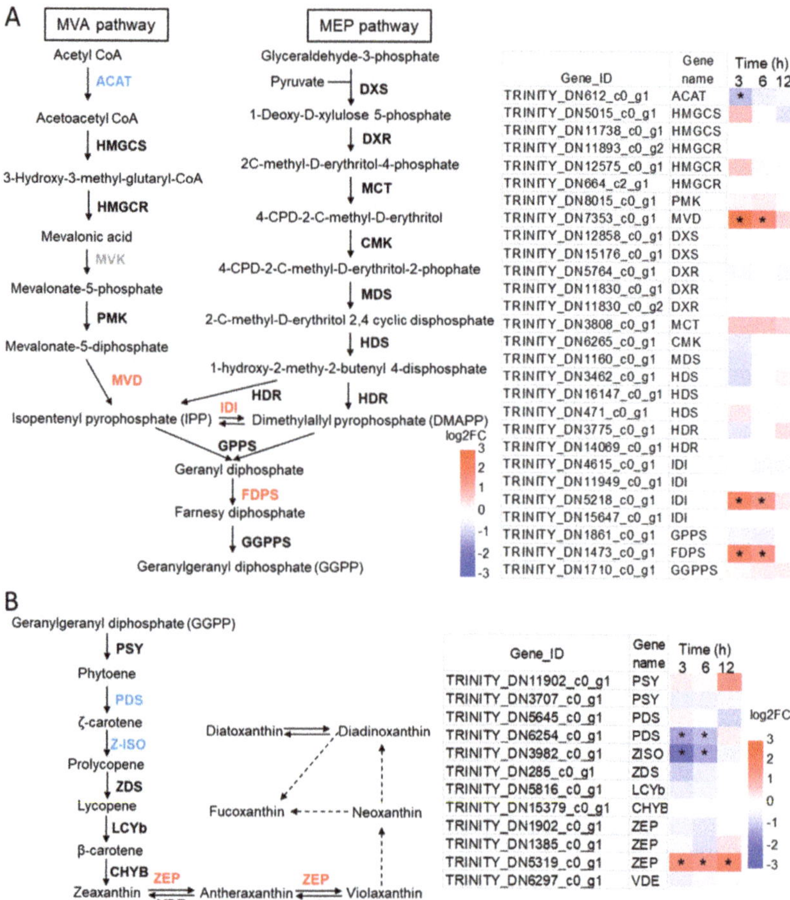

Figure 4. Expression patterns of genes in the carotenoid biosynthesis pathway. (**A**) Terpenoid backbone biosynthesis, (**B**) Carotenoid biosynthesis. ACAT, acetyl-CoA C-acetyltransferase; HMGCS, hydroxymethylglutaryl-CoA synthase; HMGCR, hydroxymethylglutaryl-CoA reductase; PMK, phosphomevalonate kinase; MVD, diphosphomevalonate decarboxylase; DXS, 1-deoxy-D-xylulose-5-phosphate synthase; DXR, 1-deoxy-D-xylulose-5-phosphate reductoisomerase; MCT, 2-C-methyl-D-erythritol 4-phosphate cytidylyltransferase; CMK, 4-diphosphocytidyl-2-C-methyl-D-erythritol kinase; MDS, 2-C-methyl-D-erythritol 2,4-cyclodiphosphate synthase; HDS, (E)-4-hydroxy-3-methylbut-2-enyl-diphosphate synthase; HDR, 4-hydroxy-3-methylbut-2-enoyl diphosphate reductase; IDI, isopentenyl-diphosphate Delta-isomerase; FDPS, farnesyl diphosphate synthase; GGPPS, geranylgeranyl diphosphate synthase; GPPS, geranyl diphosphate synthase; ZDS, zeta-carotene desaturase; PSY, phytoene synthase; PDS, phytoene desaturase; LCYb, lycopene beta-cyclase; ZEP, zeaxanthin epoxidase; VDE, violaxanthin de-epoxidase; ZISO, zeta-carotene isomerase; CHYB, beta-carotene 3-hydroxylase. * indicates the expression level in +G group was significantly differential compared with control group ($|\log_2\text{FC (Fold change)}| \geq 1.00$ and padj < 0.05). The red color of genes indicates upregulation and blue color indicates downregulation.

In the carotenoid biosynthesis pathway, GGPP is condensed to the first 40-carbon carotene, phytoene, catalyzed by phytoene synthase (PSY). The phytoene is converted to lycopene after several desaturation and isomerization steps, mediated by phytoene desaturase (PDS), zeta-carotene isomerase (ZISO), and zeta-carotene desaturase (ZDS).

The transcripts of PDS and ZISO was reduced upon glucose addition, and that of PSY and ZDS remained unchanged (Figure 4B). From the genetic engineering perspective, the reduction of PDS and ZISO correlated well the decrease of fucoxanthin content, thus the two enzymes might catalyze the rate-limiting steps in the pathway. Lycopene β-cyclase (lcyB) converts lycopene to β-carotene, and beta-carotene 3-hydroxylase (CHYB) catalyzes the hydroxylation of β-carotene to form zeaxanthin. In green algae, lycopene represents the branching point for α-carotene and β-carotene, and lycopene ε-cyclase (LCYe) diverts the carotenoid flux towards lutein instead of zeaxanthin [34]. However, in *N. laevis* the ortholog of LCYe is absent, which is consistent with previous studies suggesting LCYe gene is absent in diatoms, explaining why α-carotene and its derivatives are not found [35]. Zeaxanthin was epoxidized to violaxanthin by zeaxanthin epoxidase (ZEP) via the intermediate antheraxanthin, and the xanthophyll violaxanthin was reversely deepoxidized to zeaxanthin by violaxanthin de-epoxidase (VDE) (violaxanthin cycle). The latter conversion was essential for the photoprotection of plants or algae when exposed to excessive light [36]. In this study, three candidate ZEP isoforms and one VDE were identified, and one of ZEP isoforms (TRINITY_DN5319_c0_g1) was upregulated upon glucose addition.

Violaxanthin is the precursor of diadinoxanthin, diatoxanthin, and fucoxanthin, the major xanthophyll molecules in diatoms, and the diadinoxanthin–diatoxanthin cycle is an important short-term photoprotective mechanism [37]. However, the research on fucoxanthin biosynthesis in diatoms is still in its infancy, as the enzymes catalyze downstream conversions of violaxanthin has yet been identified in diatoms [38]. The structural studies on fucoxanthin and chlorophyll a/c binding proteins (FCPs) revealed that fucoxanthin-FCPs pigment-protein complex play both light-harvesting and photoprotection roles [23,24]. In *N. laevis*, 18 genes were annotated as putative genes encoding FCPs (Table S5), among which one gene was downregulated at 3 h, two genes were downregulated at 6 h and 12 h, while 1 gene was upregulated at 12 h.

2.5. Alterations to FA Metabolism during the Transition from Autotrophy to Mixotrophy

The first committing step of de novo fatty acid synthesis is the carboxylation of acetyl CoA to produce malonyl CoA catalyzed by acetyl-CoA carboxylase (ACC), which is a key enzyme in the de novo fatty acid synthesis [39,40]. The malonyl CoA is transferred to an acyl carrier protein (ACP) by the malonyl-CoA:acyl carrier protein transacylase (MAT) leading to the formation of malonyl-ACP for fatty acid synthesis. In this study, the transcription level of neither ACP or MAT genes had significant change between +G and −G group in *N. laevis* (Figure 5).

The subsequent condensation reactions from malonyl-ACP to C16 and/or C18 fatty acids involves a set of fatty acid synthases including 3-ketoacyl-ACP synthase (KAS, fabF and fabH), 3-ketoacyl-ACP reductase (KAR, fabG), 3-hydroxyacyl-ACP dehydratase (HAD, fabZ), and enoyl-ACP reductase (ENR, fabI, and fabK) [39]. In *N. laevis*, the transcription level of these enzymes (fabF, fabH, fabG2, fabZ) were overall upregulated in +G group at 3 h (Figure 5). Despite the fact that FabG1 was downregulated, its transcript abundance was much lower than FabG2, indicating the FabG2 rather than FabG1 was the primary functional isoform catalyzing the conversion of 3-Ketoacyl ACP to 3-Hydroxyacyl ACP.

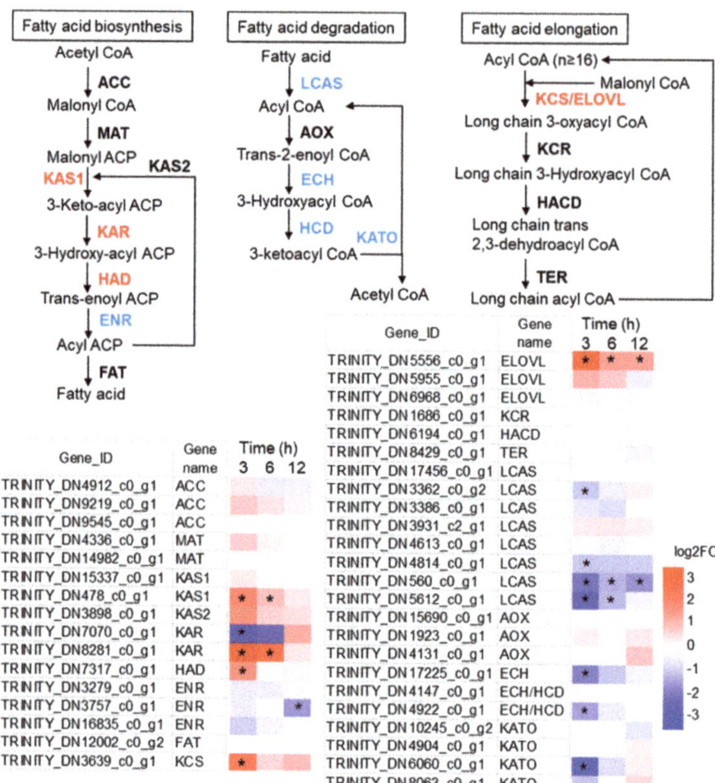

Figure 5. Expression patterns of genes involved in fatty acid metabolism. ACC, acetyl-CoA carboxylase; ACP, acyl carrier protein; MAT, malonyl-CoA:acyl carrier protein transacylase; KAS, 3-ketoacyl-ACP synthase; KAR, 3-ketoacyl-ACP reductase; HAD, 3-hydroxyacyl-ACP dehydratase; ENR, enoyl-ACP reductase; FAT, acyl-ACP thioesterase; LCAS, long-chain acyl-CoA synthetase; AOX, acyl-CoA oxidase; ECH, enoyl-CoA hydratase; HCD, 3-hydroxyacyl-CoA dehydrogenase; KATO, 3-ketoacyl-CoA thiolase; ELOVL, elongation of very long chain fatty acids protein; KCS, 3-ketoacyl-CoA synthase; KCR, very-long-chain 3-ketoacyl-CoA reductase; HACD, very-long-chain-3-hydroxyacyl-CoA dehydratase; TER, very-long-chain enoyl-CoA reductase. * indicates the expression level in +G group was significantly differential compared with control group ($|\log_2 FC| \geq 1.00$ and padj < 0.05). The red color of genes indicates upregulation and blue color indicates downregulation.

The de novo synthesized fatty acids in the form of acyl-ACP can be released as free fatty acids by an acyl-ACP thioesterase (FAT), and ligated to CoA via a long-chain acyl-CoA synthetase (LCAS). The fatty acids in the form of acyl-CoA can be further elongated or oxidated. In microalgae, isoforms of LCAS played different roles involving in TAG biosynthesis or in fatty acid β-oxidation [41]. Among the eight putative LCAS genes in *N. laevis*, four genes were downregulated by glucose addition (Figure 5), in spite that their roles in lipid metabolism needed more detailed research in future. Fatty acids degradation involves a set of enzymes including LACS, acyl-CoA oxidase (AOX), enoyl-CoA hydratase (ECH), 3-hydroxyacyl-CoA dehydrogenase (HCD) and 3-ketoacyl-CoA thiolase (KATO). Overall, the expression of these genes involved in FA degradation were strongly repressed upon glucose addition, indicative of attenuated degradation of fatty acids under mixotrophic condition.

Fatty acids are de novo synthesized in the stroma, then converted into very long chain polyunsaturated fatty acids at the endoplasmic reticulum (ER). In FA elongation pathway, the transcription of two isoforms of Elongation of Very Long Chain Fatty Acid Proteins (ELOVL) were dramatically increased (TRINITY_DN5556_c0_g1 and TRINITY_DN5955_c0_g1), and 3-ketoacyl-CoA synthase (KCS) was 4.6-fold upregulated under mixotrophic condition. These two enzymes were known as responsible for extending palmitoyl-CoA and stearoyl-CoA to very-long-chain acyl CoAs. These two upregulated genes might be crucial for EPA biosynthesis; therefore, they are the potential targets for gene engineering to enhance EPA production.

2.6. Weighted Gene Co-Expression Network Analysis (WGCNA)

To determine regulatory modules and the hub genes regulating the physiological responses related to glucose supplementation, we further carried out the WGCNA analysis using RNA-seq data of *N. laevis* (Figure 6). A total of 6 co-expressed gene modules were identified via hierarchical clustering based on gene transcript levels (Figure 6A). The correlation coefficients between group physiological traits and the identified gene modules were calculated to identify glucose regulatory modules (Figure 6B). The correlation efficient of the turquoise module and +G group is up to 0.65 with a *p*-value of 0.003, suggesting that the turquoise module was likely to be positively associated with glucose regulation. In contrary, the green module with a correlation efficient of -0.825 and a *p*-value of 2.51×10^{-5} was suggested likely to be negatively associated with glucose regulation. Therefore, the turquoise module and green module were choosing to carry out further analysis of hub genes, renamed as glucose-associated positive regulatory module (GPRM) and glucose-associated negative regulatory module (GNRM), respectively. The hierarchical cluster analysis of GPRM and GNRM showed that most genes in GPRM were strongly repressed upon glucose addition while most genes in GNRM were upregulated (Figure S4). Then, the hub genes analysis was performed, and 30 hub genes were identified from each module, respectively (Figure 6C,D). NR annotation showed that the function of most of these genes were unknown. In GPRM, there was one gene (TRIN-ITY_DN2078_c0_g1) with high sequence homology to Cell Division Cycle 20-like protein 1 (CDC20) from Fistulifera solaris, and CDC20 is essential for the anaphase onset as a cofactor of the anaphase promoting complex/cyclosome (APC/C) controlling progression through mitosis and the G1 phase of the cell cycle [42,43]. Moreover, another gene (TRIN-ITY_DN2784_c0_g1) from GPRM has the homology with a DUF111-domain-containing protein, which was likely associated to carbohydrate transport and metabolism [44]. In GNRM, several genes were identified as hub genes with putative function, including an adenylosuccinate synthetase (ADSS, TRINITY_DN5258_c0_g1), acetaldehyde dehydrogenase (ALDH, TRINITY_DN9750_c0_g1), a thiamin diphosphate-binding protein (ThDP-BP, TRINITY_DN4715_c0_g1), a formate C-acetyltransferase (TRINITY_DN7137_c0_g1), a pyruvate dehydrogenase E1 component alpha subunit (TRINITY_DN7132_c0_g1), a Phosphoethanolamine n-methyltransferase (TRINITY_DN2972_c0_g1) and a delta12 desaturase (D12D, TRINITY_DN4872_c0_g1). Among these genes, ADSS is an enzyme that plays an important role in purine biosynthesis and ALDH catalyzes the conversion of acetaldehyde to acetyl-CoA. Formate C-acetyltransferase catalyzes the reversible conversion of pyruvate and CoA into formate and acetyl-CoA, and is involved in the regulation of anaerobic glucose metabolism. The pyruvate dehydrogenase (PDH) complex is a mitochondrial multienzyme complex that catalyzes the overall conversion of pyruvate to acetyl-CoA and provides the primary link between glycolysis and the tricarboxylic acid (TCA) cycle. Phosphoethanolamine n-methyltransferase participates glycerophospholipid metabolism, and it is essential to phosphatidylcholine (PC) synthesis in Arabidopsis [45]. D12D is a key enzyme involved in the synthesis of omega-3 and omega-6 FAs, responsible for the conversion of oleic acid (OA, 18:1 n−9) to linoleic acid (LA) [46].

The assessment of RNA concentration and quality was performed by Nano-Drop (ND1000 UV–Vis spectrophotometer; NanoDrop Technologies, Wilmington, Delaware, USA) and Agilent 2100 Bio analyzer (Agilent Technologies, Inc., Palo Alto, CA, USA). After purification, poly-A containing mRNA molecules were fragmented and reverse-transcribed into cDNA, then adaptor sequences were ligated to cDNA molecules. The products were purified and enriched with PCR amplification. The PCR yield was quantified by QuantiFluor® dsDNA System, and samples were pooled together to form the final library. The pooled samples were subjected to cluster generation and sequenced using an Illumina HiSeq. Raw reads were trimmed and clipped with SeqPrep (https://github.com/jstjohn/SeqPrep, accessed on 20 January 2021) and Sickle (https://github.com/najoshi/sickle, accessed on 20 January 2021) software by filtering the low-quality reads, reads with adaptors and reads with unknown bases (N bases more than 5%) to get clean reads. The RNA-seq data were deposited in the Gene Expression Omnibus under accession number GSE168959.

De novo transcriptome assembly was performed with Trinity software (Version v2.8.5, https://github.com/trinityrnaseq/trinityrnaseq, accessed on 20 January 2021). The assembled transcriptome sequences were annotated using major databases (Non-Redundant Protein Sequence Database (NR), Swiss-Prot, Pfam protein family database, Clusters of Orthologous Groups (COG), Gene Ontology database (GO) and Kyoto Encyclopedia of Genes and Genomes (KEGG)), and statistics on its annotations were carried out in each database. RSEM software was used to calculate gene expression levels from RNA-seq data, and the expression profiles for all genes were normalized with the standardized transcript abundance measure "transcripts per million" (TPM) [48]. R studio (Version 4.0.2., Lucent Technologies Inc. (formerly AT&T Bell Laboratories), Murray Hill, NJ, USA) software was used to perform statistics: Pearson correlation between all samples was calculated using cor function, and PCA analysis was performed with all samples using princomp function. Finally, differential expressed genes (DEGs) were identified between samples and clustered with DEseq2 (Version 1.24.0, http://bioconductor.org/packages/stats/bioc/DESeq2/, accessed on 16 March 2021) software using the parameters as $|\log_2 FC|$ (Fold Change) ≥ 1.00 and adjusted p value (padj) < 0.05.

3.6. Weighted Gene Co-Expression Network (WGCNA)

For investigating correlations between genes and glucose-induced trophic transition, a global weighted gene co-expression network (WGCNA) were constructed as follows (https://horvath.genetics.ucla.edu/html/CoexpressionNetwork/Rpackages/WGCNA/, accessed on 16 March 2021): (1) genes and samples were first filtered by the variation efficient and hierarchical clustering analysis; (2) co-expressed gene modules were identified from the network using the function blockwise Modules; (3) modules with higher Pearson correlation coefficient were selected to perform the hub gene analysis.

3.7. Metabolomics Analysis

The cells were collected by centrifugation and quenched rapidly with liquid nitrogen. Cells were homogenized by a tissue lyser at 30 Hz for 1 min, and the adapter set of the tissue lyser was pre-chilled at -80 °C overnight before use. The metabolites were extracted by pre-chilled methanol: acetonitrile: H_2O (2:1:1, *v/v/v*) after homogenized using a tissue lyser. The supernatant was collected and vacuum-dried using Centri Vap benchtop vacuum concentrator and re-suspended in methanol: H_2O (1:9, *v/v*). Twenty microliters of each sample was taken and mixed into a quality control (QC) sample for evaluation of reproducibility and stability of the analysis.

Metabolomics analysis was performed with an UPLC (waters, USA) in-line with Q Exactive mass spectrometer (Thermo Fisher Scientific, Waltham, MA, USA) equipping with a BEH C18 column (1.7 μm, 2.1 mm × 100 mm, Waters, USA). The eluents were H_2O containing 0.1% (*v/v*) formic acid (A) and methanol containing 0.1% (*v/v*) formic acid (B) for positive ion mode (ESI+), and H_2O containing 10 mM ammonium formate (A) and 95% methanol containing 10 mM ammonium formate (B) for negative ion mode (ESI-).

The linear gradient was as follow: 0–1 min, 98% A; 1–9 min, 98% A to 2% A; 9–12 min, 2% A; 12–12.1 min, 2% A to 98% A and maintained until 15 min. The flow rate was at 350 µL/min, and injection volume was 5 µL. MS data was recorded over the m/z range 70–1050. The flow rate of sheath gas was 40, and aux gas flow rate was 10. The spray voltage was 3.8 kV for positive ion mode and 3.2 kV for negative ion mode. The capillary temperature was 320 °C and the aux gas heater temperature was 350 °C. Data reprocessing was performed with Compound Discoverer 3.0 (Thermo Fisher Scientific, USA) and MetaX (http://metax.genomics.cn, accessed on 16 March 2021) [49].

3.8. Statistical Analysis

All the experiments were conducted in at least three biological replicates. Experimental results were expressed as mean value ± SD. The statistical significance of the results was tested by *t* test.

4. Conclusions

In conclusion, glucose addition affected the biochemical composition of *N. laevis*. The fucoxanthin and chlorophyll contents were significantly decreased, while EPA was increased. Transcriptomic and metabolomic analysis suggested that the impact of glucose-induced trophic transition include upregulation of enzymes associated with the pyruvate metabolism, amino acid biosynthesis, and fatty acid elongation. ELOVL and KCS are likely the main contributors to EPA accumulation, and the downregulation of PDS and ZISO is likely to be responsible for the decrease of fucoxanthin content. Moreover, hub genes were identified from two glucose regulatory gene module, GPRM and GNRM respectively, and their functions were associated with cell cycle, carbohydrate metabolism, purine biosynthesis, and lipid metabolism.

Supplementary Materials: The following are available online at https://www.mdpi.com/article/10.3390/md19080426/s1, Figure S1: Transcriptome analysis overview during the transition from autotrophy to mixotrophy. (A) Principal component analysis (PCA) (B) The box plot of the gene expression level (C) Sequence similarity annotation in NR (Non-Redundant Protein Sequence) database. Figure S2: Hierarchical cluster and KEGG classification of most up-regulated and down-regulated differential expressed genes (DEGs) at 3 h. (A) hierarchical cluster analysis of the top 100 most up-regulated DEGs (B) Hierarchical cluster analysis of the top 100 most down-regulated DEGs (C) KEGG classification of the top 100 most up-regulated DEGs (D) KEGG classification of the top 100 most down-regulated. Figure S3: Metabolome analysis overview during the transition from autotrophy to mixotrophy. (A) The volcano plots of metabolites under positive ion mode (ESI+) (B) The volcano plots of metabolites under negative ion mode (ESI-) (C) KEGG classification of all metabolites. Figure S4: The hierarchical cluster analysis of glucose-associated positive regulatory module (GPRM) (A) and glucose-associated negative regulatory module (GNRM) (B). Table S1: The gene expression levels of all unigenes. Table S2: The Pearson correlation coefficients for all gene expression levels. Table S3: KEGG pathway classification of most up-regulated and down-regulated DEGs. Table S4: All peaks of Metabolomics identified and quantified. Table S5 Genes related in carotenoids biosynthesis and fatty acids metabolism.

Author Contributions: Conceptualization, X.M., B.L. and F.C.; methodology, X.M. and B.L.; formal analysis, X.M., M.G. and X.W.; investigation, X.M., M.G. and X.W.; data curation, X.M., X.W., X.L., J.Y. and B.L.; writing—original draft preparation, X.M.; writing—review and editing, X.L., J.Y., B.L. and F.C.; visualization, X.M.; supervision, B.L. and F.C.; funding acquisition, B.L. and F.C. All authors have read and agreed to the published version of the manuscript.

Funding: This research was funded by National Natural Science Foundation of China, grant number 31901624; Natural Science Foundation of Shandong Province, grant number ZR2019BC068; the Key-Area Research and Development Program of Guangdong Province, grant number 2018B020206001; Guangdong Province Zhujiang Talent Program, grant number 2019ZT08H476; Shenzhen Science and Technology Program, grant number KQTD20180412181334790; and the Innovation Team Project of Universities in Guangdong Province, grant number 2020KCXTD023.

Institutional Review Board Statement: Not applicable.

Informed Consent Statement: Not applicable.

Data Availability Statement: Not applicable.

Conflicts of Interest: The authors declare no conflict of interest.

References

1. Armbrust, E.V. The life of diatoms in the world's oceans. *Nature* **2009**, *459*, 185–192. [CrossRef] [PubMed]
2. Falkowski, P.G.; Barber, R.T.; Smetacek, V.V. Biogeochemical Controls and Feedbacks on Ocean Primary Production. *Science* **1998**, *281*, 200–207. [CrossRef]
3. Ragueneau, O.; Schultes, S.; Bidle, K.; Claquin, P.; Moriceau, B. Si and C interactions in the world ocean: Importance of ecological processes and implications for the role of diatoms in the biological pump. *Glob. Biogeochem. Cycles* **2006**, *20*. [CrossRef]
4. Bhattacharjya, R.; Kiran Marella, T.; Tiwari, A.; Saxena, A.; Kumar Singh, P.; Mishra, B. Bioprospecting of marine diatoms *Thalassiosira*, *Skeletonema* and *Chaetoceros* for lipids and other value-added products. *Bioresour. Technol.* **2020**, *318*, 124073. [CrossRef] [PubMed]
5. Marella, T.K.; Tiwari, A. Marine diatom *Thalassiosira weissflogii* based biorefinery for co-production of eicosapentaenoic acid and fucoxanthin. *Bioresour. Technol.* **2020**, *307*, 123245. [CrossRef] [PubMed]
6. Peng, J.; Yuan, J.P.; Wu, C.F.; Wang, J.H. Fucoxanthin, a marine carotenoid present in brown seaweeds and diatoms: Metabolism and bioactivities relevant to human health. *Mar. Drugs* **2011**, *9*, 1806–1828. [CrossRef] [PubMed]
7. Chuyen, H.V.; Eun, J.B. Marine carotenoids: Bioactivities and potential benefits to human health. *Crit. Rev. Food Sci.* **2017**, *57*, 2600–2610. [CrossRef]
8. Endo, J.; Arita, M. Cardioprotective mechanism of omega-3 polyunsaturated fatty acids. *J. Cardiol.* **2016**, *67*, 22–27. [CrossRef]
9. Kusaikin, M.I.; Ermakova, S.P.; Shevchenko, N.M.; Isakov, V.V.; Gorshkov, A.G.; Vereshchagin, A.L.; Grachev, M.A.; Zvyagintseva, T.N. Structural characteristics and antitumor activity of a new chrysolaminaran from the diatom alga *Synedra acus*. *Chem. Nat. Compd.* **2010**, *46*, 1–4. [CrossRef]
10. D'Ippolito, G.; Sardo, A.; Paris, D.; Vella, F.M.; Adelfi, M.G.; Botte, P.; Gallo, C.; Fontana, A. Potential of lipid metabolism in marine diatoms for biofuel production. *Biotechnol. Biofuels* **2015**, *8*, 28. [CrossRef]
11. Smith, S.R.; Abbriano, R.M.; Hildebrand, M. Comparative analysis of diatom genomes reveals substantial differences in the organization of carbon partitioning pathways. *Algal Res.* **2012**, *1*, 2–16. [CrossRef]
12. Huang, W.C.; Daboussi, F. Genetic and metabolic engineering in diatoms. *Philos. Trans. R. Soc. B* **2017**, *372*, 20160411. [CrossRef] [PubMed]
13. Falciatore, A.; Jaubert, M.; Bouly, J.P.; Bailleul, B.; Mock, T. Diatom Molecular Research Comes of Age: Model Species for Studying Phytoplankton Biology and Diversity. *Plant Cell* **2020**, *32*, 547–572. [CrossRef] [PubMed]
14. Bowler, C.; Allen, A.E.; Badger, J.H.; Grimwood, J.; Jabbari, K.; Kuo, A.; Maheswari, U.; Martens, C.; Maumus, F.; Otillar, R.P.; et al. The *Phaeodactylum* genome reveals the evolutionary history of diatom genomes. *Nature* **2008**, *456*, 239–244. [CrossRef]
15. Remmers, I.M.; D'Adamo, S.; Martens, D.E.; de Vos, R.C.H.; Mumm, R.; America, A.H.P.; Cordewener, J.H.G.; Bakker, L.V.; Peters, S.A.; Wijffels, R.H.; et al. Orchestration of transcriptome, proteome and metabolome in the diatom *Phaeodactylum tricornutum* during nitrogen limitation. *Algal Res.* **2018**, *35*, 33–49. [CrossRef]
16. Matthijs, M.; Fabris, M.; Obata, T.; Foubert, I.; Franco-Zorrilla, J.M.; Solano, R.; Fernie, A.R.; Vyverman, W.; Goossens, A. The transcription factor bZIP14 regulates the TCA cycle in the diatom *Phaeodactylum tricornutum*. *EMBO J.* **2017**, *36*, 1559–1576. [CrossRef]
17. Zaslavskaia, L.A.; Lippmeier, J.C.; Shih, C.; Ehrhardt, D.; Grossman, A.R.; Apt, K.E. Trophic obligate conversion of a photoautotrophic organism through metabolic engineering. *Science* **2001**, *292*, 2073–2075. [CrossRef] [PubMed]
18. Lu, X.; Sun, H.; Zhao, W.; Cheng, K.-W.; Chen, F.; Liu, B. A Hetero-Photoautotrophic Two-Stage Cultivation Process for Production of Fucoxanthin by the Marine Diatom *Nitzschia laevis*. *Mar. Drugs* **2018**, *16*, 219. [CrossRef]
19. Wen, Z.Y.; Chen, F. Heterotrophic production of eicosapentaenoid acid by the diatom *Nitzschia laevis*: Effects of silicate and glucose. *J. Ind. Microbiol. Biotechnol.* **2000**, *25*, 218–224. [CrossRef]
20. Lu, X.; Liu, B.; He, Y.; Guo, B.; Sun, H.; Chen, F. Novel insights into mixotrophic cultivation of *Nitzschia laevis* for co-production of fucoxanthin and eicosapentaenoic acid. *Bioresour. Technol.* **2019**, *294*, 122145. [CrossRef]
21. Rolland, F.; Baena-Gonzalez, E.; Sheen, J. Sugar sensing and signaling in plants: Conserved and novel mechanisms. *Annu. Rev. Plant Biol.* **2006**, *57*, 675–709. [CrossRef]
22. Roth, M.S.; Gallaher, S.D.; Westcott, D.J.; Iwai, M.; Louie, K.B.; Mueller, M.; Walter, A.; Foflonker, F.; Bowen, B.P.; Ataii, N.N.; et al. Regulation of Oxygenic Photosynthesis during Trophic Transitions in the Green Alga *Chromochloris zofingiensis*. *Plant Cell* **2019**, *31*, 579–601. [CrossRef]
23. Matthias, J.; Claudia, B. Properties of photosystem I antenna protein complexes of the diatom *Cyclotella meneghiniana*. *J. Exp. Bot.* **2012**, *63*, 3673–3681.
24. Wang, W.; Yu, L.J.; Xu, C.; Tomizaki, T.; Zhao, S.; Umena, Y.; Chen, X.; Qin, X.; Xin, Y.; Suga, M.; et al. Structural basis for blue-green light harvesting and energy dissipation in diatoms. *Science* **2019**, *363*, eaav0365. [CrossRef] [PubMed]

25. Markou, G.; Angelidaki, I.; Georgakakis, D. Microalgal carbohydrates: An overview of the factors influencing carbohydrates production, and of main bioconversion technologies for production of biofuels. *Appl. Microbiol. Biotechnol.* **2012**, *96*, 631–645. [CrossRef]

26. Kroth, P.G.; Chiovitti, A.; Gruber, A.; Martin-Jezequel, V.; Mock, T.; Parker, M.S.; Stanley, M.S.; Kaplan, A.; Caron, L.; Weber, T.; et al. A Model for Carbohydrate Metabolism in the Diatom *Phaeodactylum tricornutum* Deduced from Comparative Whole Genome Analysis. *PLoS ONE* **2008**, *3*, e1426. [CrossRef]

27. Wen, Z.Y.; Chen, F. Production potential of eicosapentaenoic acid by the diatom *Nitzschia laevis*. *Biotechnol. Lett.* **2000**, *22*, 727–733. [CrossRef]

28. Kathuria, S.V.; Chan, Y.H.; Nobrega, R.P.; Ozen, A.; Matthews, C.R. Clusters of isoleucine, leucine, and valine side chains define cores of stability in high-energy states of globular proteins: Sequence determinants of structure and stability. *Protein Sci.* **2016**, *25*, 662–675. [CrossRef]

29. Lohr, M.; Im, C.S.; Grossman, A.R. Genome-based examination of chlorophyll and carotenoid biosynthesis in *Chlamydomonas reinhardtii*. *Plant Physiol.* **2005**, *138*, 490–515. [CrossRef]

30. Shi, Y.; Liu, M.J.; Ding, W.; Liu, J. Novel Insights into Phosphorus Deprivation Boosted Lipid Synthesis in the Marine Alga *Nannochloropsis oceanica* without Compromising Biomass Production. *J. Agric. Food Chem.* **2020**, *68*, 11488–11502. [CrossRef]

31. Lichtenthaler, H.K. The 1-deoxy-D-xylulose-5-phosphate pathway of isoprenoid biosynthesis in plants. *Annu. Rev. Plant Phys.* **1999**, *50*, 47–65. [CrossRef]

32. Zhan, X.R.; Liao, X.Y.; Luo, X.J.; Zhu, Y.J.; Feng, S.G.; Yu, C.N.; Lu, J.J.; Shen, C.J.; Wang, H.Z. Comparative Metabolomic and Proteomic Analyses Reveal the Regulation Mechanism Underlying MeJA-Induced Bioactive Compound Accumulation in Cutleaf Groundcherry (*Physalis angulata* L.) Hairy Roots. *J. Agric. Food Chem.* **2018**, *66*, 6336–6347. [CrossRef] [PubMed]

33. Ajikumar, P.K.; Tyo, K.; Carlsen, S.; Mucha, O.; Phon, T.H.; Stephanopoulos, G. Terpenoids: Opportunities for biosynthesis of natural product drugs using engineered microorganisms. *Mol. Pharm.* **2008**, *5*, 167–190. [CrossRef]

34. Mao, X.; Lao, Y.; Sun, H.; Li, X.; Yu, J.; Chen, F. Time-resolved transcriptome analysis during transitions of sulfur nutritional status provides insight into triacylglycerol (TAG) and astaxanthin accumulation in the green alga *Chromochloris zofingiensis*. *Biotechnol. Biofuels* **2020**, *13*, 128. [CrossRef]

35. Coesel, S.; Obornik, M.; Varela, J.; Falciatore, A.; Bowler, C. Evolutionary Origins and Functions of the Carotenoid Biosynthetic Pathway in Marine Diatoms. *PLoS ONE* **2008**, *3*, e2896. [CrossRef]

36. Murchie, E.H.; Ruban, A.V. Dynamic non-photochemical quenching in plants: From molecular mechanism to productivity. *Plant J.* **2020**, *101*, 885–896. [CrossRef] [PubMed]

37. Bertrand, M. Carotenoid biosynthesis in diatoms. *Photosynth. Res.* **2010**, *106*, 89–102. [CrossRef] [PubMed]

38. Athanasakoglou, A.; Kampranis, S.C. Diatom isoprenoids: Advances and biotechnological potential. *Biotechnol. Adv.* **2019**, *37*, 107417. [CrossRef]

39. Chen, H.H.; Jiang, J.G. Lipid Accumulation Mechanisms in Auto-and Heterotrophic Microalgae. *J. Agric. Food Chem.* **2017**, *65*, 8099–8110. [CrossRef]

40. Garay, L.A.; Boundy-Mills, K.L.; German, J.B. Accumulation of High-Value Lipids in Single-Cell Microorganisms: A Mechanistic Approach and Future Perspectives. *J. Agric. Food Chem.* **2014**, *62*, 2709–2727. [CrossRef]

41. Wu, T.; Fu, Y.L.; Shi, Y.; Li, Y.L.; Kou, Y.P.; Mao, X.M.; Liu, J. Functional Characterization of Long Chain Acyl CoA Synthetase Gene Family from the Oleaginous Alga *Chromochloris zofingiensis*. *J. Agric. Food Chem.* **2020**, *68*, 4473–4484. [CrossRef]

42. Kevei, Z.; Baloban, M.; Da Ines, O.; Tiricz, H.; Kroll, A.; Regulski, K.; Mergaert, P.; Kondorosi, E. Conserved CDC20 cell cycle functions are carried out by two of the five isoforms in *Arabidopsis thaliana*. *PLoS ONE* **2011**, *6*, e20618. [CrossRef]

43. Eguren, M.; Porlan, E.; Manchado, E.; Garcia-Higuera, I.; Canamero, M.; Farinas, I.; Malumbres, M. The APC/C cofactor Cdh1 prevents replicative stress and p53-dependent cell death in neural progenitors. *Nat. Commun.* **2013**, *4*, 2880. [CrossRef]

44. Buttigieg, P.L.; Hankeln, W.; Kostadinov, I.; Kottmann, R.; Yilmaz, P.; Duhaime, M.B.; Glockner, F.O. Ecogenomic perspectives on domains of unknown function: Correlation-based exploration of marine metagenomes. *PLoS ONE* **2013**, *8*, e50869. [CrossRef]

45. Chen, W.H.; Taylor, M.C.; Barrow, R.A.; Croyal, M.; Masle, J. Loss of Phosphoethanolamine N-Methyltransferases Abolishes Phosphatidylcholine Synthesis and Is Lethal. *Plant Physiol.* **2019**, *179*, 124–142. [CrossRef] [PubMed]

46. Czumaj, A.; Sledzinski, T. Biological Role of Unsaturated Fatty Acid Desaturases in Health and Disease. *Nutrients* **2020**, *12*, 356. [CrossRef] [PubMed]

47. Jeffrey, S.W.; Humphrey, G.F. New Spectrophotometric Equations for Determining Chlorophylls a, B, C1 and C2 in Higher-Plants, Algae and Natural Phytoplankton. *Biochem. Physiol. Pflanz.* **1975**, *167*, 191–194. [CrossRef]

48. Wagner, G.P.; Kin, K.; Lynch, V.J. Measurement of mRNA abundance using RNA-seq data: RPKM measure is inconsistent among samples. *Theory Biosci.* **2012**, *131*, 281–285. [CrossRef]

49. Wen, B.; Mei, Z.L.; Zeng, C.W.; Liu, S.Q. metaX: A flexible and comprehensive software for processing metabolomics data. *BMC Bioinform.* **2017**, *18*, 183. [CrossRef]

Communication

A Method of Solubilizing and Concentrating Astaxanthin and Other Carotenoids

Kiyotaka Y. Hara [1,2,*,†], Shuwa Yagi [1,†], Yoko Hirono-Hara [2] and Hiroshi Kikukawa [1,2]

[1] Laboratory of Environmental Bioengineering, Graduate Division of Nutritional and Environmental Sciences, University of Shizuoka, 52-1 Yada, Suruga-ku, Shizuoka 422-8526, Japan; yc383537@bk2.so-net.ne.jp (S.Y.); kikukawa@u-shizuoka-ken.ac.jp (H.K.)

[2] Department of Environmental and Life Sciences, School of Food and Nutritional Sciences, University of Shizuoka, 52-1 Yada, Suruga-ku, Shizuoka 422-8526, Japan; y-hirono@u-shizuoka-ken.ac.jp

* Correspondence: k-hara@u-shizuoka-ken.ac.jp

† Both the authors contribute equally to this paper.

Abstract: The valuable marine carotenoid, astaxanthin, is used in supplements, medicines and cosmetics. In this study, crustacyanin, an astaxanthin-binding protein, was used to solubilize and concentrate astaxanthin. The recombinant crustacyanin of European lobster spontaneously formed an inclusion body when it was over-expressed in *Escherichia coli*. In this study, fusing the NusA-tag to the crustacyanin subunits made it possible to express in a soluble fraction and solubilize astaxanthin in aqueous solution. By cutting off the NusA-tag, the crustacyanin subunits generated the pure insoluble form, and captured and concentrated astaxanthin. Overall, the attaching and releasing NusA-tag method has the potential to supply solubilized carotenoids in aqueous solution and concentrated carotenoids, respectively.

Keywords: astaxanthin; marine carotenoid; crustacyanin; European lobster; NusA-tag; *Escherichia coli*

Citation: Hara, K.Y.; Yagi, S.; Hirono-Hara, Y.; Kikukawa, H. A Method of Solubilizing and Concentrating Astaxanthin and Other Carotenoids. *Mar. Drugs* **2021**, *19*, 462. https://doi.org/10.3390/md19080462

Academic Editors: Masashi Hosokawa and Hayato Maeda

Received: 23 July 2021
Accepted: 13 August 2021
Published: 16 August 2021

Publisher's Note: MDPI stays neutral with regard to jurisdictional claims in published maps and institutional affiliations.

1. Introduction

Carotenoids are used as red or orange pigments for foods, supplements, pharmaceuticals and cosmetics. Carotenoids are found as natural compounds in microorganisms, algae, plants and animals [1–6]. A valuable carotenoid, astaxanthin ($3,3'$-dihydroxy-β, β-carotene-$4,4'$-dione; $C_{40}H_{52}O_4$), acts as a protective agent against oxidative damage to cells in vivo because it has high antioxidant activity [7–9]. Astaxanthin for human consumption is often produced by extraction from the microalgae, *Haematococcus pluvialis* [10]; generally extracted from *H. pluvialis* using super-critical CO_2, a method which is not harmful with easy removal of the CO_2 [11,12]. However, the efficiency of extracting astaxanthin from *H. pluvialis* cells has been reported to be less than 50%, even when carried out under high pressure (30 MPa) [13,14]. Repeating the super-critical extraction of astaxanthin several times also leads to very high running costs [15–18]. In addition, astaxanthin is generally extracted with organic solvents and can be concentrated by evaporation [19]. For a sustainable society in the future, environmentally-negative solvent discharge and energy-intensive evaporation processes are not desirable. Therefore, there is a need to develop an environmentally friendly astaxanthin concentrating method. Furthermore, providing astaxanthin in the form of an aqueous solution is important for developing applications of this valuable product, such as cosmetics and nutritional supplement drinks.

In the cells of the shells of crustaceans, astaxanthin exists with its binding protein, crustacyanin. Wald et al. first extracted crustacyanin from the shells of lobsters [20]. Crustacyanin subunits can be classified into two groups according to the composition of their amino acids, molecular size and peptide mapping. One group consists of the A1, C1 and C2 subunits, and the other consists of the A2 and A3 subunits [21]. Heterologous expression of the crustacyanin subunits in *E. coli* has also been carried out to clarify their three-dimensional structures [22]. One possible method of concentrating carotenoids is to use astaxanthin-binding proteins. However, the expression levels of these crustacyanin subunits are lower than those required for its application.

The current study has developed a simple method for solubilizing and concentrating astaxanthin using an easily-purified single crustacyanin protein. Using the attaching and releasing NusA-tag method, the crustacyanin protein was solubilized with the NusA-tag, followed by cleavage of the tag, which easily induces a pure precipitate for concentrating astaxanthin. It has been demonstrated that the method developed in this study can be used not only for astaxanthin, the original substrate of crustacyanin, but also for α-carotene and β-carotene.

2. Results and Discussion

2.1. Optimization of Crustacyanin Preparation

For the first preparation step, native European lobster crustacyanin was produced in *E. coli*. The expression levels of A2 and C1 were higher than those of the other subunits [21], and A2C1 was a main subcomplex which forms *H. gammarus* crustacyanin [22]. Therefore, genes artificially synthesized with codon optimization for *E. coli* which encode the native *H. gammarus* crustacyanin A2 and C1 proteins were introduced into *E. coli*. Figure 1A shows that almost all the A2 (23 kDa) and C1 (26 kDa) proteins were expressed as insoluble inclusion bodies in *E. coli*. Therefore, their expressions in the inclusion bodies made the application of these proteins difficult. The NusA-tag is a protein known to increase the solubility of its conjugated target protein. In the present study, the NusA-tag (molecular weight calculated as 55 kDa and shown as 69 kDa in SDS-PAGE) [23] was fused to the N-terminal of the A2 and C1 proteins. As a result, the A2 and C1 proteins tagged with NusA were expressed in the soluble fraction at almost the same level as that of the inclusion bodies (Figure 1B).

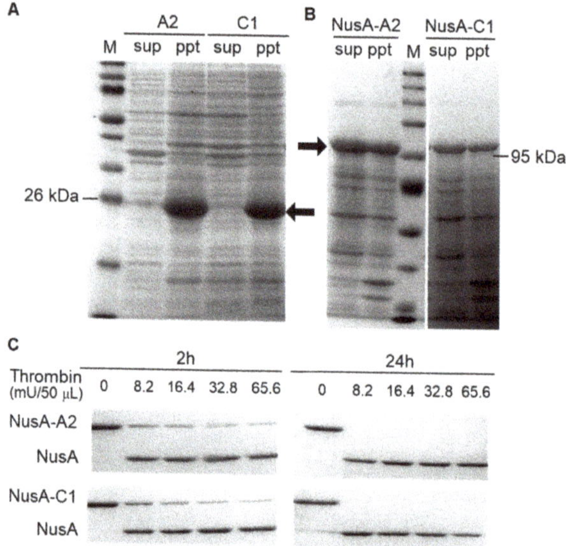

Figure 1. Expression and digestion pattern of crustacyanin. (**A**) Native crustacyanin A2 and C1 are represented by the arrows; (**B**) NusA-tagged crustacyanin A2 and C1 proteins are represented by the arrows. The abbreviations "sup", "ppt" and "M" represent supernatant, pellet and molecular marker, respectively; (**C**) Cleaving the NusA-tag from the A2 and C1 proteins tagged with NusA. Biotinylated thrombin was added to NusA-A2 and NusA-C1 crustacyanin in a 50 µL reaction mixture at different concentrations shown for 2 or 24 h. Concentration of acrylamide used in SDS-PAGE is 8.0%.

In the second step, to construct a simple method to purify the crustacyanin proteins, the NusA-tag was removed from the NusA-tagged crustacyanin proteins. The protease, thrombin, was used to digest the spacer region between the NusA-tag and the crustacyanin proteins. To optimize the conditions for digestion, the NusA-A2 and NusA-C1 proteins (0.2 µg/mL) were digested using the biotinylated thrombin at different concentrations (0, 8.2, 16.4, 32.8 and 65.6 mU) at 4 °C for different reaction times (2, 4, 8 or 24 h). After this digestion reaction in the presence of 65.6 mU thrombin protease for 2 h, both the NusA-A2 and NusA-C1 proteins were almost completely digested to generate the A2 and C1 proteins (Figure 1C). In contrast, a reaction time of 24 h using less than 8.2 mU thrombin protease was sufficient for the complete digestion of NusA from the NusA-A2 and NusA-C1 proteins.

2.2. Carotenoid Recovery by Insoluble Crustacyanin

The crustacyanin A2C1 complex has been reported to have the ability to bind to astaxanthin [21]. The binding ability of A2C1 was compared when tagged with or without NusA. The NusA-A2C1 and NusA-free A2C1 complexes were constructed by mixing the NusA-A2/NusA-C1 proteins and NusA-free A2/C1 proteins, respectively. The astaxanthin was solubilized in buffer with the NusA-A2C1 complex and no precipitation was observed after centrifugation (Figure 2), which indicated that the astaxanthin had solubilized in the aqueous phase.

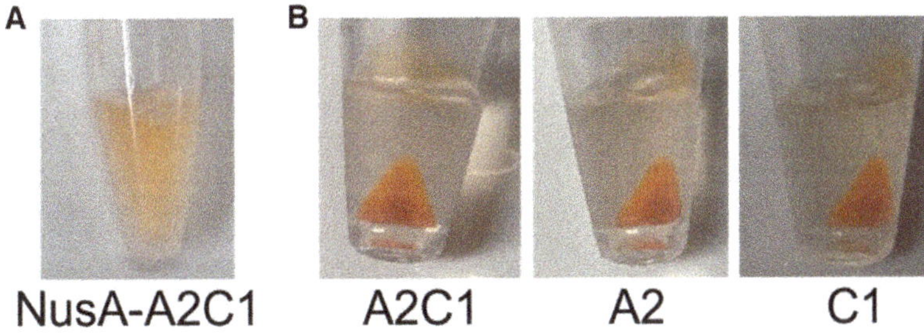

Figure 2. Solubilization and concentration of astaxanthin by crustacyanin subunits. (**A**) Addition of astaxanthin to crustacyanin A2 and C1 subunits attaching the NusA-tag; (**B**) Addition of astaxanthin to crustacyanin A2C1, A2 and C1 after digesting the NusA-tag.

This method of solubilizing astaxanthin in aqueous solution could be applied in various industries, particularly for cosmetics and supplemental nutrition drinks. In contrast, the digestion of the NusA-tag from the NusA-A2 and NusA-C1 proteins and NusA-A2C1 complex constructed from the resulting NusA-free proteins generated a red-colored precipitate after binding to astaxanthin and centrifugation (Figure 2). These precipitates indicate the ability of the method to capture and concentrate astaxanthin of a high purity as the precipitated form of crustacyanin.

To evaluate whether the crustacyanin subunits could be used to concentrate carotenoids other than astaxanthin, the abilities of the A2 or C1 protein and their A2C1 complex to bind α-carotene and β-carotene were compared (Figure 3). The crustacyanin A2 protein individually bound all three carotenoids at a recovery rate (>75%) higher than that for the C1 protein, and almost equal to that for the A2C1 complex. These results indicated that the expression of both the A2 and C1 proteins and the formation of their complex is not required for the efficient capture of carotenoids. The crustacyanin A2 subunit alone provided a simple method for concentrating carotenoids. Figure 3 also shows no differences in the recovery rate for astaxanthin, α-carotene and β-carotene from the mixture. This wide-ranging ability to capture carotenoids indicates the potential of this method for application to any other type of carotenoid.

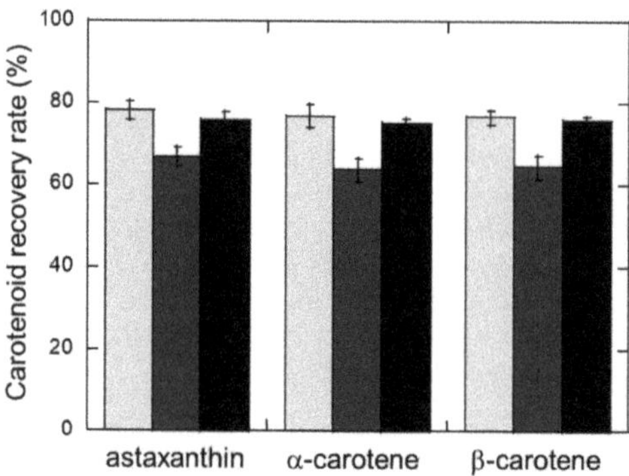

Figure 3. Capturing carotenoids from a mixture using the crustacyanin subunits. The crustacyanin subunits were reacted with a mixture of carotenoids containing astaxanthin, α-carotene and β-carotene, each at a concentration of 20 μg/mL. The amount of each carotenoid captured in the crustacyanin was measured by high-performance liquid chromatography and their recovery rates were calculated from the amounts added. The light gray, dark gray and black bars represent the crustacyanin A2, C1 subunits and A2C1 complex, respectively. Error bars represent the standard deviations.

3. Materials and Methods

3.1. Strains, Plasmids and Transformation

Genes encoding the European lobster *Homarus gammarus* crustacyanin A2 and C1 subunits (Gene bank accession No. LC598211 and No. LC598212, respectively) were artificially synthesized with codon optimization for *Escherichia coli* (Eurofins Genomic, Tokyo, Japan). These genes were amplified using PCR, then subcloned into the pET-21d vector (Merck KGaA, Darmstadt, Germany) to construct pET21d-A2 (the forward primer and reverse primer used in this amplification were 5′-GTGCGGCCGCAAGCTTCGCGCGATACACGC ATTCG-3′ and 5′-CGACAAGAGTCCGGGAGCTCTGGACGGCATTCCGTCCTTTG-3′, respectively) and pET21d-C1 (the forward primer and reverse primer used in this amplification were 5′-GTGCGGCCGCAAGCTTCAGCGTTTTCTGGGTATCATACGG-3′ and 5′-CGACAAGAGTCCGGGAGCTCTGGACAAGATTCCCGATTTTGTTG-3′, respectively). *E. coli* Rosetta2 (Merck KGaA) was then transformed using pET21d-A2 or pET21d-C1 to obtain *E. coli* strains which could over-express native A2 or C1 subunits induced by IPTG. The PCR fragments including the A2 or C1 genes were also subcloned into the pET43.1a vector (Merck KGaA) to construct pET43.1a-A2 and pET43.1a-C1. *E. coli* Origami B (Merck KGaA) was transformed by pET43.1a-A2 or pET43.1a-C1 to obtain *E. coli* strains which could over-express the A2 or C1 subunits with the NusA-tag (Nus·Tag) at their N-terminus. These *E. coli* Origami B mutant strains were then transformed using the pTF16 vector (Takara Bio, Shiga, Japan), which contains the gene-encoding Tig chaperone induced by L-arabinose, to construct *E. coli* Origami B (pET-43.1a-A2/pTF-16) and Origami B (pET-43.1a-C1/pTF-16).

3.2. Preparation of Soluble NusA-Crustacyanin

The *E. coli* Rosetta 2 and Origami B mutant strains were cultured in LB medium (10 g/L tryptone, 5 g/L yeast extract and 5 g/L sodium chloride) supplemented with 0.5 mg/mL L-arabinose and 100 μg/mL ampicillin in an Erlenmeyer flask, stirred at 100 rpm at 37 °C. When the cell concentration reached an OD_{600} value of 0.5, 1 mM IPTG was added and then cultured at 100 rpm for a further 3 h at 30 °C. After centrifugation ($3000\times$ *g*, 5 min, 4 °C), the supernatant was removed and the cell pellet was suspended in an appropriate volume of 50 mM sodium phosphate buffer (pH 7.0), then sonicated for a total of 45 s (on 5 s/off 5 s × 9 cycles) using an ultrasonic disruptor (Vibra Cell, Sonics & Materials Inc, Newtown, CT, USA). After centrifugation ($1060\times$ *g*, 5 min, 4 °C), ammonium sulfate (30% saturated concentration) was rapidly added to the disrupted cell extract, then incubated for 30 min at 4 °C. Fifteen milliliters of the resulting solution was centrifuged (1060× g, 5 min, 4 °C), then the supernatant was removed. The remaining pellet was resuspended in 3 mL of 50 mM sodium phosphate buffer (pH 7.0), dialyzed in the same buffer overnight, followed by replacement with 3 mL of fresh buffer and additional dialysis for 4 h. The dialyzed NusA-A2 and NusA-C1 proteins were centrifuged (1060× g, 5 min, 4 °C), then the concentration of each solubilized protein in the supernatant was measured using a Protein Assay BCA Kit (FUJIFILM Wako Pure Chemical Corp., Osaka, Japan). Equal volumes of the NusA-A2 and NusA-C1 solutions were mixed to form a NusA-A2C1 complex as described previously [22]. The chemicals were purchased from Nacalai Tesque (Kyoto, Japan) or FUJIFILM Wako Pure Chemical Corp.

3.3. Preparation of NusA-free Crustacyanin

NusA was digested from the NusA-A2 or NusA-C1 protein as described in the manual of the Novagen® Thrombin kits (Merk KGaA, Darmstadt, Germany). The biotinylated thrombin was used at different concentrations at 4 °C for different reaction times. Ten microliters of the reaction mixture were applied to SDS-PAGE to check the efficiency of digesting NusA from the NusA-A2 and NusA-C1 proteins. Under the optimized digestion conditions, large amounts (0.4 mg/mL) of NusA-A2 and NusA-C1 proteins were separately digested with 320 mU biotinylated thrombin at 4 °C for 3 days to prepare enough amounts of NusA-free crustacyanin for the next experiment described in Section 3.4. The NusA-free A2 and C1 proteins were collected through precipitation by centrifugation (1060× g, 5 min, 4 °C) and removing the supernatant. The proteins were concentrated by resuspension in the appropriate volume of 50 mM sodium phosphate buffer (pH 7.0).

3.4. Concentrating Carotenoids Using NusA-free Crustacyanin

Sixty microliters of a carotenoid mixture containing 20 µg/mL each of astaxanthin, α-carotene and β-carotene were dissolved in acetone, then gently mixed with 0.1 mL of each of the single NusA-free A2 or NusA-free C1 proteins (all at a concentration of 1.0 mg/mL) and 50 µL of their A2C1 complex generated from A2 and C1 proteins (at a concentration of 1.0 µg/µL). After centrifugation, the carotenoids precipitated with the A2 or C1 proteins or the A2C1 complex were suspended in 0.4 mL of acetone, vortexed and centrifuged (16,000× g, 5 min, 4 °C) to obtain the first supernatant for recovering carotenoids. To recover the remaining carotenoids from each protein, the precipitated protein was resuspended in 0.2 mL of acetone, vortexed and centrifuged (16,000× g, 5 min, 4 °C), and then the resulting second supernatant was mixed with the first supernatant (total 0.6 mL). The carotenoid concentrations were determined as described previously [24,25] using a high-performance liquid chromatography system (Shimadzu, Kyoto, Japan) equipped with a COSMOSIL Cholester Packed Column (4.6 i.d. × 250 mm, Nacalai Tesque, Kyoto, Japan). The operating conditions were: column temperature, 35 °C; mobile phase, methanol/tetrahydrofuran (80/20 (v/v)); flow rate, 1.0 mL/min; and detection of carotenoids at 470 nm using a UV detector (SPD-20AV, Shimadzu).

4. Conclusions

In this study, we developed an efficient method for producing crustacyanin subunits and used them to solubilize and concentrate carotenoids. Crustacyanin subunits with a NusA-tag attached to their N-terminus achieved soluble expression in *E. coli* and solubilized astaxanthin in aqueous solution. Astaxanthin could be concentrated by cleaving the NusA-tag from the crustacyanin subunit followed by its precipitation with astaxanthin. Furthermore, this method also could be applied to α-carotene and β-carotene. These results should indicate the potential of the attaching and releasing NusA-tag method to supply a solubilization process by attaching the NusA-tag, and a concentration process by releasing the NusA-tag, for the production of a broad range of carotenoids.

Author Contributions: Conceptualization, K.Y.H. and S.Y.; Methodology, K.Y.H., S.Y., Y.H.-H. and H.K.; Validation, K.Y.H. and S.Y.; Formal Analysis, S.Y.; Investigation, S.Y.; Data Curation, K.Y.H. and

S.Y.; Writing—Original Draft Preparation, K.Y.H. and S.Y.; Writing—Review and Editing, Y.H.-H. and H.K.; Visualization, K.Y.H. and S.Y.; Project Administration, K.Y.H.; Funding Acquisition, K.Y.H. All authors have read and agreed to the published version of the manuscript.

Funding: This work was supported by JSPS KAKENHI Grant Numbers JP21H03645 and JP16K00616. This work was also supported by JST-Mirai Program Grant Number JPMJMI17EJ.

Institutional Review Board Statement: Not applicable.

Informed Consent Statement: Not applicable.

Data Availability Statement: The article contains all the data produced in this study.

Acknowledgments: We thank N. Tanzawa for her technical assistance.

Conflicts of Interest: The authors declare no financial or commercial conflict of interest.

Abbreviations

IPTG, isopropyl-β-D-thiogalactopyranoside; PCR, polymerase chain reaction; SDS-PAGE, sodium dodecyl sulfate-polyacrylamide gel electrophoresis.

References

1. Vachali, P.; Bhosale, P.; Bernstein, P.S. Microbial carotenoids. *Methods Mol. Biol.* **2012**, *898*, 41–59.
2. Mata-Gómez, L.C.; Montañez, J.C.; Méndez-Zavala, A.; Aguilar, C.N. Biotechnological production of carotenoids by yeasts: An overview. *Microb. Cell Fact.* **2014**, *13*, 12. [CrossRef]
3. Yamamoto, K.; Hara, K.Y.; Morita, T.; Nishimura, A.; Sasaki, D.; Ishii, J.; Ogino, C.; Kizaki, N.; Kondo, A. Enhancement of astaxanthin production in *Xanthophyllomyces dendrorhous* by efficient method for the complete deletion of genes. *Microb. Cell Fact.* **2016**, *15*, 155. [CrossRef]
4. Gervasi, T.; Santini, A.; Daliu, P.; Salem, A.Z.; Gervasi, C.; Pellizzeri, A.V.; Salem, Z.M.; Gervasi, C.; Pellizzeri, V.; Barrega, L.; et al. Astaxanthin production by *Xanthophyllomyces dendrorhous* growing on a low cost substrate. *Agroforest Syst.* **2020**, *94*, 1229–1234. [CrossRef]
5. Villegas-Méndez, M.Á.; Aguilar-Machado, D.E.; Balagurusamy, N.; Montañez, J.; Morales-Oyervides, L. Agro-industrial wastes for the synthesis of carotenoids by *Xanthophyllomyces dendrorhous*: Mesquite pods-based medium design and optimization. *Biochem. Eng. J.* **2019**, *150*, 107260. [CrossRef]
6. Korumilli, T.; Mishra, S.; Korukonda, J.R. Production of astaxanthin by *Xanthophyllomyces dendrorhous* on fruit waste extract and optimization of key parameters using Taguchi method. *J. Biochem. Technol.* **2020**, *11*, 25.
7. Satoh, A.; Tsuji, S.; Okada, Y.; Murakami, N.; Urami, M.; Nakagawa, K.; Ishikura, M.; Katagiri, M.; Koga, Y.; Shirasawa, T. Preliminary clinical evaluation of toxicity and efficacy of a new astaxanthin-rich *Haematococcus pluvialis* extract. *J. Clin. Biochem. Nutr.* **2009**, *44*, 280–284. [CrossRef] [PubMed]
8. Tripathi, D.N.; Jena, G.B. Intervention of astaxanthin against cyclophosphamide-induced oxidative stress and DNA damage: A study in mice. *Chem.-Biol. Interact.* **2009**, *180*, 398–406. [CrossRef] [PubMed]
9. Hara, K.Y.; Morita, T.; Endo, Y.; Mochizuki, M.; Araki, M.; Kondo, A. Evaluation and screening of efficient promoters to improve astaxanthin production in *Xanthophyllomyces dendrorhous*. *Appl. Microbiol. Biotechnol.* **2014**, *98*, 6787–6793. [CrossRef] [PubMed]
10. Panis, G.; Rosales, C.J. Commercial astaxanthin production derived by green alga *Haematococcus pluvialis*: A microalgae process model and a techno-economic assessment all through production line. *Algal. Res.* **2016**, *18*, 175–190. [CrossRef]
11. Brunner, G.J. Supercritical fluids: Technology and application to food processing. *Food Eng.* **2005**, *67*, 21–33. [CrossRef]
12. da Silva, R.P.F.F.; Rocha-Santos, T.A.P.; Duarte, A.C. Supercritical fluid extraction of bioactive compounds. *TrAC Trends Anal. Chem.* **2016**, *76*, 40–51. [CrossRef]
13. Valderrama, J.O.; Perrut, M.; Majewski, W. Extraction of astaxantine and phycocyanine from microalgae with supercritical carbon dioxide. *J. Chem. Eng. Data* **2003**, *48*, 827–830. [CrossRef]
14. Nobre, B.; Marcelo, F.; Passos, R.; Beiro, L.; Palavra, A.; Gouveia, L.; Mendes, R. Supercritical carbon dioxide extraction of astaxanthin and other carotenoids from the microalga *Haematococcus pluvialis*. *Eur. Food Res. Technol.* **2006**, *223*, 787–790. [CrossRef]
15. Perrut, M. Supercritical fluid applications: Industrial developments and economic issues. *Ind. Eng. Chem. Res.* **2000**, *39*, 4531–4535. [CrossRef]
16. Pan, J.L.; Wang, H.M.; Chen, C.Y.; Chang, J.S. Extraction of astaxanthin from *Haematococcus pluvialis* by supercritical carbon dioxide fluid with ethanol modifier. *Eng. Life Sci.* **2012**, *12*, 638–647. [CrossRef]
17. Todd, R.; Baroutian, S.J. A techno-economic comparison of subcritical water, supercritical CO_2 and organic solvent extraction of bioactives from grape marc. *J. Clean. Prod.* **2017**, *158*, 349–358. [CrossRef]

18. Khoo, K.S.; Lee, S.Y.; Ooi, C.W.; Fu, X.; Miao, X.; Ling, T.C.; Show, P.L. Recent advances in biorefinery of astaxanthin from *Haematococcus pluvialis*. *Bioresour. Technol.* **2019**, *288*, 121606. [CrossRef]
19. Park, J.Y.; Oh, Y.K.; Choi, S.A.; Kim, M.C. Recovery of astaxanthin-containing oil from *Haematococcus pluvialis* by nano-dispersion and oil partitioning. *Appl. Biochem. Biotechnol.* **2020**, *190*, 1304–1318. [CrossRef] [PubMed]
20. Wald, G.; Nathanson, N. Crustacyanin, the blue carotenoid-protein of the lobster shell. *Biol. Bull.* **1948**, *95*, 249.
21. Chayen, N.E.; Cianci, M.; Grossmann, J.G.; Habash, J.; Helliwell, J.R.; Nneji, G.A.; Raftery, J.; Rizkallah, P.J.; Zagalsky, P.F. Unravelling the structural chemistry of the colouration mechanism in lobster shell. *Acta Crystallogr.* **2003**, *59*, 2072–2082. [CrossRef] [PubMed]
22. Ferrari, M.; Folli, C.; Pincolini, E.; McClintock, T.S.; Rössle, M.; Berni, R.; Cianci, M. Structural characterization of recombinant crustacyanin subunits from the lobster *Homarus americanus. Acta Crystallogr.* **2012**, *68*, 846–853.
23. Davis, G.D.; Elisee, C.; Newham, D.M.; Harrison, R.G. New fusion protein systems designed to give soluble expression in *Escherichia coli. Biotechnol. Bioeng.* **1999**, *65*, 382–388. [CrossRef]
24. Hara, K.Y.; Kageyama, Y.; Tanzawa, N.; Hirono-Hara, Y.; Kikukawa, H.; Wakabayashi, K. Development of astaxanthin production from citrus peel extract using *Xanthophyllomyces dendrorhous. Environ. Sci. Pollut. Res.* **2021**, *28*, 12640–12647. [CrossRef] [PubMed]
25. Kikukawa, H.; Okaya, T.; Maoka, T.; Miyazaki, M.; Murofushi, K.; Kato, T.; Hirono-Hara, Y.; Katsumata, M.; Miyahara, S.; Hara, K.Y. Carotenoid Nostoxanthin production by *Sphingomonas* sp. SG73 isolated from deep sea sediment. *Mar. Drugs* **2021**, *19*, 274. [CrossRef] [PubMed]

marine drugs

MDPI

Review

Astaxanthin Delivery Systems for Skin Application: A Review

Sarah Giovanna Montenegro Lima [1], Marjorie Caroline Liberato Cavalcanti Freire [2], Verônica da Silva Oliveira [1], Carlo Solisio [3], Attilio Converti [3] and Ádley Antonini Neves de Lima [1,*]

[1] Department of Pharmacy, Federal University of Rio Grande do Norte, Natal 59012-570, RN, Brazil; sarahmontenegrolima@gmail.com (S.G.M.L.); veronicasoliver47@gmail.com (V.d.S.O.)
[2] Physics Institute of São Carlos, University of São Paulo, São Carlos 13566-590, SP, Brazil; marjorie_freire_@hotmail.com
[3] Department of Civil, Chemical and Environment Engineering, Pole of Chemical Engineering, University of Genoa, I-16145 Genoa, Italy; solisio@unige.it (C.S.); converti@unige.it (A.C.)
* Correspondence: adleyantonini@yahoo.com.br; Tel.: +55-(84)99928-8864

Abstract: Astaxanthin (AST) is a biomolecule known for its powerful antioxidant effect, which is considered of great importance in biochemical research and has great potential for application in cosmetics, as well as food products that are beneficial to human health and medicines. Unfortunately, its poor solubility in water, chemical instability, and low oral bioavailability make its applications in the cosmetic and pharmaceutical field a major challenge for the development of new products. To favor the search for alternatives to enhance and make possible the use of AST in formulations, this article aimed to review the scientific data on its application in delivery systems. The search was made in databases without time restriction, using keywords such as astaxanthin, delivery systems, skin, cosmetic, topical, and dermal. All delivery systems found, such as liposomes, particulate systems, inclusion complexes, emulsions, and films, presented peculiar advantages able to enhance AST properties, among which are stability, antioxidant potential, biological activities, and drug release. This survey showed that further studies are needed for the industrial development of new AST-containing cosmetics and topical formulations.

Keywords: astaxanthin; delivery system; skin; cosmetics; drugs; drug release

Citation: Lima, S.G.M.; Freire, M.C.L.C.; Oliveira, V.d.S.; Solisio, C.; Converti, A.; de Lima, Á.A.N. Astaxanthin Delivery Systems for Skin Application: A Review. *Mar. Drugs* **2021**, *19*, 511. https://doi.org/10.3390/md19090511

Academic Editors: Masashi Hosokawa and Hayato Maeda

Received: 17 August 2021
Accepted: 1 September 2021
Published: 9 September 2021

1. Introduction

Astaxanthin (AST) is a xanthophyll carotenoid that was first isolated from lobster by Kuhn and Sorensen and was commercialized as a pigmentation agent for feed in the aquatic farm industry [1,2]. The AST, or 3,3′-dihydroxy-β,β-carotene-4,4′-dione, is a tetraterpene composed of 40 carbon atoms (Figure 1A), and its molecular formula is $C_{40}H_{52}O_4$ (molecular mass 596.85 g·mol^{-1}) [3,4]. This reddish-orange pigment is solid at room temperature, is fat-soluble, and its log P (octanol/water partition) is 13.27 [4]. Furthermore, its chemical structure is composed of 13 conjugated double bonds that have the ability to neutralize free radicals, conferring the strong antioxidant activity of AST [3].

In recent years, AST has gained visibility and attracted attention for cosmetic and dermatological applications, thanks to its remarkable antioxidant properties, which are much stronger than those of tocopherol, and to its positive effects on skin health and protection against UV radiation, which may suggest promising applications in anti-aging products [5,6].

Regarding industrial applications, AST has been produced synthetically through cost-effective methods for large-scale production. However, the green microalgae *Haematococcus pluvialis*, due to its ability to accumulate AST at high levels, is the main source for human consumption, besides being the most promising source for its industrial biological production [1,7]. It can be found either in plants, animals, yeasts, or other algae species. Nowadays, it finds many applications in aquaculture, cosmetics, foods, nutraceuticals, and pharmaceuticals [1] (Figure 1B).

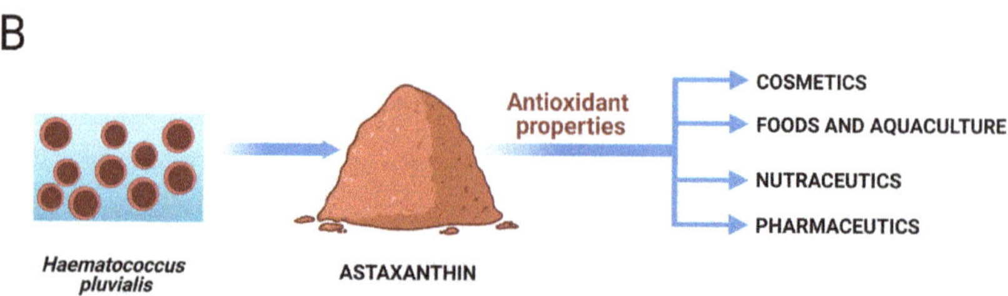

Figure 1. (**A**) Astaxanthin (AST) chemical structure. (**B**) The microalgae *Haematococcus pluvialis* is a promising source for AST industrial biological production and AST applications. Diagram created using BioRender.com (accessed on 17 August 2021).

In the foodstuffs field, AST has been widely used as a supplement and healthy functional food, due to its neuroprotective properties and its ability to scavenge the free radicals produced by physical exercise, improve human immunity, resist fatigue, delay aging and prevent a number of diseases that cause organ aging. In addition, it can be used as a food colorant and antioxidant, to enhance the sensory quality and nutritional values of foods [8]. Oral consumption of AST-containing supplements seems to have positive effects on the skin, such as facial skin rejuvenation [9], an increase in flap viability [10], protection against photoaging caused by UV irradiation [11,12], and improvements in fine lines/wrinkles, elasticity, moisture, age spot size and texture [13,14].

Topical AST application has been reported to have several skin health benefits, including antioxidant and anti-aging effects [13,15–19], protection against UV irradiation [16,20], anti-wrinkle [14,16,21], hydration [21], wound healing [22,23], anti-cancer properties [17], and anti-eczema effects [13,24].

However, the low bioavailability and solubility of AST limit its use in topical formulations. The development of delivery systems to improve AST's application for skin purposes is a promising way to develop new cosmetics and pharmaceutical products. With this goal in mind, this article aimed to review the delivery systems described in the literature for enhancing AST's properties, discuss the main results of the developed formulations and contribute to its applications in cosmetics and topical formulations.

2. Astaxanthin Delivery Systems for Skin Application

2.1. Vesicular Systems

Liposomes

Liposomes are colloidal and vesicular delivery systems, composed of at least one bilayer amphiphilic lipid membrane and a hydrophilic core [25]. These systems have many advantages, including the capability of encapsulating hydrophilic, hydrophobic, or amphiphilic components, target potential, slow-release properties, and biocompatibility due to the similarity with cell membranes, as well as biodegradability, low toxicity, ease of preparation and the ability to extend product shelf-life [25,26]. The use of liposomes as a drug delivery system allows partially overcoming the problems related to the poor stability, water-solubility, and bioavailability of AST [27] (Figure 2). For skin treatment purposes, liposomes are an excellent option for drug delivery because they can be absorbed by endocytosis, particularly by cells of the reticuloendothelial system, and release entrapped drugs. They are also able to fuse with cells or exchange lipids with the cell membrane [28]. However, in aqueous solutions, liposomes have reduced vesicle flexibility, which is a problem in the preparation of efficient liposomal formulations. In addition, absorption from topical application to the skin requires that it reaches the stratum corneum, which poses a challenge for the development of new alternatives to increase efficiency. The loaded-AST liposomes studies found are summarized in Table 1.

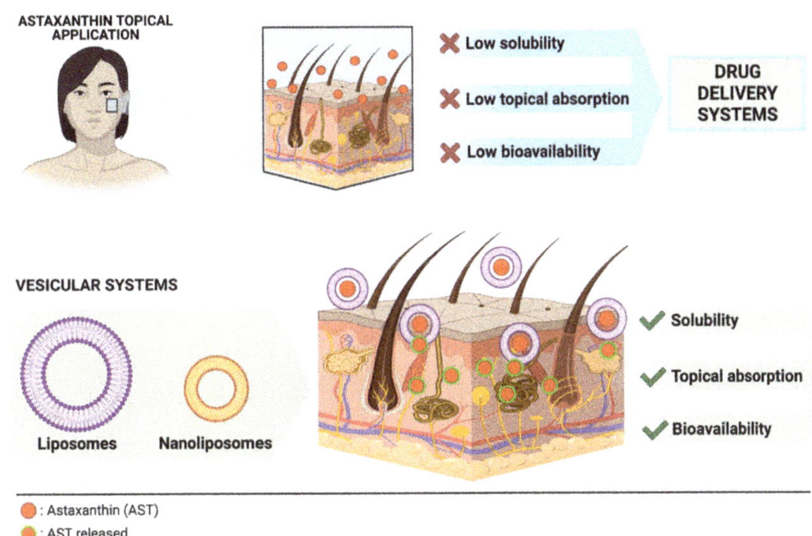

Figure 2. Disadvantages of the topical administration of AST and the use of delivery systems to improve the properties of the molecule, favoring skin absorption. Below: a schematic representation of vesicular systems (liposomes and nanoliposomes) acting as delivery systems for AST. Created with BioRender.com (accessed on 17 August 2021).

Dopierała et al. [29] have recently investigated the interactions between AST and the lipid membrane, in terms of the thermodynamic, morphological, and viscoelastic behavior and surface charge density, to understand how this molecule can affect liposomal formation and characteristics. The study showed that AST can be incorporated into the lipid monolayer and form a stable film, which regulates membrane fluidity and enhances the stability of the liposomal delivery system. Through steady-state fluorescence measurements, these authors observed that the interaction between AST and the lipid membrane reduced membrane fluidity but increased its micropolarity within a certain range of AST concentration, giving more stability to the saturated lipid bilayer. This system appears to be a promising way to improve lipid-based drug formulations, designing a dosage form that may help overcome the problem of unstable bioactive compounds. In addition to AST's antioxidant and biological activities, it is possible to enhance the stability of other formulations that are acting as adjuvants. Previously, Goto et al. [30] suggested that the two terminal rings of AST are likely to interact with the hydrophilic polar region of membrane phospholipids, and intermolecular hydrogen bonds could be formed between polar ends of the hydroxyl-ketocarotene and polar groups of phospholipids, thus regulating the fluidity of lipid membranes.

As an alternative liposomal delivery system, Lee et al. [31] developed liposomes with solid-supported lipid bilayers to overcome the flexibility issues of vesicles in an aqueous solution. Silicified liposomes were synthesized via silicification with tetraethyl orthosilicate on the hydrophilic regions of lecithin vesicles and assembled with boron nitride to enhance their stability. The resulting AST-loaded silicified-phospholipids boron nitride complex showed a release pattern and an antioxidant activity according to the 2,2-diphenyl-1-picrylhydrazyl (DPPH) free-radical scavenging method that is compatible with the literature. In addition, it presents a simple and low-cost method for AST applications in cosmetics.

Table 1. Summary of AST vesicular delivery systems, describing the preparation technique, liposome type, characterization and stability data, and assays (in vitro/in vivo) that were performed for each system.

Preparation Technique	Liposome Type	Characterization	Storage and Stability Data	Assays	References
Dissolution of hydrogenated lecithin and treatment by a high-pressure homogenizer to form nanoemulsions and tetraethyl orthosilicate addition to promote silification	Lecithin silicified liposomes	Brauner–Emmett–Teller isotherm, field emission scanning electron microscopy, Fourier transform infrared spectroscopy, UV–visible spectrophotometry	-	In vitro: DPPH free radical scavenging activity and drug release profile	[31]
Film dispersion-ultrasonic technique	Soybean phosphatidylcholine nanoliposomes	Dynamic light scattering, transmission electron microscopy, X-ray diffraction, differential scanning calorimetry, thermal gravimetric analysis, and dissolution study.	Thermal stability enhanced after encapsulation	In vitro: drug release profile	[27]
Lipid hydration method	Egg phosphatidylcholine liposomes	Dynamic light scattering	-	In vitro: antioxidant activity In vivo: UV treatment of mouse dorsal skin and effect of iontophoretic transdermal delivery	[18]
Lipid hydration method	Egg phosphatidylcholine liposomes	-	-	In vitro antioxidant activity by scavenging hydroxyl radical, and protective effect against cytotoxicity induced by hydroxyl radical	[32]

However, regular liposomes have some disadvantages, due to their large particle size (around 1–100 µm). As an alternative way to enhance the use of this delivery system, Pan et al. [27] developed nanoliposomes with a lower particle size that exhibited better penetration and new targeting properties. Using a film-dispersion ultrasonic technique and soybean phosphatidylcholine as the lipid, AST-incorporating nanoliposomes (AST-LN) were prepared, characterized, and evaluated for their in vitro release, compared with the pure molecule. Nanoliposomes ensured higher encapsulation efficiency (97.5 ± 0.3%) than in previous reports, and an average particle size of 80 ± 2 nm, which was confirmed by transmission electron microscopy (TEM). AST-LN was 17 times more soluble than pure AST and had good water dispersibility, without insoluble particles or precipitation. X-ray diffraction (XRD) analyses suggested that this improvement of water solubility could be due to the change of AST's crystalline natural form into nanoliposomes. To evaluate the release of AST-LN, an in vitro assay was carried out with pure AST solution and nanoliposomes, using a dialysis bag technique, which stopped after given intervals. The AST release from nanoliposomes after 24 h was much slower (28.74%) than from the pure solution (95.27%), which supports the potential of continuous release from nanoliposomes.

Hama et al. [18] developed liposomal topical AST formulations to assess their action against skin damage that has been induced by UV radiation. The lipid hydration method was used to prepare either neutral liposome formulations (Asx-EPC-lipo) using egg phosphatidylcholine (EPC) liposomes or 1,2-dioleoyl-3-trimethylammonium-propane-based cationic liposomal formulations (Asx-DOTAP-lipo), intended to reach the skin stratum corneum by exploiting iontophoretic delivery. The diameter of the neutral vesicles varied from 170 μm to over 300 μm, and that of the cationic ones was around 170 ± 40 μm. The in vitro test of antioxidant activity with singlet oxygen generation and chemiluminescence detection showed that liposomal formulations did not interfere with AST antioxidant activity compared with the free solution, which increased in a dose-dependent manner. The results of in vivo tests performed to protect the skin from UV radiation showed that pretreatment of the skin surface with Asx-EPC-lipo prevented morphological skin changes, which suggests that liposomal AST formulation effectively prevented stratum corneum thickening by scavenging reactive oxygen species (ROS). The effect of iontophoretic transdermal delivery with cationic Asx-DOTAP-lipo pretreatment resulted in the significant inhibition of UV-induced melanin production in the basal laminae region. This result suggests that this formulation prevented melanocyte activation by efficiently scavenging ROS in deep regions of skin exposed to UV irradiation, acting as a potential whitening agent.

In another article, Hama et al. [32] compared the hydroxyl radical scavenging activity in an aqueous solution of Asx-EPC-lipo, prepared by the method of lipid hydration using EPC as the lipid base with those of β-carotene and α-tocopherol liposomes. The average diameter of EPC-lipo increased from 150 μm to over 300 μm, alongside increasing the amount of entrapped AST. Chemiluminescence intensity, which was assessed by the Fenton reaction to check the scavenging potential of AST and its encapsulated formulations, decreased in a dose-dependent manner with respect to AST in EPC-lipo. In addition, AST-containing liposomes were more powerful than either EPC-lipo-encapsulated β-carotene or α-tocopherol. An in vitro cytotoxicity test was performed to evaluate if Asx-EPC-lipo could protect a cultured mouse skin fibroblast cell line (NIH3T3 cells) against hydroxyl radicals damage. Results showed that the formulation prevented hydroxyl radical cytotoxicity in a dose-dependent manner since liposomes could suffer from cell endocytosis, and suggested that this effect was probably due to the action on the cell surface of AST molecules distributed in the plasma membranes.

2.2. Emulsions

2.2.1. Microemulsions

Microemulsions are thermodynamically stable transparent delivery systems formed by droplets dispersed on a liquid phase, which are capable of forming spontaneously and have low interfacial energy and size, ranging from about 10 nm to 100 nm (Figure 3A) [33,34]. Zhou et al. [35] developed AST oil/water microemulsions and checked the effect of different antioxidants either alone or in combination as additives to improve their stability. The microemulsion prepared with AST alone, with Tween 80 as an emulsifier and ethanol in buffer solution, retarded AST degradation, thereby proving to be an excellent alternative to improve emulsions for AST delivery.

2.2.2. Nanoemulsions

Nanoemulsions are thermodynamically and kinetically stable systems with nanoscale droplet sizes (100 nm to 400 nm) [33], uniform size distribution, and physicochemical and biological properties different from those of other emulsions (> 500 nm) (Figure 3A) [36]. The small droplet size, the scarce probability of coalescence and flocculation, effective delivery of active ingredients, rapid penetration, long-lasting effects, and uniform deposition onto the skin make them suited for use in the personal care, cosmetic, and health science fields [37]. Articles on micro- and nanoemulsions are described in Table 2.

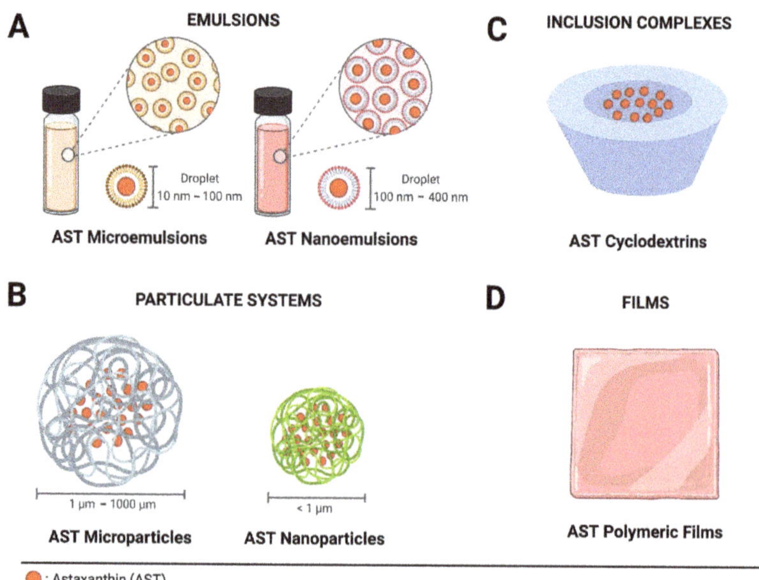

Figure 3. Schematic representation of AST delivery systems. (**A**) Emulsions: AST microemulsions and nanoemulsions; (**B**) particulate systems: AST microparticles and nanoparticles; (**C**) inclusion complexes: AST cyclodextrins (CDs); (**D**) films: AST polymeric films. Created with BioRender.com (accessed on 17 August 2021).

To obtain stabilized cosmetic products with anti-wrinkle, anti-aging, and humectant properties, Kim et al. [37] prepared oil/water AST nanoemulsions using glycerol esters such as glyceryl citrate, lactate, linoleate and oleate as alternative emulsifiers to the traditional hydrogenated lecithin. Nanoemulsions were studied in terms of their physicochemical properties, such as the emulsifier type and concentration, preparation conditions, and AST concentration, and were characterized by a freeze-fracture scanning electron microscope (FF-SEM), TEM, and high-performance liquid chromatography (HPLC). Nanoemulsions, with zeta potential in the range from -10 to -57 mV and an average particle size of 170 nm, were progressively more unstable with the increase in particle size when hydrogenated lecithin was used.

Hong et al. [38] prepared AST nanoemulsions that were functionalized with carboxymethyl chitosan, to investigate its effects on the droplet size, stability, skin permeability and cytotoxicity of the formulation. For this purpose, the low-energy emulsion phase inversion method was used, which, in addition to preventing AST degradation during nanoemulsion preparation, has the advantages of a low cost, high energy efficiency, simplicity of production and easy scale-up. Results showed improvement in chemical stability and skin permeability, while the small droplet size, satisfactory physical stability, and low cytotoxicity were not affected by the functionalized nanoemulsion.

In a comparative study, Shanmugapriya et al. [39] prepared uniform and stable oil/water nanoemulsions containing AST or α-tocopherol by the spontaneous and ultrasonication emulsification methods, using an experimental design for optimization. The AST-containing nanoemulsions displayed anticancer, wound healing and antimicrobial effects, which suggests their use in formulations for the treatment of skin cancer and wound healing, for example via incorporation into films. The same research group [39] used AST and α-tocopherol nanoemulsions with κ-carrageenan to verify topical wound healing effects in vitro and in vivo. Results suggested the treatment as a good alternative for wounds in diabetic cases with higher activity in a shorter time.

Table 2. Summary of AST emulsions delivery systems, describing the preparation technique, emulsion type, characterization and stability data, and assays (in vitro/in vivo) that were performed for each system.

Preparation Technique	Emulsion Type	Characterization	Storage and Stability Data	Assays (In Vitro, In Vivo)	References
High-pressure homogenization	Oil/water nanoemulsion, glyceryl ester and hydrogenated lecithin as emulsifiers	Dynamic light scattering and transmission electron microscopy	Stability maintained for one month of storage	-	[37]
Low-energy emulsion phase inversion method	Oil/water nanoemulsion functionalized carboxymethyl chitosan	Droplet size, zeta potential and transmission electron microscopy	Stability without alteration for three months	In vitro: skin permeation studies, Cell viability assays on L929 cells, Cell culture and cytotoxicity assays	[38]
Spontaneous and ultrasonication emulsification methods	Oil/water nanoemulsion	Dynamic light scattering and transmission electron microscopy	Interference of storage conditions	In vitro: cytotoxicity (MTT assay), antimicrobial activity and scratch wound healing assay	[39]
Spontaneous and ultrasonication emulsification methods	Oil/water nanoemulsion	Dynamic light scattering and transmission electron microscopy, Fourier transform infrared spectroscopy, differential scanning calorimetry, X-ray diffraction, thermal gravimetric analysis, and scanning electron microscopy	-	In vitro: cytotoxicity (MTT assay), scratch wound-healing assay. In vivo: wound healing in nondiabetic and diabetic mice	[40]
Oil phase dispersed with AST in ethyl butyrate and homogenizing with aqueous phase in a high-speed blender and high-pressure microfluidizer	Oil/water microemulsions	Dynamic light scattering and UV-visible spectrophotometry	-	-	[35]

2.3. Particulate Systems

Particulate systems are delivery systems that include nano/microspheres and nano/microcapsules (Figure 3B) [25]. Whereas nano/microspheres are dispersions of active ingredients in polymeric matrices, nano/microcapsules are reservoirs where distinct domains of core and wall material are present. These formulations create a compatible environment for susceptible molecules and protect them from light, oxygen, pH, heat, enzymatic degradation and other external factors that can affect their stability [41]. Microencapsulation technologies are widely used in cosmetic products to increase stability, protect against degradation, promote safe administration and allow controlled and targeted release [42]. Particle size varies from around one micron to a few millimeters, providing a large surface area available to develop active ingredients. Nanoparticles involve particulate systems in the nano range, providing a larger surface area that is available for adsorption and desorption sites, chemical reactions, light scattering, etc. [42]. Studies on which particulate systems have been used are summarized in Table 3.

Table 3. Summary of AST particulate delivery systems, describing the preparation technique, system type, characterization and stability data, and assays (in vitro/in vivo) that were performed for each system.

Preparation Technique	System Type	Characterization	Storage and Stability Data	Assays	References
AST microencapsulation by response surface methodology	Oil bodies (isolated from mature seeds) microcapsules	Fourier transform infrared spectroscopy (FT-IR), flow cytometry and microscopy	Oxidative stability, double half-life compared to free AST	In vitro: absorption assay	[43]
Multiple emulsion/solvent evaporation	Chitosan matrix cross-linked with glutaraldehyde microparticles	AST extract analysis by high-performance liquid chromatography (HPLC)	Pigment quantity during microcapsules storage at 25, 35 and 45 °C	In vitro: storage stability evaluation	[44]
Extrusion	Calcium alginate microparticles	Analysis of AST content by HPLC	Various environmental conditions: light, temperature and nitrogen gas	In vitro: assay of AST content	[45]
Supercritical anti-solvent	Poly(L-lactic acid) microspheres	Scanning electron microscopy (SEM), transmission electron microscopy (TEM), FT-IR, X-ray diffraction (XRD), thermal gravimetric analysis (TGA), differential scanning calorimetry (DSC), UV-visible spectrophotometry	6-Month measurements by UV–vis spectrophotometry	In vitro: assay of AST content and AST release profile	[46]
Emulsion solvent evaporation	Poly(lactic-co-glycolic acid) (PLGA) copolymer nanoparticles	Dynamic light scattering (DLS), SEM, TEM, FT-IR, XRD, TGA, DSC	-	In vitro: anti-photodamage effect in HaCaT cells	[47]
Hot homogenization	Nanostructured lipid carriers	DLS, atomic force microscopy, SEM	Samples stored at 4 °C, protected from light for 1 month	In vitro: antioxidant activity by the α-tocopherol equivalent antioxidant capacity assay	[48]
Macromolecular co-assembly combined with solvent evaporation	Natural DNA and chitosan nanocarriers	DLS, TEM, field emission SEM, HPLC (AST content)	-	In vitro: oxidative stress, cytotoxicity (MTT assay) and cell uptake assay	[49]
Antisolvent precipitation method combined with electrostatic deposition method	PLGA and chitosan oligosaccharides nanoparticles	DLS, SEM, TEM, FT-IR, XRD, DSC	72 h of storage at room temperature	In vitro: cytotoxicity and AST release profile	[50]
Solvent displacement process	Ethylcellulose, Poly(ethylene oxide) 4-methoycinnamoyl-phthaloylchitosan and poly(vinylalcohol-covinyl-4-methoxycinnamate nanospheres	SEM, TEM	Thermal stability	In vitro: AST release profile	[51]

2.3.1. Microparticles

Higuera-Ciapara et al. [44] developed non-homogeneous-sized AST microcapsules with a diameter of 5–50 μm, using a chitosan matrix cross-linked with glutaraldehyde by the multiple emulsion/solvent evaporation method. The system stability was evaluated based on retained pigment quantity during microcapsule storage at 25, 35 and 45 °C, which was quantified weekly by HPLC. For any of the treatments, the results did not show a marked decrease in AST concentration in the microcapsules, and AST was maintained in stable conditions, as evidenced by poor isomerization and pigment degradation.

Oil bodies (OB) are minute plant organelles similar to liposomes, with 0.5–2.0 μm diameter, consisting of an oil core surrounded by a phospholipid monolayer with a proteinaceous membrane, and these are used to deliver phytohormones and other hydrophobic compounds in plants. These structures, which have been isolated from rapeseeds, have been shown to constitute a novel type of microcapsule that is suitable for the extraction of hydrophobic organic compounds from aqueous environments [52]. Hydrophobic compounds like AST that are used in cosmetics have been loaded into these systems, to protect them from oxidation. Acevedo et al. [43] developed AST microcapsules (AST-M) with OB extracted from *Brassica napus* seeds with high microencapsulation efficiency (>99%) and used a response surface methodology to optimize the microencapsulation conditions. The AST-M were examined by optical microscopy, which evidenced morphological stability, autofluorescence, spherical structures and an AST presence in the core. Their larger mean diameter (3.4 ± 0.5 mm) compared to those containing OB alone (1.56 ± 0.06 mm) was likely due to the AST intercalation into microcapsule monolayer, as previously described by other authors. Stability studies showed high stability of microcapsules to aggregation and coalescence, as well as a double half-life in the presence of air and light exposure compared with free AST, thereby highlighting the protective role of OB against AST degradation. After 2 h of cell incubation (CRL1730 endothelial cell line), antioxidant assays demonstrated the higher antioxidant power of AST-M in comparison with free AST, which was dose- and time-dependent. The cell viability assay did not show any cell toxicity, and microencapsulated AST displayed higher oxidative stability than its free counterpart. The authors suggested that the use of OB as a new delivery system is promising for the cosmetic field because it joins once in contact with the skin, releasing the antioxidant safely, and offers a new and natural carrier to deliver stable AST.

Lin et al. [45] searched for the best conditions to prepare AST that is encapsulated in sodium alginate beads, varying the concentrations of calcium chloride solution, medium, sodium alginate solution as the encapsulation agent, and Tween 20 as a surfactant, and selecting the average yield weight, microencapsulation yield, the average size of beads and loading efficiency as the responses. It seemed that the higher the alginate concentration, the higher the average weight and size of the beads. Different conditions of three-week storage were tested for free AST, AST encapsulated in various formulations, and a commercial product with 10% AST, in terms of bioactive content. The percentage of AST amount retained after storage was higher (90%) for beads compared to the controls without encapsulation, which suggests that the matrix covering the molecule protected it from thermal degradation and oxygen attack.

AST microspheres were developed by Liu et al. [46] using the supercritical anti-solvent (SAS) process and poly (L-lactic acid) as the polymeric carrier. The authors believe that this alternative method has great potential for preparing encapsulation systems, thanks to its single-pass process and mild operating conditions, to prevent the degradation of AST and other sensitive molecules. SAS operating conditions were varied according to an orthogonal experimental design to elucidate encapsulation circumstances. Optimal conditions ensured an encapsulation efficiency of 91.5% and a mean particle size of 954.6 nm. Characterization assays proved the formation of uniform particles and the amorphous state of AST encapsulated in the matrix, while 6-month storage tests at 40 °C showed the enhancement of its stability.

2.3.2. Nanoparticles

Poly(lactic-co-glycolic acid) (PLGA) is a copolymer recently proposed for use in topical formulations to prevent and treat photodamage, one that is non-toxic and can be hydrolyzed in vivo into its biodegradable monomers, i.e., lactic and glycolic acids [53]. PLGA-AST nanoparticles (AST-PLGA NP) were developed by Hu et al. [47] in an optimization study, based on an experimental design, aiming to maximize encapsulation efficiency (96.42 ± 0.73%) and drug-loading capacity (7.19 ± 0.12%), and to simultaneously minimize particle size (154.4 ± 0.35 nm). Fourier transform infrared (FT-IR) spectroscopy, differential scanning calorimetry and thermogravimetric analyses confirmed AST encapsulation within PLGA nanoparticles, while XRD analysis suggested its amorphous state, and scanning electron microscopy (SEM) and TEM evidenced globular nanoparticles without drug crystals. This would contribute to easier penetration compared to pure AST, and would afford good size distribution. To evaluate the capacity of cellular uptake, an in vitro model with a fluorescent probe was performed with HaCaT cells, an aneuploid immortal keratinocyte cell line. The results showed cellular uptake of the nanoparticles that increased in a time-dependent way. In vitro cytotoxicity, assessed by the MTT assay, showed no significant decrease in cell viability, while AST-PLGA NP increased cell viability and had stronger antioxidant activity, compared to pure AST. Protection by scavenging free radicals was performed on the same cell line, and pure AST and AST-PLGA NP displayed similar antioxidant properties.

Chitosan nanoparticles are promising as drug delivery systems, due to their biocompatibility, biodegradability, atoxicity, bioactivity, and large target options triggered by their cationic character [54]. Chitosan and natural DNA were used by Wang et al. [49] as oppositely charged biomaterials, to develop a nanoparticle system to deliver and release AST. Thus, AST-loaded DNA/chitosan (ADC) and empty nanocarrier (DNA/CS) were prepared by macromolecular co-assembly combined with the solvent evaporation method, and their antioxidant activities and cellular uptake were assessed. Both showed a positive surface charge, due to the predominant cationic characteristic of chitosan, and very close zeta potential (35.3 and 32.2 mV, respectively), which suggests that AST incorporation did not influence this parameter. The larger average particle size of ADC (211 ± 5 nm) compared to DNA/CS, and its PDI of 0.29 ± 0.01, confirmed its nano scale, good dispersity, and AST entrapment. The AST content in ADC was higher than in ethanol solution, confirming enhancement of its solubility. The MTT assay, performed to assess the cytoprotective effect, showed that ADC prevented the deleterious effect of H_2O_2 on cell viability more than vitamin C (positive control) and free AST, in spite of a vitamin C concentration that was three times higher than that of AST, likely due to the facilitated endocytosis of ADC nanoparticles. The ROS scavenging efficiency of ADC nanoparticles was twice that of free AST at the same bioactive concentration.

Liu et al. [50] prepared an innovative type of nanoparticle to improve solubility and stability, using chitosan oligosaccharides (COS) to coat PLGA AST nanoparticles (Ax-PLGA@COS NP). To prepare AST-loaded PLGA nanoparticles, a two-step process was used, which consisted of antisolvent precipitation, followed by coating nanoparticles with COS by electrostatic deposition. The formulation was characterized in terms of encapsulation efficiency, morphology, PDI, zeta potential and particle size, while the stability of nanoparticles was evaluated in terms of changes in their color and particle size during 72 h of storage at room temperature. Ax-PLGA@COS NP suspensions had a uniform orange color that did not change appreciably during storage. Furthermore, no particle aggregation or sedimentation was observed in these systems, and the formulation showed no cytotoxicity to Caco-2 cells.

Due to AST's susceptibility to heat, Tachaprutinun et al. [51] aimed to assess the resistance to thermal degradation of AST nanoparticles prepared by the solvent displacement process, using three different polymers, namely poly(ethylene oxide)-4-methoxy cinnamoyl phthaloyl-chitosan (PCPLC), poly(vinylalcohol-co-vinyl-4-methoxycinnamate) (PB4) and ethylcellulose (EC). While EC was inefficient and PB4 poorly efficient for encapsulating the bioactive compound, PCPLC allowed for an encapsulation efficiency of 98%, a loading of ~40%, and a 312 ± 5.83 nm particle size. In thermal degradation tests performed at 70 °C for two hours, most of the free AST was degraded, while the nanoparticle formulation was able to protect the molecule from degradation.

Solid lipid nanoparticles (SLN) are colloidal particles prepared from solid lipids, surfactants, the active ingredient and water. This method has shown several advantages, including the use of biocompatible lipids, high in vivo stability and a wide application spectrum. However, they have limitations, such as a low drug-carrying capacity and drug leakage during storage. Thus, nanostructured lipid carriers (NLC) emerge as second-generation lipid nanoparticles to overcome these deficiencies [55].

NLC are obtained using lipids (solids and liquids) and emulsifiers, forming a solid lipid matrix at body temperature that is capable of incorporating hydrophobic and hydrophilic molecules [25]. This matrix has promising potential for the pharmaceutical and cosmetics industry due to its beneficial effects, such as skin hydration, occlusion, greater bioavailability, and targeting application to the skin [55].

In this sense, Rodriguez-Ruiz et al. [48] developed an innovative possible solution with a green chemistry process formulation of AST-loaded nanostructured lipid carriers (NLC). The compounds used to synthesize NLC by the hot homogenization method were sunflower oil as the liquid lipid, glyceryl palmitostearate as the solid lipid, and Poloxamer 407 and Tween 80 as the surfactants. AST-loaded NLC and AST-free NLC were characterized by dynamic light scattering (DLS), atomic force microscopy (AFM) and SEM. DLS analysis showed that for the former particle size of 60 ± 7 nm, a polydispersity index (PDI) of around 0.33 ± 0.09 and zeta potential of -25.5 ± 0.7 mV were achieved, which is indicative of stability. The AFM and SEM analyses demonstrated nanoparticle spherical shape and nanoscale. To evaluate the stability, samples were stored for one month, protected from light, at low temperatures. Results showed no significant change in the size, PDI and zeta potential of both preparations. No less than $90 \pm 5\%$ of the starting AST content in NLC was observed after this time, and the additional effect of antioxidants present in sunflower oil seemed to protect the bioactive compound from oxidation. The antioxidant potential determined by the α-tocopherol equivalent antioxidant capacity assay demonstrated an inhibition curve slope that was almost double for AST-loaded NLC compared to the control, confirming the high antioxidant activity of AST [48].

2.4. Inclusion Complexes

Cyclodextrin

Cyclodextrins (CDs) are a family of cyclic polysaccharides used to form inclusion complexes with a wide variety of substances used in pharmaceuticals, drug delivery systems, cosmetics, and in the food and chemical industries [56]. Their molecular structure is composed of a cavity size, which is determined by the number of glucose units, where the space inside the cyclodextrin molecules allows the formation of inclusion complexes with poorly soluble compounds (Figure 3C) [56]. The inclusion of guest molecules into CDs can change their physical and chemical properties, as well as increasing their water solubility and stability [57]. The CDs are an excellent alternative for the inclusion of a variety of natural compounds, such as oils [58,59] and other compounds [60,61]. The AST inclusion complexes with cyclodextrin are summarized in Table 4.

One of the first studies on the inclusion of AST in CDs to enhance its solubility for topical applications was developed by Lockwood et al. [62]. When used in proportions from 0 to 60% (*w/v*), a sulfobutyl ether β-cyclodextrin was shown to complex with crystalline AST. At 60%, AST water solubility has increased by more than 50 times, and the implementation of a pre-solubilization process could increase it by 71 times over the parent compound in water.

Kim et al. [63] prepared AST inclusion complexes with various types of CDs in different ratios, characterized each formulation by HPLC, SEM and FT-IR, and evaluated their 28-day stability and water solubility under different conditions of pH, light, temperature and oxidation. To minimize the costs, a β-cyclodextrin (β-CD) that is widely used in food and cosmetics applications was used for comparison. Inclusion complexes formed at the AST ratio of 1:200, and the host molecule showed a uniform shape and particle size. β-CD was proved to incorporate AST, with an inclusion yield higher than 90%, and the solubility of the resulting AST-loaded inclusion complex was 13-fold at 25 °C and about 100-fold that of free AST at a pH of 6.5. In the stability study, the yield of the inclusion complex remained above 80% after 21 days of UV irradiation, while the free AST was completely degraded. In addition, it proved stable against oxidation, was favored by acidic conditions and exhibited greater temperature resistance for industrial processing.

Table 4. Summary of AST cyclodextrins (CDs) delivery systems, describing the preparation characterization, storage and stability data, and assays (in vitro/in vivo) that were performed for each system.

CDs	Characterization	Storage and Stability Data	Assays	References
β-cyclodextrin (β-CD)	High-performance liquid chromatography (HPLC), scanning electron microscopy and Fourier transform infrared spectroscopy (FT-IR)	Stability enhanced by over 7–9 folds under various storage conditions such as pH, temperature, ultraviolet irradiation, and presence of oxygen	In vitro: water solubility	[63]
Sulfobutyl ether β-CD	UV-visible spectrophotometry	-	In vitro: water solubility	[62]
β-CD	HPLC	Storage at 4, 30, 57 °C and under light (light intensity of 1500 lux)	In vitro water solubility	[57]
Hydroxypropyl-β-cyclodextrin (HP-β-CD)	Thermogravimetry, UV-visible spectrophotometry, FT-IR, molecular modeling, nucleic magnetic resonance	Stability under oxygen and light at 4, 25 and 50 °C, storage at 4 and 25 °C in dark incubators	In vitro: water solubility, antioxidant capacity by reducing power, DPPH free radical scavenging activity and hydroxyl radical scavenging activity	[64–66]
HP-β-CD	FT-IR, UV-visible spectrophotometry	Storage at 6 °C under light protection for 6 months	In vitro cytoprotective activity of HP-β-CD complex. Direct biological evaluation of HP-β-CD antioxidant capacity Indirect HP-β-CD antioxidant protection against reactive oxygen species	[67]

Chen et al. [57] prepared an AST β-cyclodextrin complex and measured its water solubility and stability to heat and light. Complex formation was checked by infrared spectroscopy and HPLC, showing an inclusion yield of 48.96%. The water solubility of AST was slightly increased, while its heat stability was greatly enhanced compared to the free bioactive compound.

Hydroxypropyl-β-cyclodextrin (HP-β-CD) is a hydroxyalkyl derivative alternative to parent CDs, which offers improved water solubility and is slightly more friendly from a toxicological standpoint [68]. Yuan et al. [64] prepared a new water-soluble formulation of AST with HP-β-CD, analyzed its thermal behavior, and investigated its stability in heat and light. The inclusion yield of the formulation was 46.5%, and no less than 200 mg/mL of it was dispersed in water. Water solubility was greatly enhanced (>1.0 mg/mL) in comparison with the previous study, due to the higher solubility of HP-β-CD compared with β-CD. On the other hand, the overall amount of AST entrapped in the inclusion complex was lower, likely because the hydroxypropyl substituent concentrated at the edge of the CD cavity made the entry of AST molecules more difficult. In 2012, the same research group studied this complex via UV-Vis, FT-IR, 1H nuclear magnetic resonance (NMR) spectroscopies and molecular modeling, to enhance knowledge about the molecule structure [65]. Storage stability at 4 and 25 °C was higher than that of free AST, while in vitro antioxidant tests showed greater antioxidant activity than ascorbic acid [66].

To prevent antioxidative stress on endothelial cells, Zuluaga et al. [67] developed a similar complex with AST and HP-β-CD. The differential of this study was the direct and indirect measurement of its antioxidant capacity by understanding the cells' molecular mechanisms involved in gene expression. Results showed that the inclusion complex formulation could protect cells by activating endogenous AST systems through the Nrf2/HO-

1/NQO1 pathway, in addition to the enhancement of solubility due to its incorporation into the cyclodextrin core.

2.5. Films

Topical film-forming systems are drug delivery systems for topical applications that are capable of adhering to the skin, forming a thin transparent film that provides delivery of the active ingredients to the body tissue (Figure 3D) [69]. These systems are composed of the drug- and film-forming agents in a volatile vehicle, which evaporates in contact with the skin [69]. Even though it is a promising option for topical drug delivery, the literature on the incorporation of AST as a way to exert antioxidant effects on the skin is scarce.

Veeruraj et al. [23] prepared films to demonstrate the wound healing properties of AST when incorporated in collagen films. In vivo assays were performed to assess tissue regeneration and drug delivery from the formulation, and in vitro assays to check the antioxidant activity. In addition to the AST collagen film, a gentamicin-incorporating collagen film was developed to assess its antibiotic effects. The filming agent and AST were extracted from the waste material of the outer skin of the squid *Doryteuthis singhalensis*, which is an innovative and sustainable alternative for the development of this delivery system. Biodegradation tests showed that film materials degraded more rapidly than the collagen matrix, suggesting the controlled degradation of collagen materials. Wound healing activity was measured by the reduction of the non-healing area in the healing process, occurring over 21 days. The untreated control exhibited the lowest wound contraction, whereas the AST collagen film showed the highest one among the experimental groups, as well as the fastest wound-healing progress, with complete healing in 15 days. Antioxidant assay by the DPPH free radical scavenging method showed the higher activity of AST collagen film compared to ascorbic acid. The main information on this article is presented in Table 5.

Table 5. Summary of AST film delivery systems, describing the preparation technique, filming agent, characterization and stability data, and assays (in vitro/in vivo) that were performed for the system.

Preparation Technique	Filming Agent	Characterization	Storage and Stability Data	Assays
Collagen solution incorporating AST and gentamicin	Biomaterials extracted from the waste material of the outer skin of the squid *Doryteuthis singhalensis*	Scanning electron microscopy, energy dispersive X-ray spectroscopy, X-ray diffraction	-	In vitro: biodegradation study and DPPH free radical scavenging activity In vivo: wound-healing activity

3. Discussion

Liposomes in themselves are relatively unstable delivery systems because of their membrane instability in aqueous solution, which can affect their bioavailability and pharmacological effect, the addition of adjuvants being necessary to bring more rigidity and stability. AST seems to enhance the liposomes' stability when on the membrane, which helps its delivery. The strategy of adding other adjuvants to improve membrane rigidity seems to be necessary to ensure stability, which should be taken into account when assessing production costs. Reducing the particle size with the development of nanoliposomes seems to be an excellent alternative method to improve their stability and consequently their penetration, solubility, and continuous release; however, it is still necessary to consider the machinery and costs involved. Another innovative alternative method to deliver AST on deep skin layers is the use of the iontophoretic technique with charged delivery systems, developed with liposomes, which makes it possible to reach the stratum corneum and act as a whitening agent.

Emulsion delivery systems, especially nanoemulsions, are an effective and widely applied form to deliver bioactive compounds in medicines and cosmetics. Due to the hydrophobic character of AST, most of the nanoemulsions prepared in the literature are oil/water systems. Studies based on experimental design allow researchers to identify the best conditions for nanoemulsion preparation, with reduced time and costs of research. In addition, nanoemulsions can be incorporated in other delivery systems, such as films, which can significantly improve their biological effects and their acceptance by the patient or customer. Unfortunately, only one article was found concerning the use of microemulsions in dermatological and cosmetics applications. More investment in research on this new product delivery system is desirable.

Particulate systems are a promising alternative way to develop innovative formulations for AST delivery, thanks to the possibility of using different matrix components and methods to form spheres and capsules at micro- or nanoscale. Similar to nanoemulsions, studies based on experimental design can be performed to optimize the conditions needed to prepare particulate systems. It is important to consider alternative methods, to prevent AST degradation. Some articles on particulate systems considered in this review presented either ecological or natural alternatives to develop delivery systems, which can aggregate to more sustainable products.

The use of CDs as systems to guest AST in inclusion complexes is able to significantly improve the solubility of hydrophobic compounds like AST. However, it would be necessary to have a greater number of inclusion complexes, in order to carry out a comparison by which to identify the most advantageous system. From the results examined in this review, it is possible to state that AST can form inclusion complexes either with natural CDs, such as β-CDs, or with modified CDs, such as hydroxypropyl β-CD. Films for the delivery of bioactive compounds on the skin are an alternative, especially for wound healing and scald treatments. The antioxidant effect accelerates the healing process, due to its anti-inflammatory activity. The AST-incorporating films may also be applied as an easy adjunct treatment, to treat skin cancer, which should be considered in the research. In addition, AST seems to act merely as an active principle, which requires a film-forming agent such as collagen, chitosan, cellulose derivatives, and others to create the film structure.

The systems that seem to be more developed and robust are the emulsions and the particulate systems, given the large number of articles and assays performed on them. However, they are not the only systems to consider. On the other hand, the limited studies carried out on other systems, many of which are not mentioned in this review article, show opportunities for the development of new and innovative formulations. Examples are solid dispersions, transdermal systems, and others.

4. Conclusions

The aim of this article was to review the literature about the development of new systems for loading AST for cosmetics and topical usage. Delivery systems are useful for the improvement of the physicochemical profile of this compound, such as stability, water-solubility, antioxidant properties, drug release, and in vivo and in vitro biological activities. Therefore, the delivery systems for loading AST described in this article create opportunities for industrial applications; however, other industrial issues for the development of new products must be evaluated. Moreover, AST-optimized products become potentially attractive for the development of further studies, due to the antioxidant properties and benefits of AST.

Funding: This research was funded by in part by the Coordenação de Aperfeiçoamento de Pessoal de Nível Superior—Brasil (CAPES)—Finance Code 001 and by the Conselho Nacional de Desenvolvimento Científico e Tecnológico (CNPq).

Acknowledgments: The authors thank CNPq and CAPES for their support and Federal University of Rio Grande do Norte.

Conflicts of Interest: The authors declare no conflict of interest. CNPq and CAPES have no role in the design of the study, in the collection, analyses or interpretation of data, in the writing of the manuscript, or in the decision to publish it.

References

1. Davinelli, S.; Nielsen, M.E.; Scapagnini, G. Astaxanthin in skin health, repair, and disease: A comprehensive review. *Nutrients* **2018**, *10*, 522. [CrossRef] [PubMed]
2. Vollmer, D.L.; West, V.A.; Lephart, E.D. Enhancing skin health: By oral administration of natural compounds and minerals with implications to the dermal microbiome. *Int. J. Mol. Sci.* **2018**, *19*, 3059. [CrossRef]
3. Higuera-Ciapara, I.; Felix-Valenzuela, L.; Goycoolea, F.M. Astaxanthin: A review of its chemistry and applications. *Crit. Rev. Food Sci. Nutr.* **2006**, *46*, 185–196. [CrossRef]
4. Jannel, S.; Caro, Y.; Bermudes, M.; Petit, T. Novel insights into the biotechnological production of *Haematococcus pluvialis*-derived astaxanthin: Advances and key challenges to allow its industrial use as novel food ingredient. *J. Mar. Sci. Eng.* **2020**, *8*, 789. [CrossRef]
5. Couteau, C.; Coiffard, L. Microalgal Application in Cosmetics. In *Microalgae in Health and Disease Prevention*; Fleurence, J., Levine, I.A., Eds.; Academic Press: Cambridge, MA, USA, 2018; pp. 317–323.
6. Sotiropoulou, G.; Zingkou, E.; Pampalakis, G. Redirecting drug repositioning to discover innovative cosmeceuticals. *Exp. Dermatol.* **2021**, *30*, 628–644. [CrossRef]
7. Fakhri, S.; Abbaszadeh, F.; Dargahi, L.; Jorjani, M. Astaxanthin: A mechanistic review on its biological activities and health benefits. *Pharmacol. Res.* **2018**, *136*, 1–20. [CrossRef]
8. Zhao, T.; Yan, X.; Sun, L.; Yang, T.; Hu, X.; He, Z.; Liu, F.; Liu, X. Research progress on extraction, biological activities and delivery systems of natural astaxanthin. *Trends Food Sci. Technol.* **2019**, *91*, 354–361. [CrossRef]
9. Chalyk, N.E.; Klochkov, V.A.; Bandaletova, T.Y.; Kyle, N.H.; Petyaev, I.M. Continuous astaxanthin intake reduces oxidative stress and reverses age-related morphological changes of residual skin surface components in middle-aged volunteers. *Nutr. Res.* **2017**, *48*, 40–48. [CrossRef]
10. Gürsoy, K.; Teymur, H.; Koca, G.; Tanas Işikçi, Ö.; Göktaş Demircan, F.B.; Kankaya, Y.; Koçer, U. The Effect of Astaxanthin on Random Pattern Skin Flaps. *Ann. Plast. Surg.* **2020**, *84*, 208–215. [CrossRef] [PubMed]
11. Faraone, I.; Sinisgalli, C.; Ostuni, A.; Armentano, M.F.; Carmosino, M.; Milella, L.; Russo, D.; Labanca, F.; Khan, H. Astaxanthin anticancer effects are mediated through multiple molecular mechanisms: A systematic review. *Pharmacol. Res.* **2020**, *155*, 104689. [CrossRef] [PubMed]
12. Ito, N.; Seki, S.; Ueda, F. The protective role of astaxanthin for UV-induced skin deterioration in healthy people—A randomized, double-blind, placebo-controlled trial. *Nutrients* **2018**, *10*, 817. [CrossRef] [PubMed]
13. Yoshihisa, Y.; Andoh, T.; Matsunaga, K.; Ur Rehman, M.; Maoka, T.; Shimizu, T. Efficacy of astaxanthin for the treatment of atopic dermatitis in a murine model. *PLoS ONE* **2016**, *11*, e0152288. [CrossRef] [PubMed]
14. Tominaga, K.; Hongo, N.; Karato, M.; Yamashita, E. Cosmetic benefits of astaxanthin on humans subjects. *Acta Biochim. Pol.* **2012**, *59*, 43–47. [CrossRef] [PubMed]
15. Eren, B.; Tuncay Tanrıverdi, S.; Aydın Köse, F.; Özer, Ö. Antioxidant properties evaluation of topical astaxanthin formulations as anti-aging products. *J. Cosmet. Dermatol.* **2019**, *18*, 242–250. [CrossRef] [PubMed]
16. Imokawa, G. The Xanthophyll Carotenoid Astaxanthin has Distinct Biological Effects to Prevent the Photoaging of the Skin Even by its Postirradiation Treatment. *Photochem. Photobiol.* **2019**, *95*, 490–500. [CrossRef]
17. Rao, A.R.; Sindhuja, H.N.; Dharmesh, S.M.; Sankar, K.U.; Sarada, R.; Ravishankar, G.A. Effective inhibition of skin cancer, tyrosinase, and antioxidative properties by astaxanthin and astaxanthin esters from the green alga haematococcus pluvialis. *J. Agric. Food Chem.* **2013**, *61*, 3842–3851. [CrossRef]
18. Hama, S.; Takahashi, K.; Inai, Y.; Shiota, K.; Sakamoto, R.; Yamada, A.; Tsuchiya, H.; Kanamura, K.; Yamashita, E.; Kogure, K. Protective Effects of Topical Application of a Poorly Soluble Antioxidant Astaxanthin Liposomal Formulation on Ultraviolet-Induced Skin Damage. *J. Pharm. Sci.* **2012**, *101*, 2271–2280. [CrossRef] [PubMed]
19. Suganuma, K.; Nakajima, H.; Ohtsuki, M.; Imokawa, G. Astaxanthin attenuates the UVA induced up-regulation of matrix-metalloproteinase-1 and skin fibroblast elastase in human dermal fibroblasts. *J. Dermatol. Sci.* **2010**, *58*, 136–142. [CrossRef]
20. Yoshihisa, Y.; Rehman, M.u.; Shimizu, T. Astaxanthin, a xanthophyll carotenoid, inhibits ultraviolet-induced apoptosis in keratinocytes. *Exp. Dermatol.* **2014**, *23*, 178–183. [CrossRef]
21. Seki, T.; Sueki, H.; Kono, H.; Suganuma, K.; Yamashita, E. Effects of astaxanthin from haematococcus pluvialis on human skin. *Fr. J.* **2001**, *12*, 98–103.
22. Meephansan, J.; Rungjang, A.; Yingmema, W.; Deenonpoe, R.; Ponnikorn, S. Effect of astaxanthin on cutaneous wound healing. *Clin. Cosmet. Investig. Dermatol.* **2017**, *10*, 259–265. [CrossRef] [PubMed]
23. Veeruraj, A.; Liu, L.; Zheng, J.; Wu, J.; Arumugam, M. Evaluation of astaxanthin incorporated collagen film developed from the outer skin waste of squid Doryteuthis singhalensis for wound healing and tissue regenerative applications. *Mater. Sci. Eng. C* **2019**, *95*, 29–42. [CrossRef] [PubMed]
24. Palareti, G.; Legnani, C.; Cosmi, B.; Antonucci, E.; Erba, N.; Poli, D.; Testa, S.; Tosetto, A. Anti-inflammatory effect of Astaxanthin in phthalic anhydride- induced atopic dermatitis animal model. *Int. J. Lab. Hematol.* **2016**, *38*, 42–49. [CrossRef] [PubMed]
25. Costa, R.; Santos, L. Delivery systems for cosmetics—From manufacturing to the skin of natural antioxidants. *Powder Technol.* **2017**, *322*, 402–416. [CrossRef]
26. Tan, C.; Zhang, Y.; Abbas, S.; Feng, B.; Zhang, X.; Xia, S. Modulation of the carotenoid bioaccessibility through liposomal encapsulation. *Colloids Surfaces B Biointerfaces* **2014**, *123*, 692–700. [CrossRef]

27. Pan, L.; Wang, H.; Gu, K. Nanoliposomes as vehicles for astaxanthin: Characterization, in vitro release evaluation and structure. *Molecules* **2018**, *23*, 2822. [CrossRef] [PubMed]

28. Schäfer-Korting, M.; Korting, H.C.; Braun-Falco, O. Liposome preparations: A step forward in topical drug therapy for skin disease? A review. *J. Am. Acad. Dermatol.* **1989**, *21*, 1271–1275. [CrossRef]

29. Dopierała, K.; Karwowska, K.; Petelska, A.D.; Prochaska, K. Thermodynamic, viscoelastic and electrical properties of lipid membranes in the presence of astaxanthin. *Biophys. Chem.* **2020**, *258*, 106318. [CrossRef]

30. Goto, S.; Kogure, K.; Abe, K.; Kimata, Y.; Kitahama, K.; Yamashita, E.; Terada, H. Efficient radical trapping at the surface and inside the phospholipid membrane is responsible for highly potent antiperoxidative activity of the carotenoid astaxanthin. *Biochim. Biophys. Acta-Biomembr.* **2001**, *1512*, 251–258. [CrossRef]

31. Lee, H.S.; Sung, D.K.; Kim, S.H.; Choi, W.I.; Hwang, E.T.; Choi, D.J.; Chang, J.H. Controlled release of astaxanthin from nanoporous silicified-phospholipids assembled boron nitride complex for cosmetic applications. *Appl. Surf. Sci.* **2017**, *424*, 15–19. [CrossRef]

32. Hama, S.; Uenishi, S.; Yamada, A.; Ohgita, T.; Tsuchiya, H.; Yamashita, E.; Kogure, K. Scavenging of Hydroxyl Radicals in Aqueous Solution by Astaxanthin Encapsulated in Liposomes. *Biol. Pharm. Bull.* **2012**, *35*, 2238–2242. [CrossRef] [PubMed]

33. Callender, S.P.; Mathews, J.A.; Kobernyk, K.; Wettig, S.D. Microemulsion utility in pharmaceuticals: Implications for multi-drug delivery. *Int. J. Pharm.* **2017**, *526*, 425–442. [CrossRef]

34. Nastiti, C.M.R.R.; Ponto, T.; Abd, E.; Grice, J.E.; Benson, H.A.E.; Roberts, M.S. Topical nano and microemulsions for skin delivery. *Pharmaceutics* **2017**, *9*, 37. [CrossRef] [PubMed]

35. Zhou, Q.; Xu, J.; Yang, S.; Xue, Y.; Zhang, T.; Wang, J.; Xue, C. The effect of various antioxidants on the degradation of o/w microemulsions containing esterified astaxanthins from *Haematococcus pluvialis*. *J. Oleo Sci.* **2015**, *64*, 515–525. [CrossRef] [PubMed]

36. Rai, V.K.; Mishra, N.; Yadav, K.S.; Yadav, N.P. Nanoemulsion as pharmaceutical carrier for dermal and transdermal drug delivery: Formulation development, stability issues, basic considerations and applications. *J. Control Release* **2018**, *270*, 203–225. [CrossRef]

37. Kim, D.M.; Hyun, S.S.; Yun, P.; Lee, C.H.; Byun, S.Y. Identification of an emulsifier and conditions for preparing stable nanoemulsions containing the antioxidant astaxanthin. *Int. J. Cosmet. Sci.* **2012**, *34*, 64–73. [CrossRef]

38. Hong, L.; Zhou, C.L.; Chen, F.P.; Han, D.; Wang, C.Y.; Li, J.X.; Chi, Z.; Liu, C.G. Development of a carboxymethyl chitosan functionalized nanoemulsion formulation for increasing aqueous solubility, stability and skin permeability of astaxanthin using low-energy method. *J. Microencapsul.* **2017**, *34*, 707–721. [CrossRef]

39. Shanmugapriya, K.; Kim, H.; Saravana, P.S.; Chun, B.S.; Kang, H.W. Astaxanthin-alpha tocopherol nanoemulsion formulation by emulsification methods: Investigation on anticancer, wound healing, and antibacterial effects. *Colloids Surf. B Biointerfaces* **2018**, *172*, 170–179. [CrossRef]

40. Shanmugapriya, K.; Kim, H.; Kang, H.W. A new alternative insight of nanoemulsion conjugated with κ-carrageenan for wound healing study in diabetic mice: In vitro and in vivo evaluation. *Eur. J. Pharm. Sci.* **2019**, *133*, 236–250. [CrossRef]

41. Nesterenko, A.; Alric, I.; Violleau, F.; Silvestre, F.; Durrieu, V. A new way of valorizing biomaterials: The use of sunflower protein for α-tocopherol microencapsulation. *Food Res. Int.* **2013**, *53*, 115–124. [CrossRef]

42. Casanova, F.; Santos, L. Encapsulation of cosmetic active ingredients for topical application-a review. *J. Microencapsul.* **2016**, *33*, 1–17. [CrossRef] [PubMed]

43. Acevedo, F.; Rubilar, M.; Jofré, I.; Villarroel, M.; Navarrete, P.; Esparza, M.; Romero, F.; Vilches, E.A.; Acevedo, V.; Shene, C. Oil bodies as a potential microencapsulation carrier for astaxanthin stabilisation and safe delivery. *J. Microencapsul.* **2014**, *31*, 488–500. [CrossRef]

44. Higuera-Ciapara, I.; Felix-Valenzuela, L.; Goycoolea, F.M.; Argüelles-Monal, W. Microencapsulation of astaxanthin in a chitosan matrix. *Carbohydr. Polym.* **2004**, *56*, 41–45. [CrossRef]

45. Lin, S.F.; Chen, Y.C.; Chen, R.N.; Chen, L.C.; Ho, H.O.; Tsung, Y.H.; Sheu, M.T.; Liu, D.Z. Improving the stability of astaxanthin by microencapsulation in calcium alginate beads. *PLoS ONE* **2016**, *11*, e0153685. [CrossRef] [PubMed]

46. Liu, G.; Hu, M.; Zhao, Z.; Lin, Q.; Wei, D.; Jiang, Y. Enhancing the stability of astaxanthin by encapsulation in poly (l-lactic acid) microspheres using a supercritical anti-solvent process. *Particuology* **2019**, *44*, 54–62. [CrossRef]

47. Hu, F.; Liu, W.; Yan, L.; Kong, F.; Wei, K. Optimization and characterization of poly(lactic-co-glycolic acid) nanoparticles loaded with astaxanthin and evaluation of anti-photodamage effect in vitro. *R. Soc. Open Sci.* **2019**, *6*, 191184. [CrossRef] [PubMed]

48. Rodriguez-Ruiz, V.; Salatti-Dorado, J.Á.; Barzegari, A.; Nicolas-Boluda, A.; Houaoui, A.; Caballo, C.; Caballero-Casero, N.; Sicilia, D.; Venegas, J.B.; Pauthe, E.; et al. Astaxanthin-loaded nanostructured lipid carriers for preservation of antioxidant activity. *Molecules* **2018**, *23*, 2601. [CrossRef]

49. Wang, Q.; Zhao, Y.; Guan, L.; Zhang, Y.; Dang, Q.; Dong, P.; Li, J.; Liang, X. Preparation of astaxanthin-loaded DNA/chitosan nanoparticles for improved cellular uptake and antioxidation capability. *Food Chem.* **2017**, *227*, 9–15. [CrossRef]

50. Liu, C.; Zhang, S.; McClements, D.J.; Wang, D.; Xu, Y. Design of Astaxanthin-Loaded Core-Shell Nanoparticles Consisting of Chitosan Oligosaccharides and Poly(lactic- co-glycolic acid): Enhancement of Water Solubility, Stability, and Bioavailability. *J. Agric. Food Chem.* **2019**, *67*, 5113–5121. [CrossRef]

51. Tachaprutinun, A.; Udomsup, T.; Luadthong, C.; Wanichwecharungruang, S. Preventing the thermal degradation of astaxanthin through nanoencapsulation. *Int. J. Pharm.* **2009**, *374*, 119–124. [CrossRef]

52. Boucher, J.; Cengelli, F.; Trumbic, D.; Marison, I.W. Sorption of hydrophobic organic compounds (HOC) in rapeseed oil bodies. *Chemosphere* **2008**, *70*, 1452–1458. [CrossRef]

53. Kumari, A.; Yadav, S.K.; Yadav, S.C. Biodegradable polymeric nanoparticles based drug delivery systems. *Colloids Surfaces B Biointerfaces* **2010**, *75*, 1–18. [CrossRef]
54. Ali, A.; Ahmed, S. A review on chitosan and its nanocomposites in drug delivery. *Int. J. Biol. Macromol.* **2018**, *109*, 273–286. [CrossRef]
55. Chauhan, I.; Yasir, M.; Verma, M.; Singh, A.P. Nanostructured lipid carriers: A groundbreaking approach for transdermal drug delivery. *Adv. Pharm. Bull.* **2020**, *10*, 150–165. [CrossRef]
56. Carneiro, S.B.; Duarte, F.Í.C.; Heimfarth, L.; Quintans, J.D.S.S.; Quintans-Júnior, L.J.; Júnior, V.F.D.V.; De Lima, Á.A.N. Cyclodextrin-drug inclusion complexes: In vivo and in vitro approaches. *Int. J. Mol. Sci.* **2019**, *20*, 642. [CrossRef] [PubMed]
57. Chen, X.; Chen, R.; Guo, Z.; Li, C.; Li, P. The preparation and stability of the inclusion complex of astaxanthin with β-cyclodextrin. *Food Chem.* **2007**, *101*, 1580–1584. [CrossRef]
58. Magalhães, T.S.S.d.A.; Macedo, P.C.d.O.; Pacheco, S.Y.K.; da Silva, S.S.; Barbosa, E.G.; Pereira, R.R.; Costa, R.M.R.; Silva Junior, J.O.C.; da Silva Ferreira, M.A.; de Almeida, J.C.; et al. Development and evaluation of antimicrobial and modulatory activity of inclusion complex of euterpe oleracea mart oil and β-cyclodextrin or HP-β-cyclodextrin. *Int. J. Mol. Sci.* **2020**, *21*, 942. [CrossRef] [PubMed]
59. Pinheiro, J.G.d.O.; Tavares, E.d.A.; Silva, S.S.D.; Félix Silva, J.; Carvalho, Y.M.B.G.D.; Ferreira, M.R.A.; Araúj, A.A.d.S.; Barbosa, E.G.; Fernandes Pedrosa, M.d.F.; Soares, L.A.L.; et al. Inclusion complexes of copaiba (Copaifera multijuga hayne) oleoresin and cyclodextrins: Physicochemical characterization and anti-inflammatory activity. *Int. J. Mol. Sci.* **2017**, *18*, 2388. [CrossRef] [PubMed]
60. da Silva Júnior, W.F.; Bezerra de Menezes, D.L.; de Oliveira, L.C.; Koester, L.S.; Oliveira de Almeida, P.D.; Lima, E.S.; de Azevedo, E.P.; da Veiga Júnior, V.F.; Neves de Lima, Á.A. Inclusion complexes of β and HPβ-cyclodextrin with α, β amyrin and in vitro anti-inflammatory activity. *Biomolecules* **2019**, *9*, 241. [CrossRef]
61. Ferreira, E.B.; da Silva Junior, W.F.; de Oliveira Pinheiro, J.G.; Da Fonseca, A.G.; Moura Lemos, T.M.A.; de Oliveira Rocha, H.A.; De Azevedo, E.P.; Mendonça Junior, F.J.B.; Neves de Lima, A.A. Characterization and antiproliferative activity of a novel 2-aminothiophene derivative-β-cyclodextrin binary system. *Molecules* **2018**, *23*, 3130. [CrossRef]
62. Lockwood, S.F.; O'Malley, S.; Mosher, G.L. Improved aqueous solubility of crystalline astaxanthin (3,3′-dihydroxy-β, β-carotene-4,4′-dione) by Captisol® (sulfobutyl ether β-cyclodextrin). *J. Pharm. Sci.* **2003**, *92*, 922–926. [CrossRef]
63. Kim, S.; Cho, E.; Yoo, J.; Cho, E.; Ju Choi, S.; Son, S.M.; Lee, J.M.; In, M.J.; Kim, D.C.; Kim, J.H.; et al. β-CD-mediated encapsulation enhanced stability and solubility of Astaxanthin. *J. Appl. Biol. Chem.* **2010**, *53*, 559–565. [CrossRef]
64. Yuan, C.; Jin, Z.; Xu, X.; Zhuang, H.; Shen, W. Preparation and stability of the inclusion complex of astaxanthin with hydroxypropyl-β-cyclodextrin. *Food Chem.* **2008**, *109*, 264–268. [CrossRef] [PubMed]
65. Yuan, C.; Jin, Z.; Xu, X. Inclusion complex of astaxanthin with hydroxypropyl-β-cyclodextrin: UV, FTIR, 1H NMR and molecular modeling studies. *Carbohydr. Polym.* **2012**, *89*, 492–496. [CrossRef] [PubMed]
66. Yuan, C.; Du, L.; Jin, Z.; Xu, X. Storage stability and antioxidant activity of complex of astaxanthin with hydroxypropyl-β-cyclodextrin. *Carbohydr. Polym.* **2013**, *91*, 385–389. [CrossRef] [PubMed]
67. Zuluaga, M.; Barzegari, A.; Letourneur, D.; Gueguen, V.; Pavon-Djavid, G. Oxidative Stress Regulation on Endothelial Cells by Hydrophilic Astaxanthin Complex: Chemical, Biological, and Molecular Antioxidant Activity Evaluation. *Oxid. Med. Cell. Longev.* **2017**, *2017*, 8073798. [CrossRef]
68. Gould, S.; Scott, R.C. 2-Hydroxypropyl-β-cyclodextrin (HP-β-CD): A toxicology review. *Food Chem. Toxicol.* **2005**, *43*, 1451–1459. [CrossRef] [PubMed]
69. Kathe, K.; Kathpalia, H. Film forming systems for topical and transdermal drug delivery. *Asian J. Pharm. Sci.* **2017**, *12*, 487–497. [CrossRef]

marine drugs

MDPI

Article

Production of Fucoxanthin from *Phaeodactylum tricornutum* Using High Performance Countercurrent Chromatography Retaining Its FOXO3 Nuclear Translocation-Inducing Effect

Daniela Bárcenas-Pérez [1,2], Antonín Střížek [1,3], Pavel Hrouzek [1], Jiří Kopecký [1], Marta Barradas [4], Arantzazu Sierra-Ramirez [4], Pablo J. Fernandez-Marcos [4] and José Cheel [1,*]

[1]   Laboratory of Algal Biotechnology—Centre ALGATECH, Institute of Microbiology of the Czech Academy of Sciences, Opatovický Mlýn, 379 81 Třeboň, Czech Republic; barcenas@alga.cz (D.B.-P.); strizek@alga.cz (A.S.); hrouzek@alga.cz (P.H.); kopecky@alga.cz (J.K.)
[2]   Faculty of Science, University of South Bohemia, Branišovská 1760, 370 05 České Budějovice, Czech Republic
[3]   Faculty of Science, Department of Ecology, Charles University, Viničná 7, 128 44 Prague 2, Czech Republic
[4]   Metabolic Syndrome Group—BIOPROMET, Madrid Institute for Advanced Studies—IMDEA Food, CEI UAM+CSIC, 28049 Madrid, Spain; marta.barradas@imdea.org (M.B.); aranzazu.sierra@imdea.org (A.S.-R.); pablojose.fernandez@imdea.org (P.J.F.-M.)
*   Correspondence: jcheel@alga.cz; Tel.: +420-384-340-465

Citation: Bárcenas-Pérez, D.; Střížek, A.; Hrouzek, P.; Kopecký, J.; Barradas, M.; Sierra-Ramirez, A.; Fernandez-Marcos, P.J.; Cheel, J. Production of Fucoxanthin from *Phaeodactylum tricornutum* Using High Performance Countercurrent Chromatography Retaining Its FOXO3 Nuclear Translocation-Inducing Effect. *Mar. Drugs* **2021**, *19*, 517. https://doi.org/10.3390/md19090517

Academic Editors: Masashi Hosokawa and Hayato Maeda

Received: 28 July 2021
Accepted: 8 September 2021
Published: 11 September 2021

Publisher's Note: MDPI stays neutral with regard to jurisdictional claims in published maps and institutional affiliations.

Abstract: *Phaeodactylum tricornutum* is a rich source of fucoxanthin, a carotenoid with several health benefits. In the present study, high performance countercurrent chromatography (HPCCC) was used to isolate fucoxanthin from an extract of *P. tricornutum*. A multiple sequential injection HPCCC method was developed combining two elution modes (reverse phase and extrusion). The lower phase of a biphasic solvent system (*n*-heptane, ethyl acetate, ethanol and water, ratio 5/5/6/3, *v/v/v/v*) was used as the mobile phase, while the upper phase was the stationary phase. Ten consecutive sample injections (240 mg of extract each) were performed leading to the separation of 38 mg fucoxanthin with purity of 97% and a recovery of 98%. The process throughput was 0.189 g/h, while the efficiency per gram of fucoxanthin was 0.003 g/h. Environmental risk and general process evaluation factors were used for assessment of the developed separation method and compared with existing fucoxanthin liquid-liquid isolation methods. The isolated fucoxanthin retained its well-described ability to induce nuclear translocation of transcription factor FOXO3. Overall, the developed isolation method may represent a useful model to produce biologically active fucoxanthin from diatom biomass.

Keywords: fucoxanthin; *Phaeodactylum tricornutum*; high performance countercurrent chromatography (HPCCC); countercurrent chromatography (CCC); centrifugal partition chromatography (CPC)

1. Introduction

Fucoxanthin (Figure 1) is an orange carotenoid found in brown seaweeds and some classes of microalgae, especially diatoms [1]. Unlike other carotenoids found in nature, fucoxanthin has a unique molecular structure composed of an unusual allenic bond, a 5,6-monoepoxide and nine conjugated double bounds [2]. This carotenoid has been mainly isolated from seaweeds and widely investigated for its biological properties [2,3]. Its benefits include anti-obesity [4–7], anti-diabetic [8], anti-inflammatory [9–11], anti-cancer [12–15] and antioxidant [3,16–20] effects. Moreover, fucoxanthin has been previously described to induce nuclear translocation of the transcription factor FOXO3, which inhibits the fibronectin and collagen IV expression as well as oxidative stress, resulting in the reduction of fibrosis in diabetic nephropathy [21,22]. Commercial products containing brown algae-sourced fucoxanthin which are mainly used for weight and fat control include Solaray Fucoxanthin from SolarayVR, fucoTHIN from Garden of Life [23] and ThinOgen Fucoxanthin from BGG [24]. Only one product (FucoVitalTM from Algatech, Ketura) containing fucoxanthin from the diatom *Phaeodactylum tricornutum* is commercially available [23].

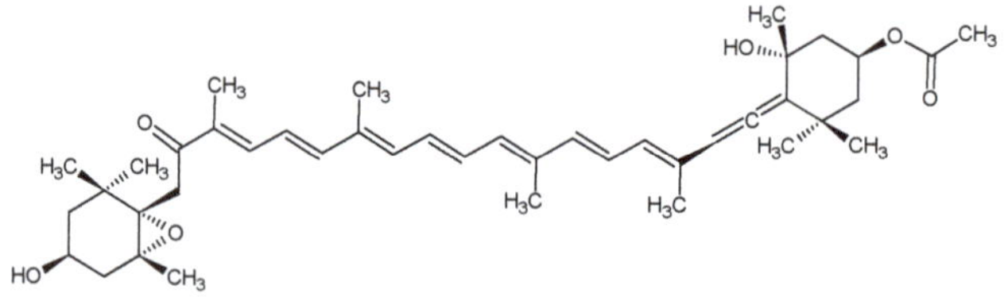

Figure 1. Chemical structure of fucoxanthin.

Diatom microalgae have gained attention as a valuable source of fucoxanthin since its biomass content can be enriched up to ten times more than for macroalgae [3]. Among diatoms, *P. tricornutum* "the model diatom" is one of the most studied species. Until now, most of the efforts devoted to exploit *P. tricornutum* as a source of fucoxanthin have been mainly focused on culture optimization, reaching fucoxanthin contents up to 16 mg [25] and 26 mg [26] per gram dry weight. Fucoxanthin has an increasing demand in nutraceutical, cosmetic and food industry sectors and its total global market size is expected to reach USD 107.4 million in the period from 2020 to 2025 [27].

So far, the methods reported for the isolation of fucoxanthin from algae have required the use of liquid-solid chromatography [25,28–33], which includes multi-step procedures. Fucoxanthin-enriched extracts have been also produced through enzyme-assisted extraction followed by co-solvent extraction [34], aqueous two-phase systems (ATPS) extraction system [35], supercritical CO_2 extraction [36], pressurized liquid extraction [37,38], pressurized subcritical extraction [26] and subcritical fluid extraction [1]. Despite the aforementioned efforts for obtaining fucoxanthin from algal biomass, it has been noted that low cost, simple and scalable isolation methods still require to be developed [23,39]. In these circumstances, liquid-liquid chromatography such as countercurrent chromatography (CCC) is able to play an essential role, as it takes advantage of its liquid stationary phase. CCC does not use any solid support and the stationary and the mobile phases are liquids [40]. The stationary phase is immobilized in the column by means of a centrifugal force field generated by the column spinning, whereas the mobile phase is pumped through the column. The separation of target compounds from a mixture is based on the difference in their partitioning between the two immiscible phases. Given that this technology lacks solid support, it has many advantages over the solid chromatographic techniques including large sample loading capacity, low risk of irreversible adsorption and sample denaturation, high sample recovery, low consumption of solvents and great operational versatility, since the roles of mobile and stationary phases can be exchanged during the chromatographic operation [40–42]. The capacity of countercurrent chromatography for obtaining valuable compounds from microalgae biomass has been widely demonstrated [42–48] and its application at industrial level is already a reality [49]. High-speed countercurrent chromatography (HSCCC) has been used for the isolation of fucoxanthin from edible brown macroalgae species such as *Laminaria japonica*, *Undaria pinnatifida* and *Sargassum fusiforme* [50] applying a single sample injection method. Centrifugal partition chromatography (CPC), another variant of liquid-liquid chromatography, was also used in two steps to isolate fucoxanthin from the macroalgae *Eisenia bicyclis* [51] and in one step followed by flash chromatography to produce fucoxanthin from the microalgae *Tisochrysis lutea* [52]. The present study reports, for the first time, an efficient and scalable HPCCC isolation method to obtain fucoxanthin from the diatom *P. tricornutum* using a multiple sequential-injection separation strategy. The developed method does not need to align with another chromatographic technique to achieve the desired purity. The biomass extraction and HPCCC isolation process were unified using the same solvent system to improve the fucoxanthin final recovery. The isolated compound was found to maintain its described bioactivity as inductor of nuclear translocation of the transcription factor FOXO3.

2. Results and Discussion

2.1. Extract Preparation

In order to achieve a high recovery of target compounds, the isolation process should be preceded by an efficient extraction method. Generally, the solvent used for obtaining biomass extracts is not the same to that used for the isolation work, since different techniques are applied. Unifying the chemical nature of the solvent used in biomass extraction and isolation could favor the recovery of the target compound, as recently described [45]. In the present study, the upper (UP1) and lower (LP1) phases used for the HPCCC isolation of fucoxanthin (Section 2.2) were tested for their capacity to produce a fucoxanthin-enriched extract from *P. tricornutum* biomass. These liquid phases were compared with other solvents generally used to generate microalgae extracts (Figure 2). Ultrasound assisted extraction (UAE) for 10 and 30 min and mortar and pestle-assisted extraction (MPE) were applied as extraction methods. The results (Figure 2) showed that the fucoxanthin yield was highly dependent on the solvent and extraction method. LP1 and 80% ethanol led to similar fucoxanthin recoveries in all experiments, but the highest values were obtained by using UAE for 30 min. In previous investigations [25,32], 100% ethanol was found to lead to the best extraction yield. To determine the extraction efficiency of LP1 in this study, the Bligh–Dyer method was used as a reference method, as described in Section 3.2. In Figure 2, the contents of fucoxanthin in the extracts produced using UAE for 30 min with LP1 and Bligh–Dyer methods were 4.84 mg/g and 2 mg/g, respectively. Therefore, LP1 showed extraction efficiency over 100%. It is conceivable to assume that the major components in the upper (UP1) and lower (LP1) phases are *n*-heptane and water; respectively, with ethanol and ethyl acetate distributed variably between the two liquid phases. Using LP1 as the solvent for biomass extraction allows a selective extraction that excludes highly lipophilic impurities and, thus, benefits its further isolation. Therefore, the extraction of *P. tricornutum* biomass with LP1 under UAE for 30 min was selected for the large-scale biomass extraction. Using these conditions, an amount of 10 g of dried *P. tricornutum* biomass was extracted with 1.2 L of LP1 affording 3.56 g of dried extract, which was used for the HPCCC isolation of fucoxanthin.

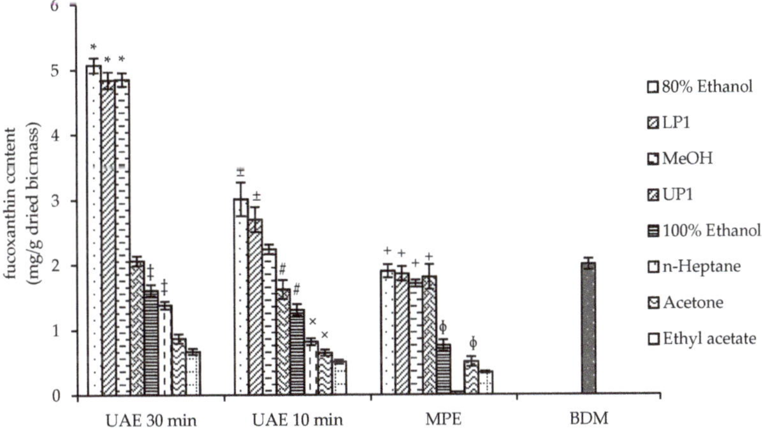

Figure 2. Extraction optimization of fucoxanthin from dried biomass of *Phaeodactylum tricornutum*. Data are means ± SD (n = 3). Bars values with the same symbols (*,‡,±,#,×,+,φ) within the same extraction treatment are not significantly different from each other (Tukey's test, $p < 0.05$). Ultrasound assisted extraction (UAE), mortar and pestle-assisted extraction (MPE), lower phase (LP1) and upper phase (UP1) of the selected biphasic solvent system, Bligh–Dyer method (BDM).

2.2. Development and Optimization of HPCCC Separation at Analytical Scale and Scale-Up to Semi-Prep Column

Different biphasic solvent systems composed of *n*-heptane, ethyl acetate, ethanol and water at different proportions were prepared and investigated for their capacity to be used in the isolation of fucoxanthin from extract of *P. tricornutum* biomass (Table 1). An ideal solvent system has to meet some basic requirements. Firstly, it has to provide a proper K value ($0.5 \leq K \leq 3.5$) [43,53] that permits the separation of the target compound between the two immiscible phases of the selected biphasic solvent system. Secondly, it has to retain enough stationary phase within the HPCCC column by providing a proper density difference (>0.08 g/mL) between its two immiscible liquid phases [46,54] and a short settling time (t < 30 s) [40]. From the Table 1, the system 1 was selected for the isolation of fucoxanthin and thus transferred to the analytical coil (24 mL) of the HPCCC equipment for the optimization of the isolation process in order to maximize the throughput and efficiency.

Table 1. The partition coefficient (K) of fucoxanthin in different biphasic solvent systems, density differences and the settling times.

Solvent Systems	Composition	Relative Proportions of Solvents (*v/v/v/v*)	Phase Volume Ratio (UP/LP)	Settling Time (s)	Density Difference (LP−UP, g/mL)	Partition Coefficient (K) of Fucoxanthin
1	*n*-Hep–EtoAc–EtOH–H$_2$O	5/5/6/3	0.58	17	0.1284	0.515
2	*n*-Hep–EtoAc–EtOH–H$_2$O	5/5/7/3	0.43	15	0.1171	0.314
3	*n*-Hep–EtoAc–EtOH–H$_2$O	5/5/8/3	0.38	18	0.1311	0.205
4	*n*-Hep–EtoAc–EtOH–H$_2$O	5/5/6/4	0.59	18	0.1301	0.897
5	*n*-Hep–EtoAc–EtOH–H$_2$O	5/5/6/5	0.59	20	0.1497	1.942
6	*n*-Hep–EtoAc–EtOH–H$_2$O	5/5/5/3	0.80	20	0.1232	0.590

LP: Lower phase. UP: Upper phase.

It has been well established that a high retention of the stationary phase leads to a good resolution in countercurrent chromatography [55]. Given that flow rate and sample loading can affect the retention of the stationary phase, these parameters were optimized in the current study. The optimization studies were performed using sample loadings from 20 to 60 mg and mobile phase flow rates of 0.5 and 1.0 mL/min using the analytical coil (24 mL) of the HPCCC equipment. The column rotational speed and loop volume were at fixed 1600 rpm and 0.5 mL, respectively, operating at 30 °C. This process was performed in reverse elution mode, which means that the lower phase of the selected biphasic solvent system was used as the mobile phase, while the upper phase was the stationary phase. It was observed that fucoxanthin was well separated with sample loadings of 20 and 40 mg at mobile phase flow rates of both 0.5 and 1 mL/min (Figure 3a,b,d,e). These operating conditions permitted a good retention of stationary phase within the column (Table 2), which was estimated using equation 1. The highest fucoxanthin purity was achieved at a flow rate of 0.5 mL/min and when 20 or 40 mg of sample was injected (Table 2). Finally, a sample loading of 40 mg was chosen as the optimal value, as it would favor the process throughput.

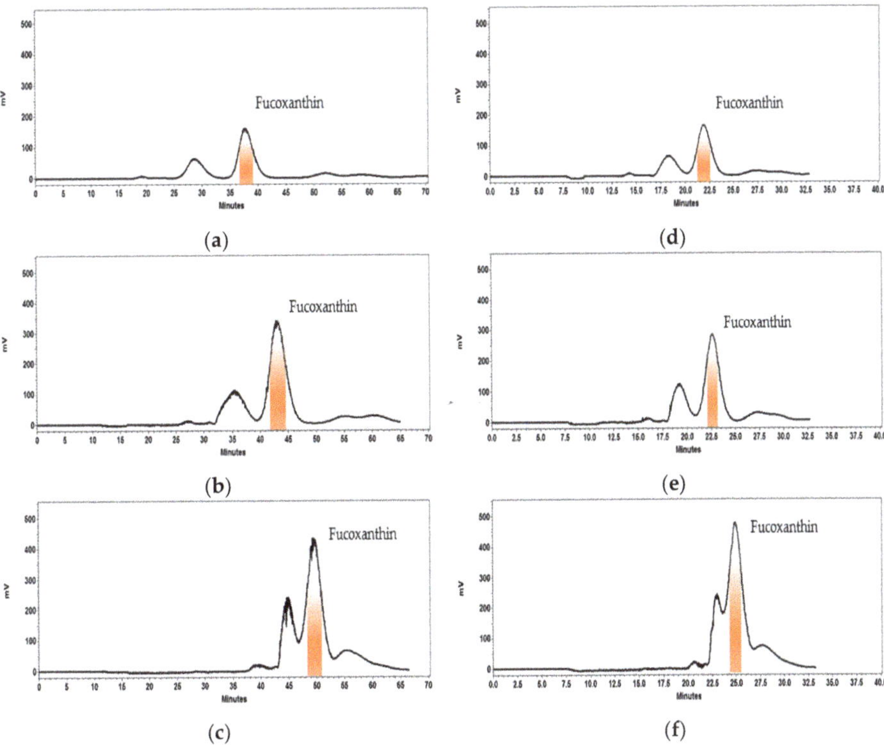

Figure 3. High performance countercurrent chromatography (HPCCC) optimization using different sample loadings and flow rates of mobile phase for obtaining fucoxanthin from *Phaeodactylum tricornutum* extract. (**a**) 20 mg at 0.5 mL/min. (**b**) 40 mg at 0.5 mL/min. (**c**) 60 mg at 0.5 mL/min. (**d**) 20 mg at 1.0 mL/min. (**e**) 40 mg at 1.0 mL/min. (**f**) 60 mg at 1.0 mL/min. The samples were dissolved in 0.5 mL of mobile phase (0.5 mL sample loop). Biphasic solvent system: System 1, mixture of *n*-heptane, ethyl acetate, ethanol and water (ratio 5:5:6:3, *v/v/v/v*). Elution mode: Reverse, the mobile phase is the lower phase of the system 1. Column temperature: 30 °C. Detection: 450 nm.

Table 2. Stationary phase retention (*Sf*) and peak resolution in response to different sample loadings and flow rates in the analytical coil (24 mL) of high performance countercurrent chromatography (HPCCC) for obtaining fucoxanthin from *Phaeodactylum tricornutum* extract. Biphasic solvent system: System 1, mixture of *n*-heptane, ethyl acetate, ethanol and water (ratio 5:5:6:3, *v/v/v/v*). Elution mode: Reverse, the mobile phase is the lower phase of the biphasic solvent system. Column temperature: 30 °C. Detection: 450 nm. Loop volume: 0.5 mL.

Optimization Experiments	Flow Rate (mL/min)	*Sf* at The Hydrodynamic Equilibrium in HPCCC (%)	Loading Per Injection (mg)	Peak Resolution (1/2)	*Sf* at The End of The HPCCC Separation Run (%)	Peak Purity (%)
a	0.5	56.25	20	2.9	52.08	98
b	0.5	56.25	40	2.1	32.25	97
c	0.5	56.25	60	1.7	10.41	70
d	1.0	50	20	2.0	29.16	96
e	1.0	50	40	1.8	20.83	94
f	1.0	50	60	1.4	4.16	55

Rt: Retention time. The peak resolution was calculated as follows: Rs = 2(Rt$_2$-Rt$_1$)/(W$_2$+W$_1$). (1) major contaminant. (2) fucoxanthin. W: the peak width at half height. a (Rt$_1$ = 29 min, Rt$_2$ = 38 min and W$_1$~3.2, W$_2$~3). b (Rt$_1$ = 35 min, Rt$_2$ = 43.4 min and W$_1$~4.6, W$_2$~3.35). c (Rt$_1$ = 44.7 min, Rt$_2$ = 49.4 min and W$_1$~2.4, W$_2$~3). d (Rt$_1$ = 18.35 min, Rt$_2$ = 21.9 min and W$_1$~1.7, W$_2$~1.9). e (Rt$_1$ = 19.1 min, Rt$_2$ = 22.5 min and W$_1$~2.1, W$_2$~1.75). f (Rt$_1$ = 23 min, Rt$_2$ = 24.9 min and W$_1$~1.0, W$_2$~1.7).

Once the operating parameters were optimized on the HPCCC analytical coil (24 mL), they were transferred to the HPCCC semi-prep coil (134 mL) to scale-up the developed method. The sample loading, flow rates and loop volume were scaled up linearly based on a principle previously established [56]. The "g" force in the analytical and semi-prep columns remained the same and the volume ratio between the analytical (24 mL) and semi-prep (134 mL) HPCCC coils was approximately 6. Accordingly, the sample loading, mobile phase flow rate and loop volume were adjusted proportionally to 240 mg (40 mg × 6), 3 mL/min (0.5 mL/min × 6) and 3 mL (0.5 mL × 6), respectively. These scale up settings led to a semi-prep chromatographic process (Figure 4a), showing a good stationary phase retention (*Sf*: 68.65%) at the hydrodynamic equilibrium stage. It was observed that fucoxanthin eluted at 37.2 min with a peak resolution of 2.5 with respect to the major contaminant and purity of 97%. However, it was noticed that a higher flow rate (4 mL/min) was still possible to produce fucoxanthin with a purity of 97% (Figure 4b) and a peak resolution of 2.3 with respect to the major contaminant. Therefore, these last conditions were selected in order to shorten the process and, thereby, increase the overall throughput and efficiency. In CCC, it is possible to predict the retention time of a given compound, once the stationary phase retention, partition coefficient value and flow rate are known. In the present study, it was calculated using the equation 2 as earlier described [57]. This information is of particular interest for calculating a priori the separation process duration and the solvent consumption. In the present study, the predicted retention time of fucoxanthin was 22.83 min, while its experimentally observed retention time was 28.45 min (Figure 4b). These values did not fully match up from each other, which may be due to a decrease in the stationary phase retention (*Sf*: 19.4%) during the separation process.

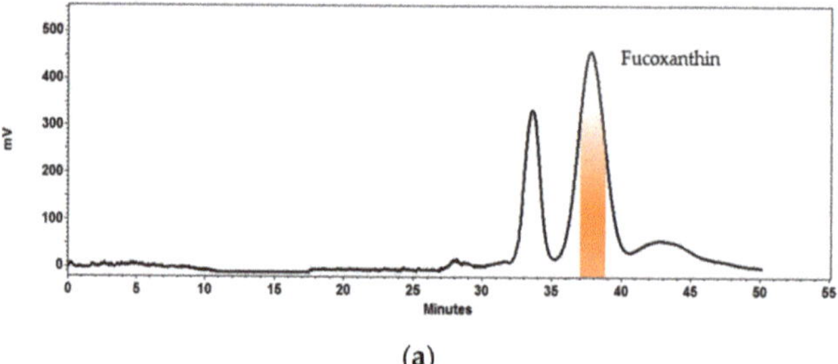

(a)

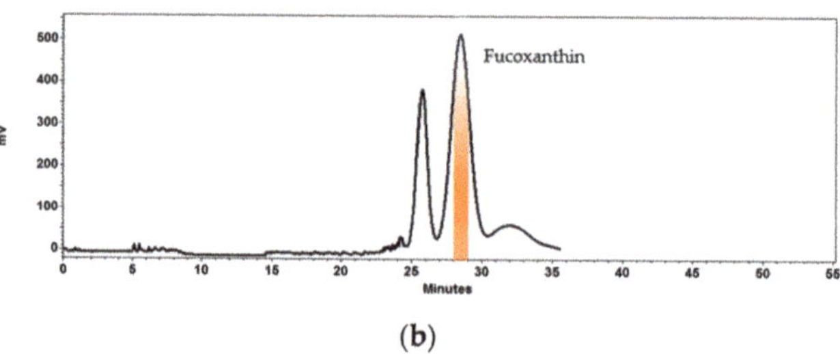

(b)

Figure 4. Scale-up of high performance countercurrent chromatography (HPCCC) method to obtain fucoxanthin from *Phaeodactylum tricornutum* extract. Biphasic solvent system: System 1, mixture of *n*-heptane, ethyl acetate, ethanol and water (ratio 5:5:6:3, *v/v/v/v*). Loading per injection: 240 mg of *Phaeodactylum tricornutum* extract dissolved in 3 mL of mobile phase (3 mL sample loop). (**a**) 3 mL/min of flow rate of the mobile phase. (**b**) 4 mL/min of flow rate of the mobile phase. Rotational speed: 1600 rpm. Column temperature: 30 °C. Detection: 450 nm.

Next, in order to increase the productivity of the developed HPCCC separation process, a multiple-sequential injection system was applied combining two elution modes. It involves reverse phase elution mode for the fucoxanthin separation followed by elution-extrusion mode to replenish the column with new stationary phase without stopping the column rotation. After that, a new hydrodynamic equilibrium was achieved for a new separation cycle of fucoxanthin (Figure 5). The combination of these two HPCCC elution modes resulted as a feasible method for sequentially obtaining fucoxanthin from *P. tricornutum* biomass.

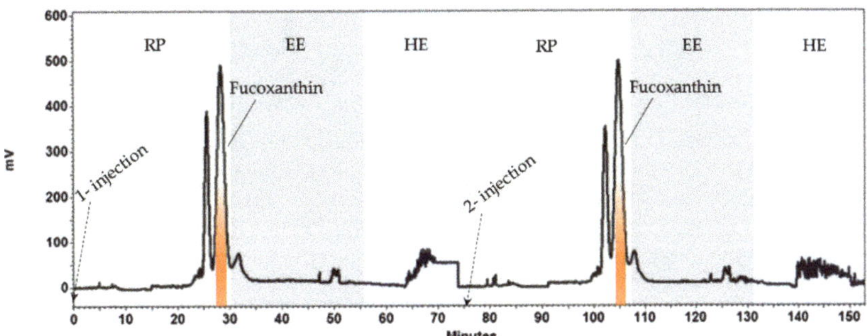

Figure 5. Development of two-injections high performance countercurrent chromatography (HPCCC) method to obtain fucoxanthin from *Phaeodactylum tricornutum* extract. Biphasic solvent system: System 1, mixture of *n*-heptane, ethyl acetate, ethanol and water (ratio 5:5:6:3, $v/v/v/v$). Elution modes: Reverse (RP) and elution-extrusion (EE). Hydrodynamic Equilibrium (HE). Loading per injection: 240 mg of *Phaeodactylum tricornutum* extract dissolved in 3 mL of mobile phase (3 mL sample loop). Runs: 2 consecutive injections. Flow rate: 4 mL/min. Rotational speed: 1600 rpm. Column temperature: 30 °C. Detection: 450 nm.

2.3. HPCCC Sequential Isolation of Fucoxanthin

Based on the previous optimized parameters and criteria, a sequential HPCCC separation process was satisfactorily performed to isolate fucoxanthin from *P. tricornutum* biomass (Figure 6). An amount of 240 mg of *P. tricornutum* extract was injected in each separation cycle. In total, 10 separation cycles were performed processing 2.4 g of *P. tricornutum* extract in 715 min. The entire HPCCC process took 762 min, which included a time period of 27 min for the first column filling with upper phase at 10 mL/min; 20 min for equilibration of the two liquid phases within the column by pumping lower phase at flow rate of 4 mL/min and followed by 715 min that comprised the ten separation cycles. Each separation cycle consisted of three steps including the separation of fucoxanthin in reversed phase elution mode (pumping lower phase at 10 mL/min for 31 min), extrusion (pumping upper phase at 10 mL/min for 25 min) and equilibration of liquid phases inside the column (pumping lower phase at 4 mL/min for 20 min). The chromatogram of the developed process is shown in Figure 6. The total separation process consumed 4.050 L of solvents, resulting in fucoxanthin (38 mg) with a purity of 97% (Figure 7b). To estimate the reproducibility of the separation process from run to run, the relative standard deviation (RSD) of the resolution values between the fucoxanthin and its major contaminant during the ten separation cycles process was calculated. It was found a RSD value of 7.54. Accordingly, the developed process showed a good reproducibility. Finally, as the content of fucoxanthin in the processed extract of *P. tricornutum* biomass was found to be 16.129 mg/g dried extract; therefore, the developed HPCCC process led to a fucoxanthin recovery of 98.16%.

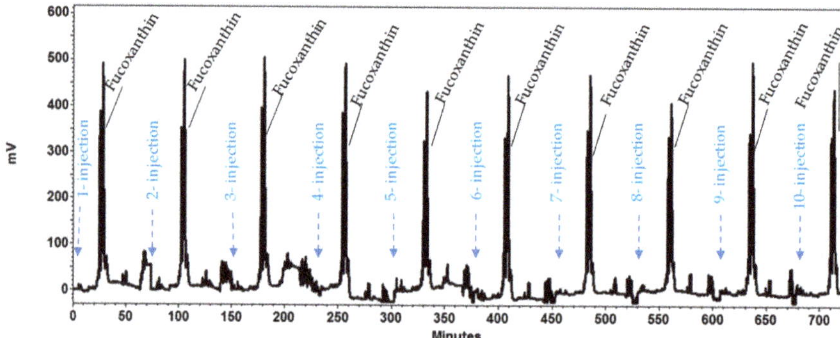

Figure 6. Multiple sequential injections—high performance countercurrent chromatography (HPCCC) method to obtain fucoxanthin from *Phaeodactylum tricornutum* extract. Biphasic solvent system: System 1, mixture of *n*-heptane, ethyl acetate, ethanol and water (ratio 5:5:6:3, *v/v/v/v*). Sample loading: 240 mg of extract dissolved in 3 mL of mobile phase (3 mL sample loop). Runs: 10 consecutive injections. Flow rate: 4 mL/min. Rotational speed: 1600 rpm. Column temperature: 30 °C. Detection: 450 nm.

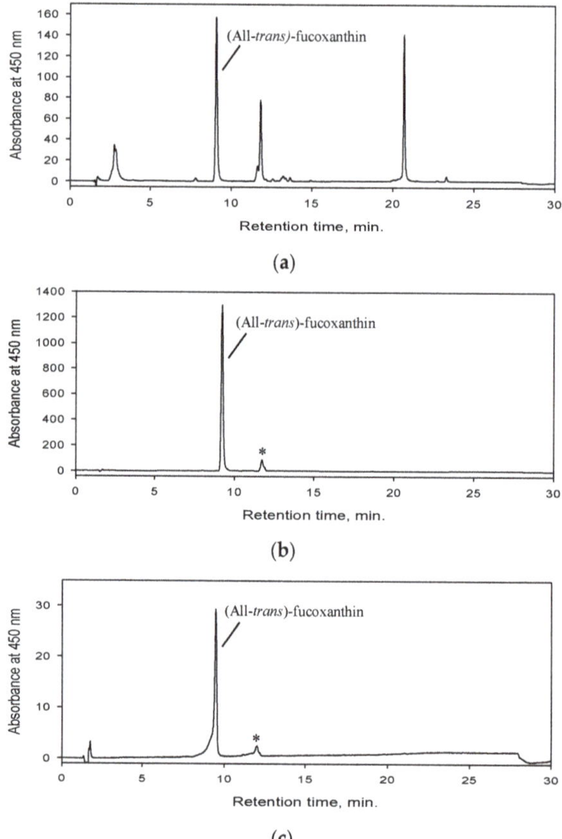

Figure 7. High performance liquid chromatography with diode array detection (HPLC-DAD) chromatograms of *Phaeodactylum tricornutum* extract (**a**), All-*trans*-fucoxanthin fraction obtained by high performance countercurrent chromatography (HPCCC) (**b**) and commercial standard of fucoxanthin from Sigma Aldrich (**c**). * (13 or 13′-*cis*)-fucoxanthin. The chromatograms were monitored at 450 nm.

2.4. Identity Confirmation of The Isolated Target Compound

The identity of the target compound was established as (All-*trans*)-fucoxanthin on the basis of its APCI-HRMS (Figure 8a) and UV-Visible (Figure 8b) spectra in comparison to the literature data [25,58]. The APCI-HRMS spectrum of the target compound peak displayed the molecular ion $[M+H]^+$ at *m/z* 659.4349; a fragment ion $[M+H-H_2O]^+$ at *m/z* 641.4246 corresponding to the cleavage of a water molecule; a fragment ion $[M+H-2H_2O]^+$ at *m/z* 623.4127 formed by the loss of two water units; the ion with second-highest relative abundance $[M+H-H_2O-AcOH]^+$ was observed at *m/z* 581.4021 corresponding to the elimination of water and acetyl group from the molecular ion. The last ion at *m/z* 563.3923 $[M+H-2H_2O-AcOH]^+$ was generated by the cleavage of two water units followed by the dissociation of an acetyl group from the molecular ion. In the Figure 7b, one minor contaminant present in the fucoxanthin fraction obtained by means of HPCCC was observed, which showed an APCI-HRMS fragmentation pattern (Figure 8c) similar to that of (All-*trans*)-fucoxanthin (Figure 8a); therefore, they could not be distinguished from each other using mass spectrometry. The All-*trans*-fucoxanthin peak (Figure 7b) showed its typical UV-VIS spectrum (λmax: 450 and 466 nm) (Figure 8b), while the minor contaminant with retention time of 11.3 min was identified as a *cis*-fucoxanthin isomer based on its UV–VIS spectra (λmax: 442 and 460 nm together with characteristic band at 332 nm) (Figure 8d). This last compound is more likely to be 13- or 13′-*cis*-fucoxanthin, in line with its hypsochromic shift peculiarity and the intensity of the *cis* peak (D_B/D_{II}: 47.0%), as earlier published [59]. The *cis* isomers of carotenoids have been shown to be commonly generated from (All-*trans*)-carotenoids by light and temperature effects and constitute no risk to human health [45,47].

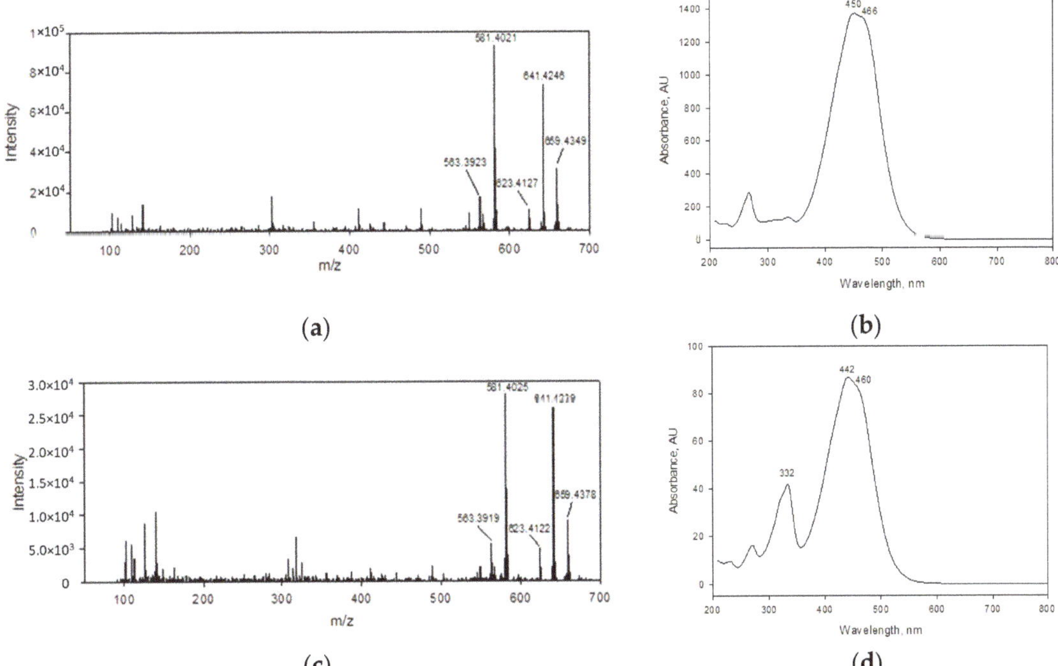

Figure 8. Atmospheric pressure chemical ionization-high resolution mass spectrometry (APCI-HRMS) of All-*trans*-fucoxanthin obtained by HPCCC (**a**) and (13 or 13′-*cis*)-fucoxanthin (**c**). Ultraviolet-visible (UV-Vis) spectra of All-*trans*-fucoxanthin obtained by HPCCC (**b**) and (13 or 13′-*cis*)-fucoxanthin (**d**).

2.5. HPCCC Process Performance

Fucoxanthin has been mostly obtained from algal biomass using liquid-solid chromatography applying multi-step procedures [25,28–33]. Efforts have been dedicated to the production of fucoxanthin from algae using liquid-liquid chromatography namely CCC or CPC [50–52]; however, little has been done to improve the efficiency of the process for obtaining microalgae-sourced fucoxanthin. Unlike solid–liquid chromatography, no expensive columns are required in countercurrent chromatography [49,60]; thus, its use may represent a significant cost saving. The present study, copes with the challenge of applying a multiple-sequential injection strategy using HPCCC (Figure 6) for obtaining fucoxanthin from a microalgae diatom. The scale up of the developed isolation process can be performed in a volumetric way [56], as displayed in Table 3. The projected scaling up production of fucoxanthin from the lab scale (134 mL semi-prep column—Spectrum) to pilot size (18 L column—Maxi), maintaining the same "g" level, would enable processing 609.31 g of algal extract in less than a week. This demonstrates the scalability of the developed multiple-sequential injection system using HPCCC, which could cope with demand for large-scale production of fucoxanthin.

Table 3. Projected throughput for semi-preparative (used in this paper), preparative and pilot scale equipment for sequential (ten injections)—high performance countercurrent chromatography (HPCCC) process for obtaining fucoxanthin.

Equipment	Column Volume	Throughput	Throughput
Spectrum	134 mL	0.189 g/h	7.56 g/week [a]
Midi	980 mL	1.382 g/h	55.28 g/week [a]
Maxi	4.6 L	6.488 g/h	155.71 g/week [b]
NSMS	8.820 L	12.440 g/h	298.56 g/week [b]
Maxi	18 L	25.388 g/h	609.31 g/week [b]

Throughput measured as mass of algal extract processed per unit time. New Spectrum modular series: NSMS. [a] Estimation assumes lab-scale equipment runs for 40 h/week. [b] Estimation assumes suitable equipment in the pilot plant and 24 h operation/week.

In this study, the developed separation process using a multiple-sequential injection strategy was aimed to enable the quick and large-scale isolation of fucoxanthin from the diatom *P. tricornutum*. Different liquid-liquid separation methods using CCC or CPC have already been reported for the separation of fucoxanthin from different algae species [50–52]. Two of these methods processed directly crude extracts using CPC for one microalgae species [52] and CCC for three macroalgae species [50]. A third reported method [51] involved a previous fractionation of macroalgal extract by solvent partitioning before CPC separation. Therefore, this last method [51] would not be useful for a comparative evaluation in this paper, as it uses a fucoxanthin-enriched fraction and would not ensure equality of experimental conditions, besides being an extensive method. Table 4 compares the process performance indicators (purity, *Pt*, *Pe*, *Er* and *Ge*) of the different methods. As shown in Table 4, most of the reported methods led to high purity fucoxanthin (90–99% pure). The purity of fucoxanthin produced in the present study was 97%. Although the different methods produced fucoxanthin with only slightly different purity values, they differed in the other process indicators. Due to the lack of some data referred to fucoxanthin separation in methods A [50] and B [52], some values had to be assumed to make a comparison possible. Studies describing the method A (it uses macroalgae species *L. japonica*, *U. pinnatifida* and *S. fusiforme*) and B (it uses microalgae *Tisochrysis lutea*) did not report solvent consumption for filling columns; therefore, we assumed two column volumes for column filling, as was also performed in the present study. The method described in the current paper used a 134 mL column volume, while methods A and B used 240 and 250 mL of column volumes, respectively. Unlike method B, method A does not report the flow rate of solvent used for column filling; thus, we assumed a flow rate of 30 mL/min, as was also applied in method B. In Table 4, the highest *Pt* values (methods A and B) were in the range from 0.22 and 12, which were higher than that obtained in this paper (0.19). However, the *Pe* value of the developed method (0.003) was about 6–31-fold higher than those of the method A and 1.15-fold higher than that in method B. The method developed in this paper showed the lowest environmental risk factor (*Er*) representing about 6.43–35.06-fold lower than the value indicated in the method A and 2.46-fold lower than that of the method B. This demonstrates the reduced environmental impact of the developed multiple-sequential injection method. It is well established that for production on an industrial scale, a higher efficiency of the process and a lower associated environmental risk factor are the main goals [61], which is achieved with as high a *Ge* as possible. In the present study, the *Ge* factor of the developed multiple-sequential injection HPCCC method was 36–1075-fold and 2.78-fold higher than that of methods A and B, respectively.

Table 4. Comparative results for different liquid-liquid separation methods.

HPCCC Process	Purity (%)	*Pt* (g/h)	*Pe* (g/h)	*Er* (L/g)	*Ge* (g^2 h^{-1} L $^{-1}$)
Method in this paper	97.0	0.189	0.003	106.578	0.000028
Method A1 [50]	94.8	12.195	0.000405	900.6024	0.0000004496
Method A2 [50]	90.2	0.732	0.0005	685.7798	0.000000775
Method A3 [50]	90.4	7.317	0.0000976	3737.500	0.0000000261
Method B [52]	99.0	0.222	0.0026	261.797	0.00001010

Pt: Process throughput. *Pe*: Process efficiency. *Er*: Process environmental risk factor. *Ge*: General process evaluation factor. Estimation of *Pe* and *Er* uses the mass of isolated fucoxanthin. Methods A1, A2 and A3 use the macroalgae species *L. japonica*, *U. pinnatifida* and *S. fusiforme*, respectively.

2.6. Induction of Nuclear Translocation of FOXO3 by Fucoxanthin

Fucoxanthin has been earlier described to induce FOXO3 nuclear translocation resulting in the reduction of oxidative stress and fibrosis in diabetic nephropathy [21,22]. A way to validate the viability of the developed multiple-sequential injection method using HPCCC for obtaining fucoxanthin is to check that the isolated compound retains its bioactivity. For these purposes, we tested the ability of the isolated fucoxanthin to act as an inductor of nuclear translocation of the transcription factor FOXO3, as already described [21,22]. In this bioassay, a well-described method based on the translocation of FOXO3 in the human osteosarcoma cell line U2OS [62–64] was used. As shown in Figure 9a,b, treatment of these cells for 5 h with the positive control (BYL-719) [65], induced a dose-dependent translocation of FOXO3 to the nucleus. Using this time of treatment, there was no evidence of viability decrease when cells were treated with fucoxanthin, as determined by cell number and morphology in the confocal assays. Previously described fucoxanthin-induced cell death in U2OS cells [66] was only apparent after a much longer treatment (48 h); and FOXO3 phosphorylation after treatment with fucoxanthin was measured 24 h after treatment [21]. Therefore, the FOXO3 nuclear translocation assay that is shown in the present study reports quick responses to treatments. Accordingly, compounds concentrations needed to elicit these quick responses (5 h) tend to be higher than those needed for long responses (24 or 48 h in previous literature). Importantly, treatment with the isolated fucoxanthin retained a clear ability to induce nuclear translocation of FOXO3 in a dose-dependent manner (Figure 9a,b). These results demonstrate that the isolated target compound not only shows good purity, but also retains bioactivity in an in vivo setting, thus proving the robustness and the chemically inert profile of the isolation method developed in this work.

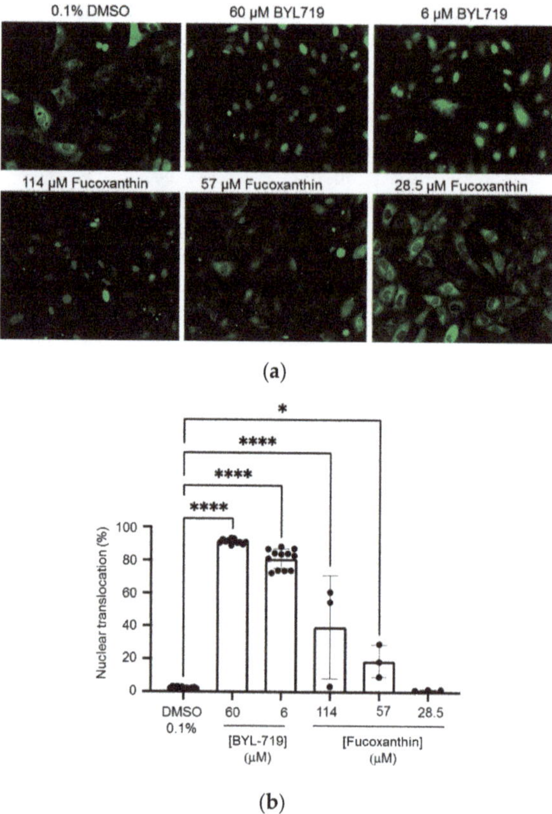

Figure 9. Induction of nuclear translocation of FOXO3 by *Phaeodactylum tricornutum*-derived fucoxanthin. (**a**) U2OS human osteosarcoma cell lines stably transfected with the chimeric construct FOXO3-GFP were treated with the isolated compound for 5 h and sub-cellular localization of the fusion protein was determined by confocal microscopy. (**b**) Quantification of the % values of nuclear translocation of the fusion protein FOXO3-GFP in the cells shown in (**a**). Bars represent the average of at least 3 replicates, indicated by dots. Error bars represent the standard deviation. Statistical significance was assessed using the one-way ANOVA with Tukey correction for multiple comparisons. *, $p < 0.05$; ****, $p < 0.0001$.

3. Materials and Methods

3.1. Biomass Production

In this study, the microalgae *P. tricornutum* strain CCAP 1055/5 (Culture Collection of Algae and Protozoa—Scottish Association for Marine Science, Scotland, UK) was phototrophically cultivated in 80 L glass tubular photobioreactor. The culture was bubbled with a mixture of air and CO_2 (98:2; v/v) at rate of 5 L/min to maintain a high turbulence in the reactor and to prevent cells sedimentation. The culture was continuously illuminated with a LED day light white lamp placed centrally inside the photobioreactor. Light intensity was set to 1500 µmol (photons) m^{-2} s^{-1} measured with a LI-250 light meter (LI-COR Biosciences, Lincoln, NE, USA) 3 cm above light body. The modified Artificial Seawater Medium [67] with added SiO3 solution (0.1 M) was used for cultivation at a temperature kept to $20 \pm 1\ °C$. The cells were harvested by centrifugation using a refrigerated centrifuge (Sigma 8KS) at $20,461 \times g$ for 10 min. The resulting biomass was frozen to $-70\ °C$ and then lyophilized using a ScanVac CoolSafe freeze dryer (LaboGene ApS, Lynge, Denmark) for 72 h. An amount of 105 g of lyophilized *P. tricornutum* biomass was obtained. The microalgae growth curve is shown in Supplementary Figure S1.

3.2. Optimization of Biomass Extraction

For optimizing the production of fucoxanthin-enriched extract from *P. tricornutum* biomass, two extraction methods were investigated including ultrasound-assisted extraction (UAE) and mortar and pestle-assisted extraction (MPE). The tested extraction solvents were absolute ethanol, 80% ethanol (AnalaR Normapur, VWR Inc., Fontenay-sous-Bois, France), acetone, methanol (HiPerSolv Chromanorm, VWR Inc., Fontenay-sous-Bois, France), ethyl acetate, *n*-heptane (HiPerSolv Chromanorm, VWR Inc., Gliwice, Poland) as well as the upper and lower phases of the selected biphasic solvent system (Table 1). Bligh–Dyer method [68] was used as a reference procedure for extraction efficiency estimation. An amount of 10 mg of dried biomass was extracted with 5 mL of the corresponding solvent. UAE was performed for 10 and 30 min employing an ultrasound bath (K6 Kraintek, s.r.o., Podhájska, Slovakia) with a frequency of 38 kHz and an intensity of 47.7707 W/cm at 25 °C. High performance liquid chromatography with diode array detection (HPLC-DAD) was used for the determination of fucoxanthin content in the resulting extracts, as shown in Section 3.4. The best extraction system was used for the large-scale production of a fucoxanthin-enriched extract, from which pure fucoxanthin was obtained using HPCCC.

3.3. High Performance Countercurrent Chromatography (HPCCC) Separation
3.3.1. HPCCC Equipment

For the isolation of fucoxanthin from *P. tricornutum* extract, a HPCCC equipment (Spectrum model, Dynamic Extractions Ltd., Slough, UK) equipped with a 134 mL column (PTFE bore tubing = 3.2 mm) was used. A speed regulator installed into the HPCCC chassis was used to control the speed of the HPCCC column. For controlling the column temperature, a H50/H150 Smart Water Chiller (LabTech Srl, Sorisole Bergamo, Italy) was used. To pump the mobile phase through the column, a Q-Grad pump (LabAlliance, State College, PA, USA) was used. The monitoring of the separation process was performed using a sapphire UV-VIS spectrophotometer (ECOM spol. s.r.o., Prague, Czech Republic) working at a wavelength of 450 nm. The separation process was simultaneously recorded by an EZChrom SI software platform (Agilent Technologies, Pleasanton CA, USA).

3.3.2. Selection of the Suitable Biphasic Solvent System for HPCCC

Several biphasic solvent systems were prepared using different volume ratios of *n*-heptane, ethyl acetate (HiPerSolv Chromanorm, VWR Inc., Gliwice, Poland), ethanol (AnalaR Normapur, VWR Inc., Fontenay-sous-Bois, France) and water. The obtained biphasic solvent systems were tested for their ability to be employed in HPCCC. A correct biphasic solvent system should provide a suitable partition coefficient (K) of the target compound [40,43], a proper density difference between its upper and lower phases and a short settling time [40,43]. The K value calculation was performed by dissolving 2 mg of the produced extract in 1 mL of each phase of the biphasic solvent systems. The obtained mixture was shaken and left to stand until the formation of two clear phases. The formed upper and lower phases were separated and used for the calculation of the K value using HPLC-DAD. The K value was calculated as the ratio between the fucoxanthin peak area in the upper phase to that of the lower phase [47,48]. The measurement of the settling time was performed as previously reported [47]. The density differences between the two immiscible phases were calculated by weighting 1 mL of each phase using a microbalance [43].

3.3.3. HPCCC Separation Process

The HPCCC separation of fucoxanthin from *P. tricornutum* biomass was carried out using the selected biphasic solvent system composed of *n*-heptane, ethyl acetate, ethanol and water. The preparation of the solvent system was performed by mixing the individual amount of each solvent in a decanting funnel. Then, the mixture was vigorously shaken and left to stand until the formation of the two immiscible phases. The upper phase was employed as stationary phase and the lower phase as mobile phase. The HPCCC column was filled with the selected stationary phase and the rotational speed of the HPCCC column was set at 1600 rpm under a controlled temperature of 30 °C. Once the column was totally filled, the mobile phase was pumped through it until reaching the hydrodynamic equilibrium between the two immiscible phases within the HPCCC column. The equilibrium is achieved when the mobile phase front has emerged from the column without the carryover of the stationary phase; thus, at this steady stage, the system is ready for sample injection. The *P. tricornutum* extract dissolved in one volume of mobile phase was the sample solution. The HPCCC fractions were manually collected and analyzed by HPLC-DAD.

The following equation was used to calculate the retention of the stationary phase (*Sf*) in the HPCCC column:

$$Sf\ (\%) = \frac{Vs}{Vc} \times 100 \tag{1}$$

where Vc is the HPCCC column volume and Vs is the stationary phase volume in the column when hydrodynamic equilibrium has been reached [57].

The following equation was used to predict the retention time (t_R) of the target compound in the HPCCC separation:

$$t_R = \frac{V_M + (K \times V_S)}{F} \tag{2}$$

where V_M is the mobile phase volume when the hydrodynamic equilibrium is reached, K is the partition coefficient of the target compound, V_S is the stationary phase volume when the hydrodynamic equilibrium has been reached and F is the mobile phase flow rate [57].

3.4. HPLC-DAD Analysis of Extract and Fractions

The *P. tricornutum* extract and HPCCC fractions were analyzed using a high performance liquid chromatography (HPLC) system (Agilent 1100 Series, Santa Clara, CA, USA) equipped with diode array detection (DAD). The chromatographic separation was performed using a reversed phase column (Luna® C8 column, 100 × 4.6 mm, 3 μm, 100 Å) at 30 °C. The mobile phase consisted of the mixture of water (A) and methanol (B) which was pumped at a flow rate of 0.8 mL/min using a gradient elution as follows: 0–20 min, 20–0% A; 20–25 min, 0% A; 25–27 min, 0–20% A; 27–30 min, 20–20% A [47]. The HPLC analysis was monitored at 450 nm. A commercial standard of fucoxanthin was used for quantification and comparison purposes (Sigma Aldrich, Darmstadt, Germany). For estimating the content of fucoxanthin in microalgae extracts, a calibration curve was generated using five concentration points of the commercial standard of fucoxanthin ranging from 0.5 to 50 μg/mL, employing sample injection volumes of 20 μL. The resulting regression line equation was y = 126.3x (R^2 = 0.9999), where x expresses fucoxanthin concentration (μg/mL) and y is the HPLC peak area. Purity of isolated fucoxanthin was determined using the same regression line equation.

3.5. Confirmation of the Chemical Identity of the Purified Target Compound

The chemical identity of the isolated target compound was confirmed through a Dionex UltiMate 3000 HPLC system (Thermo Scientific, Carlsbad, CA, USA) connected to a high-resolution tandem mass spectrometry (HRMS/MS) detector with atmospheric pressure chemical ionization (APCI) source (Impact HD mass spectrometer Bruker, Billerica, MA, USA) (HPLC-APCI-HRMS) operated in positive ionization mode. Aiming to improve ionization efficiency, a formic acid (0.1%) solution was put in both solvents A and B. The MS operation parameters were set as follows: capillary voltage (2500 V), drying gas flow (5 L/min), drying gas temperature (350 °C), vaporizer temperature (450 °C) and nebulizer pressure (20 psi). The scanning of mass range between 100 and 2000 *m/z* was used for recording full-scan mass spectra. For the fragmentation of fucoxanthin, the collision energy was set to 35 eV and the collision gas was nitrogen. The identity of the target compound was confirmed in comparison with data published in literature. The conditions for the chromatographic separation are described in Section 3.4.

3.6. High Performance Countercurrent Chromatography (HPCCC) Process Performance

The HPCCC process performance was evaluated as previously reported [61]. Four performance indicators were evaluated including process throughput (Pt), process efficiency (Pe), process environmental risk factor (Er) and general process evaluation factor (Ge).

The process throughput (Pt), which measures the mass processed per unit time, was calculated using the following equation:

$$Pt = \frac{M_c}{t} \tag{3}$$

where M_c is the mass of the algal extract injected per separation process and t is the time per separation process.

The process efficiency (Pe), which shows the rate of production of one mass unit of the isolated compound per unit time, was estimated using the following equation:

$$Pe = \frac{M_t}{t} \tag{4}$$

where M_t is the mass of the isolated target compound and t is the time consumption.

Process environmental risk factor (Er), which shows the volume of the waste solvent generated in the production of one mass unit of isolated target compound, was estimated using the next equation:

$$Er = \frac{V}{M_t} \tag{5}$$

where V is the total volume of solvent used in the process and Mt is the mass of the isolated target compound.

General process evaluation factor (Ge), which represents the process efficiency (Pe) relative to its environmental influence (Er), was determined using the following equation:

$$Ge - \frac{Pe}{Er} \tag{6}$$

The indicators Pt and Pe were used to determine the potential productivity of the developed HPCCC process. The Er factor provided information on the environmental influence of the HPCCC process, and the Ge factor was used to show the overall performance of the entire HPCCC separation process in relation to its environmental impact. The higher the Ge factor, the more beneficial it will be.

3.7. Induction of Nuclear Translocation of FOXO3

Induction of FOXO3 nuclear translocation was measured as previously described [62]. Briefly, U2OS human osteosarcoma-derived cells were stably transfected with the chimeric construct FOXO3-GFP [63]. Cells with the highest fluorescent intensity were sorted using an INFLUX™ cell sorter (BD Biosciences, San Jose, CA, USA) and plated in 384-well plates. One day later, cells were treated with the indicated concentrations of compound and after 5 h, cells were fixed in 4% paraformaldehyde and stained with 1 μg/mL 40,6-Diamidino-2-phenylindole dihydrochloride (DAPI). Green fluorescence was measured using the high-throughput confocal microscope Opera LX (Perkin Elmer, Boston, MA, USA). Nuclear translocation was quantified using Acapella v2.0 (Perkin Elmer, Boston, MA, USA).

3.8. Statistical Analysis

The one-way ANOVA statistical test ($p < 0.05$) was used to determine difference among means followed by a Tukey's multiple comparison test ($p < 0.05$) to perform pairwise comparisons. Relative standard deviation (RSD) of resolution values between the target compounds across the high performance countercurrent chromatography (HPCCC) separation was estimated. The data were analyzed using the Statistical Package S-Plus 2000.

Supplementary Materials: The following are available online at https://www.mdpi.com/article/10.3390/md19090517/s1, Figure S1: Growth curve of *Phaeodactylum tricornutum*.

Author Contributions: Conceptualization, J.C. and D.B.-P.; methodology, J.C. and D.B.-P.; formal analysis, D.B.-P., P.H., J.K., A.S., P.J.F.-M., M.B. and A.S.-R.; investigation, D.B.-P.; writing—original draft preparation, D.B.-P. and J.C.; writing—review and editing, P.H., J.K. and A.S.; supervision, J.C.; funding acquisition, P.H. All authors have read and agreed to the published version of the manuscript.

Funding: This research was funded by the Technology Agency of the Czech Republic (NCK grant TN010000048/03, J.C., D.B-P., P.H. and A.S.) and the National Programme of Sustainability I of the Ministry of Education, Youth and Sports of the Czech Republic (ID: LO1416, JC, PH, JK). Biotests were funded by the IMDEA Food Institute, the Ramón Areces Foundation (CIVP18A3891) and a Ramon y Cajal Fellowship (MICINN, RYC-2017-22335).

Acknowledgments: Daniela Bárcenas-Pérez gratefully acknowledges the research supervision of José Cheel (Centre Algatech–Czech Academy of Sciences) during the doctoral study as well as the University of South Bohemia for a doctoral scholarship and the grant GAJU 017/2019/P. The authors thank Alastair T. Gardiner for English proofreading of the article and Wolfgang Link for kindly providing the FOXO3-GFP construct.

Conflicts of Interest: The authors declare no conflict of interest.

References

1. Guler, B.A.; Deniz, I.; Demirel, Z.; Yesil-Celiktas, O.; Imamoglu, E. A novel subcritical fucoxanthin extraction with a biorefinery approach. *Biochem. Eng. J.* **2020**, *153*, 107403. [CrossRef]
2. Zhang, H.; Tang, Y.; Zhang, Y.; Zhang, S.; Qu, J.; Wang, X.; Kong, R.; Han, C.; Liu, Z. Fucoxanthin: A promising medicinal and nutritional ingredient. *Evid. Based Complement. Alternat. Med.* **2015**, 723515. [CrossRef]
3. Neumann, U.; Derwenskus, F.; Flaiz Flister, V.; Schmid-Staiger, U.; Hirth, T.; Bischoff, S.C. Fucoxanthin, a carotenoid derived from *Phaeodactylum tricornutum* exerts antiproliferative and antioxidant activities in vitro. *Antioxidants* **2019**, *8*, 183. [CrossRef] [PubMed]
4. Maeda, H.; Hosokawa, M.; Sashima, T.; Funayama, K.; Miyashita, K. Fucoxanthin from edible seaweed, *Undaria pinnatifida*, shows antiobesity effect through UCP1 expression in white adipose tissues. *Biochem. Biophys. Res. Commun.* **2005**, *332*, 392–397. [CrossRef] [PubMed]
5. Jeon, S.M.; Kim, H.J.; Woo, M.N.; Lee, M.K.; Shin, Y.C.; Park, Y.B.; Choi, M.S. Fucoxanthin-rich seaweed extract suppresses body weight gain and improves lipid metabolism in high-fat-fed C57BL/6J mice. *Biotechnol. J.* **2010**, *5*, 961–969. [CrossRef] [PubMed]
6. Woo, M.N.; Jeon, S.M.; Shin, Y.C.; Lee, M.K.; Kang, M.A.; Choi, M.S. Anti-obese property of fucoxanthin is partly mediated by altering lipid-regulating enzymes and uncoupling proteins of visceral adipose tissue in mice. *Mol. Nutr. Food Res.* **2009**, *53*, 1603–1611. [CrossRef] [PubMed]
7. Hosokawa, M.; Miyashita, T.; Nishikawa, S.; Emi, S.; Tsukui, T.; Beppu, F.; Okada, T.; Miyashita, K. Fucoxanthin regulates adipocytokine mRNA expression in white adipose tissue of diabetic/obese KK-Ay mice. *Arch. Biochem. Biophys.* **2010**, *1*, 17–25. [CrossRef] [PubMed]
8. Maeda, H.; Hosokawa, M.; Sashima, T.; Murakami-Funayama, K.; Miyashita, K. Anti-obesity and anti-diabetic effects of fucoxanthin on diet-induced obesity conditions in a murine model. *Mol. Med. Rep.* **2009**, *2*, 897–902. [CrossRef]
9. Heo, S.J.; Yoon, W.J.; Kim, K.N.; Ahn, G.N.; Kang, S.M.; Kang, D.H.; Affan, A.; Oh, C.; Jung, W.K.; Jeon, J.Y. Evaluation of anti-inflammatory effect of fucoxanthin isolated from brown algae in lipopolysaccharide-stimulated RAW 264.7 macrophages. *Food Chem. Toxicol.* **2010**, *48*, 2045–2051. [CrossRef]
10. Kim, K.N.; Heo, S.J.; Yoon, W.J.; Kang, S.M.; Ahn, G.; Yi, T.H.; Jeon, Y.J. Fucoxanthin inhibits the inflammatory response by suppressing the activation of NF-κB and MAPKs in lipopolysaccharide-induced RAW 264.7 macrophages. *Eur. J. Pharmacol.* **2010**, *649*, 369–375. [CrossRef]
11. Su, J.; Guo, K.; Huang, M.; Liu, Y.; Zhang, J.; Sun, L.; Li, D.; Pang, K.; Wang, G.; Chen, L.; et al. Fucoxanthin, a marine xanthophyll isolated from *Conticribra weissflogii* ND-8: Preventive anti-inflammatory effect in a mouse model of sepsis. *Front. Pharmacol.* **2019**, *10*, 906. [CrossRef]

12. Moghadamtousi, S.Z.; Karimian, H.; Khanabdali, R.; Razavi, M.; Firoozinia, M.; Zandi, K.; Kadir, H.A. Anticancer and antitumor potential of fucoidan and fucoxanthin, two main metabolites isolated from brown algae. *Sci. World J.* **2014**, 768323. [CrossRef]
13. Ishikawa, C.; Tafuku, S.; Kadekaru, T.; Sawada, S.; Tomita, M.; Okudaira, T.; Nakazato, T.; Toda, T.; Uchihara, J.N.; Taira, N.; et al. Antiadult T-cell leukemia effects of brown algae fucoxanthin and its deacetylated product, fucoxanthinol. *Int. J. Cancer* **2008**, *123*, 2702–2712. [CrossRef]
14. Das, S.K.; Hashimoto, T.; Kanazawa, K. Growth inhibition of human hepatic carcinoma HepG2 cells by fucoxanthin is associated with down-regulation of cyclin D. *Biochim. Biophys.* **2008**, *1780*, 743–749. [CrossRef] [PubMed]
15. Zhang, Z.; Zhang, P.; Hamada, M.; Takahashi, S.; Xing, G.; Liu, J.; Sugiura, N. Potential chemoprevention effect of dietary fucoxanthin on urinary bladder cancer EJ-1 cell line. *Oncol. Rep.* **2008**, *20*, 1099–1103. [CrossRef]
16. Sachindra, N.M.; Sato, E.; Maeda, H.; Hosokawa, M.; Niwano, Y.; Kohno, M.; Miyashita, K. Radical scavenging and singlet oxygen quenching activity of marine carotenoid fucoxanthin and its metabolites. *J. Agric. Food Chem.* **2007**, *55*, 8516–8522. [CrossRef] [PubMed]
17. Nomura, T.; Kikuchi, M.; Kubodera, A.; Kawakami, Y. Proton-donative antioxidant activity of fucoxanthin with 1,1-diphenyl-2-picrylhydrazyl (DPPH). *Biochem. Mol. Biol. Int.* **1997**, *42*, 361–370. [CrossRef]
18. Yan, X.; Chuda, Y.; Suzuki, M.; Nagata, T. Fucoxanthin as the major antioxidant in *Hijikia fusiformis*, a common edible seaweed. *Biosci. Biotechnol. Biochem.* **1999**, *63*, 605–607. [CrossRef] [PubMed]
19. Mise, T.; Ueda, M.; Yasumoto, T. Production of fucoxanthin-rich powder from *Cladosiphon okamuranus*. *Adv. J. Food Sci. Technol.* **2011**, *3*, 73–76.
20. Sangeetha, R.K.; Bhaskar, N.; Baskaran, V. Comparative effects of beta-carotene and fucoxanthin on retinol deficiency induced oxidative stress in rats. *Mol. Cell Biochem.* **2009**, *331*, 59–67. [CrossRef]
21. Yang, G.; Jin, L.; Zheng, D.; Tang, X.; Yang, J.; Fan, L.; Xie, X. Fucoxanthin Alleviates Oxidative Stress through Akt/Sirt1/FoxO3α Signaling to Inhibit HG-Induced Renal Fibrosis in GMCs. *Mar. Drugs* **2019**, *17*, 702. [CrossRef]
22. Xiao, H.; Zhao, J.; Fang, C.; Cao, Q.; Xing, M.; Li, X.; Hou, J.; Ji, A.; Song, S. Advances in Studies on the Pharmacological Activities of Fucoxanthin. *Mar. Drugs* **2020**, *18*, 634. [CrossRef] [PubMed]
23. Yang, R.; Wei, D.; Xie, J. Diatoms as cell factories for high-value products: Chrysolaminarin, eicosapentaenoic acid, and fucoxanthin. *Crit. Rev. Biotechnol.* **2020**, *40*, 993–1009. [CrossRef] [PubMed]
24. BGG World. Available online: https://bggworld.com/thinogentm-fucoxanthin/ (accessed on 10 June 2021).
25. Kim, S.M.; Jung, Y.J.; Kwon, O.N.; Cha, K.H.; Um, B.H.; Chung, D.; Pan, C.H. A potential commercial source of fucoxanthin extracted from the microalga *Phaeodactylum tricornutum*. *Appl. Biochem. Biotechnol.* **2012**, *166*, 1843–1855. [CrossRef] [PubMed]
26. Derwenskus, F.; Metz, F.; Gille, A.; Schmid-Staiger, U.; Briviba, K.; Schließmann, U.; Hirth, T. Pressurized extraction of unsaturated fatty acids and carotenoids from wet *Chlorella vulgaris* and *Phaeodactylum tricornutum* biomass using subcritical liquids. *Glob. Bioenergy* **2019**, *11*, 335–344. [CrossRef]
27. Global Fucoxanthin (CAS 3351-86-8) Market 2020 by Manufacturers, Regions, Type and Application, Forecast to 2025. Available online: https://www.360researchreports.com/global-fucoxanthin-cas-3351-86-8-market-16507769 (accessed on 15 January 2021).
28. Noviendri, D.; Jaswir, I.; Salleh, H.M.; Taher, M.; Miyashita, K.; Ramli, N. Fucoxanthin extraction and fatty acid analysis of *Sargassum bindery* and *S. dulicatum*. *J. Med. Plants Res.* **2011**, *5*, 2405–2412.
29. Jaswir, I.; Noviendri, D.; Salleh, H.M.; Muhammad, T.; Miyashita, K. Isolation of fucoxanthin and fatty acids analysis of *Padina australis* and cytotoxic effect of fucoxanthin on human lung cancer (H1299) cell lines. *Afr. J. Biotechnol.* **2011**, *10*, 18855–18862. [CrossRef]
30. Xia, S.; Wang, K.; Wan, L.; Li, A.; Hu, Q.; Zhang, C. Production, characterization, and antioxidant activity of fucoxanthin from the marine diatom *Odontella aurita*. *Mar. Drugs* **2013**, *11*, 2667–2681. [CrossRef]
31. Jaswir, I.; Noviendri, D.; Salleh, H.M.; Taher, M.; Miyashita, K.; Ramli, N. Analysis of fucoxanthin content and purification of all-trans-fucoxanthin from *Turbinaria turbinate* and *Sargassum plagyophyllum* by SiO₂ open column chromatography and reversed phase-HPLC. *J. Liq. Chromatogr. Relat. Technol.* **2013**, *36*, 1340–1354. [CrossRef]
32. Zhang, W.; Wang, F.; Gao, B.; Huang, L.; Zhang, C. An integrated biorefinery process: Stepwise extraction of fucoxanthin, eicosapentaenoic acid and chrysolaminarin from the same *Phaeodactylum tricornutum* biomass. *Algal Res.* **2018**, *32*, 193–200. [CrossRef]
33. Sun, P.; Wong, C.C.; Li, Y.; He, Y.; Mao, X.; Wu, T.; Ren, Y.; Chen, F. A novel strategy for isolation and purification of fucoxanthinol and fucoxanthin from the diatom *Nitzschia laevis*. *Food Chem.* **2019**, *30*, 566–572. [CrossRef]
34. Billakanti, M.J.; Catchpole, O.; Fenton, T.; Mitchell, K. Enzyme-assisted extraction of fucoxanthin and lipids containing polyunsaturated fatty acids from *Undaria pinnatifida* using dimethyl ether and ethanol. *Process. Biochem.* **2013**, *48*, 1999–2008. [CrossRef]
35. Gómez-Loredo, A.; Benavides, J.; Rito-Palomares, M. Partition behavior of fucoxanthin in ethanol-potassium phosphate two-phase systems. *J. Chem. Technol. Biotechnol.* **2014**, *89*, 1637–1645. [CrossRef]
36. Roh, M.K.; Uddin, M.S.; Chun, B.S. Extraction of fucoxanthin and polyphenol from *Undaria pinnatifida* using supercritical carbon dioxide with co-solvent. *Biotechnol Bioprocess. Eng.* **2008**, *13*, 724–729. [CrossRef]
37. Shang, Y.F.; Kim, S.M.; Lee, W.J.; Um, B.H. Pressurized liquid method for fucoxanthin extraction from *Eisenia bicyclis* (Kjellman) setchell. *J. Biosci Bioeng.* **2011**, *111*, 237–241. [CrossRef]
38. Gilbert-López, B.; Barranco, A.; Herrero, M.; Cifuentes, A.; Ibáñez, E. Development of new green processes for the recovery of bioactives from *Phaeodactylum tricornutum*. *Food Res. Int.* **2017**, *99*, 1056–1065. [CrossRef] [PubMed]

39. Kuczynska, P.; Jemiola-Rzeminska, M.; Strzalka, K. Photosynthetic pigments in diatoms. *Mar. Drugs* **2015**, *13*, 5847–5881. [CrossRef] [PubMed]

40. Ito, Y. Golden rules and pitfalls in selecting optimum conditions for high-speed counter-current chromatography. *J. Chromatogr. A* **2005**, *1065*, 145–168. [CrossRef] [PubMed]

41. Michel, T.; Destandau, E.; Elfakir, C. New advances in countercurrent chromatography and centrifugal partition chromatography: Focus on coupling strategy. *Anal. Bioanal. Chem.* **2014**, *406*, 957–969. [CrossRef] [PubMed]

42. Bárcenas-Pérez, D.; Lukeš, M.; Hrouzek, P.; Kubáč, D.; Kopecký, J.; Kaštánek, P.; Cheel, J. A biorefinery approach to obtain docosahexaenoic acid and docosapentaenoic acid n-6 from *Schizochytrium* using high performance countercurrent chromatography. *Algal Res.* **2021**, *55*, 102241. [CrossRef]

43. Cheel, J.; Urajová, P.; Hájek, J.; Hrouzek, P.; Kuzma, M.; Bouju, E.; Faure, K.; Kopecký, J. Separation of cyclic lipopeptide puwainaphycins from cyanobacteria by countercurrent chromatography combined with polymeric resins and HPLC. *Anal. Bioanal. Chem.* **2017**, *409*, 917–930. [CrossRef]

44. Cheel, J.; Hájek, J.; Kuzma, M.; Saurav, K.; Smýkalová, I.; Ondráčková, E.; Urajová, P.; Vu, D.L.; Faure, K.; Kopecký, J.; et al. Application of HPCCC combined with polymeric resins and HPLC for the separation of cyclic lipopeptides muscotoxins A–C and their antimicrobial activity. *Molecules* **2018**, *23*, 2653. [CrossRef]

45. Fábryová, T.; Cheel, J.; Kubac, D.; Hrouzek, P.; Vu, D.L.; Tůmová, L.; Kopecký, J. Purification of lutein from the green microalgae *Chlorella vulgaris* by integrated use of a new extraction protocol and a multi-injection high performance counter-current chromatography (HPCCC). *Algal Res.* **2019**, *41*, 101574. [CrossRef]

46. Fábryová, T.; Tůmová, L.; da Silva, D.C.; Pereira, D.M.; Andrade, P.B.; Valentão, P.; Hrouzek, P.; Kopecký, J.; Cheel, J. Isolation of astaxanthin monoesters from the microalgae *Haematococcus pluvialis* by high performance countercurrent chromatography (HPCCC) combined with high performance liquid chromatography (HPLC). *Algal Res.* **2020**, *49*, 101947. [CrossRef]

47. Nováková, M.; Fábryová, T.; Vokurková, D.; Dolečková, I.; Kopecký, J.; Hrouzek, P.; Tůmová, L.; Cheel, J. Separation of the glycosylated carotenoid myxoxanthophyll from *Synechocystis salina* by HPCCC and evaluation of its antioxidant, tyrosinase inhibitory and immune-stimulating properties. *Separations* **2020**, *7*, 73. [CrossRef]

48. Fábryová, T.; Kubáč, D.; Kuzma, M.; Hrouzek, P.; Kopecký, J.; Tůmová, L.; Cheel, J. High-performance countercurrent chromatography for lutein production from a chlorophyll-deficient strain of the microalgae *Parachlorella kessleri* HY1. *J. Appl. Phycol.* **2021**. [CrossRef]

49. Sutherland, I.; Thickitt, C.; Douillet, N.; Freebairn, K.; Johns, D.; Mountain, C.; Wood, P.; Edwards, N.; Rooke, D.; Harris, G.; et al. Scalable technology for the extraction of pharmaceutics: Outcomes from a 3 year collaborative industry/academia research programme. *J. Chromatogr. A* **2013**, *1282*, 84–94. [CrossRef]

50. Xiao, X.; Si, X.; Yuan, Z.; Xu, X.; Li, G. Isolation of fucoxanthin from edible brown algae by microwave-assisted extraction coupled with high-speed countercurrent chromatography. *J. Sep. Sci.* **2012**, *35*, 2313–2317. [CrossRef]

51. Kim, S.M.; Shang, Y.F.; Um, B.H. A preparative method for isolation of fucoxanthin from *Eisenia bicyclis* by centrifugal partition chromatography. *Phytochem Anal.* **2011**, *22*, 322–329. [CrossRef]

52. Gonçalves de Oliveira-Júnior, R.; Grougnet, R.; Bodet, P.E.; Bonnet, A.; Nicolau, E.; Jebali, A.; Rumin, J.; Picot, L. Updated pigment composition of *Tisochrysis lutea* and purification of fucoxanthin using centrifugal partition chromatography coupled to flash chromatography for the chemosensitization of melanoma cells. *Algal Res.* **2020**, *51*, 102035. [CrossRef]

53. Cheel, J.; Minceva, M.; Urajová, P.; Aslam, R.; Hrouzek, P.; Kopecký, J. Separation of Aeruginosin -865 from cultivated soil cyanobacterium (Nostoc sp.) by centrifugal partitition chromatography combined with gel permeation chromatography. *Nat. Prod. Commun.* **2015**, *10*, 1719–1722. [CrossRef]

54. Ito, Y.; Conway, W.D. Experimental observations of the hydrodynamic behavior of solvent systems in high-speed counter-current chromatography. III. Effects of physical properties of the solvent systems and operating temperature on the distribution of two-phase solvent systems. *J. Chromatogr. A* **1984**, *301*, 405–414. [CrossRef]

55. Berthod, A.; Maryutina, T.; Spivakov, B.; Shpigun, O.; Sutherland, I. Countercurrent chromatography in analytical chemistry (IUPAC Technical Report). *Pure Appl. Chem.* **2009**, *81*, 355–387. [CrossRef]

56. Wood, P.; Ignatova, S.; Janaway, L.; Keay, D.; Hawes, D.; Garrard, I.; Sutherland, I.A. Counter-current chromatography separation scaled up from an analytical column to a production column. *J. Chromatogr. A* **2007**, *1151*, 25–30. [CrossRef] [PubMed]

57. Sutherland, I.A. Liquid stationary phase retention and resolution in hydrodynamic CCC. In *Comprehensive Analytical Chemistry*; Berthod, A., Ed.; Elsevier Science B.V.: Amsterdam, The Netherlands, 2002; Volume 38, pp. 159–176.

58. Crupi, P.; Toci, A.T.; Mangini, S.; Wrubl, F.; Rodolfi, L.; Tredici, M.R.; Coletta, A.; Antonacci, D. Determination of fucoxanthin isomers in microalgae (*Isochrysis* sp.) by high-performance liquid chromatography coupled with diode-array detector multistage mass spectrometry coupled with positive electrospray ionization. *Rapid Commun. Mass Spectrom.* **2013**, *15*, 1027–1035. [CrossRef]

59. Haugan, J.A.; Englert, G.; Glinz, E.; Liaaen-Jensen, S. Algal Carotenoids. 48. Structural Assignments of Geometrical Isomers of Fucoxanthin. *Acta Chem. Scand.* **1992**, *46*, 389–395. [CrossRef]

60. DeAmicis, C.; Edwards, N.A.; Giles, M.B.; Harris, G.H.; Hewitson, P.; Janaway, L.; Ignatova, S. Comparison of preparative reversed phase liquid chromatography and countercurrent chromatography for the kilogram scale purification of crude spinetoram insecticide. *J. Chromatogr. A* **2011**, *1218*, 6122–6127. [CrossRef] [PubMed]

61. Zhang, M.; Ignatova, S.; Liang, Q.; Wu Jun, F.; Sutherland, I.; Wang, Y.; Luo, G. Rapid and high-throughput purification of salvianolic acid B from *Salvia miltiorrhiza* Bunge by high-performance counter-current chromatography. *J. Chromatogr. A* **2009**, *18*, 3869–3873. [CrossRef] [PubMed]

62. Barradas, M.; Link, W.; Megias, D.; Fernandez-Marcos, P.J. High-Throughput Image-Based Screening to Identify Chemical Compounds Capable of Activating FOXO. *Methods Mol. Biol.* **2019**, *1890*, 151–161. [CrossRef]

63. Zanella, F.; Rosado, A.; Garcia, B.; Carnero, A.; Link, W. Using multiplexed regulation of luciferase activity and GFP translocation to screen for FOXO modulators. *BMC Cell Biol.* **2009**, *10*, 14. [CrossRef]

64. Zanella, F.; Rosado, A.; García, B.; Carnero, A.; Link, W. Chemical genetic analysis of FOXO nuclear-cytoplasmic shuttling by using image-based cell screening. *Chembiochem* **2008**, *22*, 2229–2237. [CrossRef]

65. Furet, P.; Guagnano, V.; Fairhurst, R.A.; Imbach-Weese, P.; Bruce, I.; Knapp, M.; Fritsch, C.; Blasco, F.; Blanz, J.; Aichholz, R.; et al. Discovery of NVP-BYL719 a potent and selective phosphatidylinositol-3 kinase alpha inhibitor selected for clinical evaluation. *Bioorg. Med. Chem. Lett.* **2013**, *23*, 3741–3748. [CrossRef]

66. Garg, S.; Afzal, S.; Elwakeel, A.; Sharma, D.; Radhakrishnan, N.; Dhanjal, J.K.; Sundar, D.; Kaul, S.C.; Wadhwa, R. Marine Carotenoid Fucoxanthin Possesses Anti-Metastasis Activity: Molecular Evidence. *Mar. Drugs* **2019**, *17*, 338. [CrossRef] [PubMed]

67. Starr, R.C.; Zeikus, J.A. UTEX–The Culture Collection of Algae at the University of Texas at Austin 1993 List of Cultures. *J. Phycol.* **1993**, *29*, 106. [CrossRef]

68. Bligh, E.G.; Dyer, W.J. A rapid method of total lipid extraction and purification. *Can. J. Biochem. Physiol.* **1959**, *37*, 911–917. [CrossRef] [PubMed]

marine drugs

MDPI

Review

The Critical Studies of Fucoxanthin Research Trends from 1928 to June 2021: A Bibliometric Review

Yam Sim Khaw [1], Fatimah Md. Yusoff [2,3,*], Hui Teng Tan [1], Nur Amirah Izyan Noor Mazli [1], Muhammad Farhan Nazarudin [1], Noor Azmi Shaharuddin [4] and Abdul Rahman Omar [5]

[1] Laboratory of Aquatic Animal Health and Therapeutics, Institute of Bioscience, Universiti Putra Malaysia, Serdang 43400, Selangor, Malaysia; yskhaw@gmail.com (Y.S.K.); huiteng.tan28@gmail.com (H.T.T.); nuramirahizyan@gmail.com (N.A.I.N.M.); m_farhannaza@upm.edu.my (M.F.N.)
[2] Department of Aquaculture, Faculty of Agriculture, Universiti Putra Malaysia, Serdang 43400, Selangor, Malaysia
[3] International Institute of Aquaculture and Aquatic Sciences, Universiti Putra Malaysia, Port Dickson 71050, Negeri Sembilan, Malaysia
[4] Department of Biochemistry, Faculty of Biotechnology and Biomolecular Sciences, Universiti Putra Malaysia, Serdang 43400, Selangor, Malaysia; noorazmi@upm.edu.my
[5] Laboratory of Vaccines and Immunotherapeutic, Institute of Bioscience, Universiti Putra Malaysia, Serdang 43400, Selangor, Malaysia; aro@upm.edu.my
* Correspondence: fatimamy@upm.edu.my; Tel.: +60-3-89408311

Abstract: Fucoxanthin is a major carotenoid in brown macroalgae and diatoms that possesses a broad spectrum of health benefits. This review evaluated the research trends of the fucoxanthin field from 1928 to June 2021 using the bibliometric method. The present findings unraveled that the fucoxanthin field has grown quickly in recent years with a total of 2080 publications. Japan was the most active country in producing fucoxanthin publications. Three Japan institutes were listed in the top ten productive institutions, with Hokkaido University being the most prominent institutional contributor in publishing fucoxanthin articles. The most relevant subject area on fucoxanthin was the agricultural and biological sciences category, while most fucoxanthin articles were published in *Marine Drugs*. A total of four research concepts emerged based on the bibliometric keywords analysis: "bioactivities", "photosynthesis", "optimization of process", and "environment". The "bioactivities" of fucoxanthin was identified as the priority in future research. The current analysis highlighted the importance of collaboration and suggested that global collaboration could be the key to valorizing and efficiently boosting the consumer acceptability of fucoxanthin. The present bibliometric analysis offers valuable insights into the research trends of fucoxanthin to construct a better future development of this treasurable carotenoid.

Keywords: microalgae; macroalgae; carotenoids; fucoxanthin; Scopus; bibliometric; applications; biosynthesis; health benefits; production

Citation: Khaw, Y.S.; Yusoff, F.M.; Tan, H.T.; Noor Mazli, N.A.I.; Nazarudin, M.F.; Shaharuddin, N.A.; Omar, A.R. The Critical Studies of Fucoxanthin Research Trends from 1928 to June 2021: A Bibliometric Review. *Mar. Drugs* **2021**, *19*, 606. https://doi.org/10.3390/md19110606

Academic Editors: Masashi Hosokawa and Hayato Maeda

Received: 24 September 2021
Accepted: 12 October 2021
Published: 26 October 2021

Publisher's Note: MDPI stays neutral with regard to jurisdictional claims in published maps and institutional affiliations.

1. Introduction

Fucoxanthin is an orange-colored pigment, which accounted for more than 10% of the estimated total carotenoids production in the nature [1]. It is the most abundant and signature carotenoid in brown macroalgae and diatoms [2]. It is a part of fucoxanthin–chlorophyll *a* protein (FCP) complexes along with proteins and chlorophyll *a* (chl *a*) that are functionally associated with the light-harvesting complex of the algae [3].

It has been shown that fucoxanthin possesses numerous biological activities and health-stimulating properties such as anti-angiogenic, anti-diabetic, antioxidant, anti-inflammatory, and antimalarial activity [4,5]. Notably, in vivo studies (mice and human) have proven that fucoxanthin effectively decreases body weight [6,7]. In addition, fucoxanthin has been demonstrated to exhibit a wide range of anticancer and anti-proliferative activity in different in vivo and in vitro studies [8]. These bioactivities of fucoxanthin are attributed to its unique molecular structure such as an allenic bond, an acetyl group, and a conjugated carbonyl with a 5,6-monoepoxide [9]. Due to its multiple health benefits, the global total fucoxanthin market is estimated to increase from $88 million in 2019 to more than $100 million over the next five years (Global Fucoxanthin Market 2020).

Bibliometric analysis is an approach that analyzes the worldwide scientific production and qualitative data. The analysis could provide information on the research trends and future directions within a specific research field [10]. Lately, a bibliometric analysis of microalgae research, microalgae-based wastewater treatment, global dinoflagellate research, and microalgae-derived pigments has been performed [10–13]. These bibliometric analyses focused on microalgae or metabolites generated from microalgae. Silva et al. (2020) [12] only reviewed four pigments, chlorophylls, phycocyanin, and β-carotene of the microalgae using a bibliometric approach. A critical examination of the research trends of fucoxanthin utilizing the bibliometric method is still lacking. Thus, this review aimed to analyze the research trends of the fucoxanthin field via bibliometric analysis. Several conclusions can be drawn through the analysis, such as evaluating essential milestones over the history of a scientific field and the scientific fads. The keyword analysis in the current review highlighted the trends and directions of fucoxanthin research. The review could serve as a basis in fortifying the future direction of fucoxanthin research.

2. Methods

Bibliometric analysis in the present study employed three-step approaches. These steps were data mining, data refinement, and data analysis.

2.1. Data Mining

A set of publications associated with fucoxanthin research was first extracted from the Elsevier Scopus database (collected on 15 July 2021). A complete search of the database was performed using [TITLE-ABS-KEY (fucoxanthin)] as the search query. This query retrieved a total of 2080 publications. Different search results will be obtained if distinct search parameters or databases were utilized.

2.2. Data Refinement

Data refinement is indispensable in bibliometric study, as it could remove the irrelevant articles. The retrieved data were polished using OpenRefine software, which excluded duplicate articles [11]. In addition, each article was also examined meticulously by manual inspection to eliminate unrelated articles.

2.3. Data Analysis

The information of publications was manually extracted. For example, journal impact factors (IFs) were retrieved from Journal Citation Reports (JRC) in 2020. In addition, the number of total citations and the h-index value were obtained from the Scopus database (https://www.scopus.com/search/form.uri?display=basic#basic; collected on 15 July 2021). On the other hand, software such as VOSviewer was utilized to perform bibliometric analyses. VOSviewer was employed to visualize the similarities among different subjects (countries and keywords) [14]. In addition, the visualization map was generated using VOSviewer to display the relationship between subjects in a research field. The research network and keyword co-occurrence network were constructed and visualized with VOSviewer. In the case of the keyword co-occurrence network, the most commonly utilized keywords related to the searched words (fucoxanthin) were screened. The minimum number of occurrences for the keyword co-occurrence network was determined as 15. Manual inspection was performed to eliminate the unrelated keywords. Moreover, the data were analyzed using the bibliometric R-package incorporated in the open-source RStudio software (www.rstudio.com; version 4.0.5, assessed on 15 July 2021) [15].

3. Publication Output

A total of 2080 fucoxanthin research articles were published from 1928 to June 2021 (Figure 1). According to the Scopus database, the first fucoxanthin research article was published in 1928. A small number of publications per year were observed (<10 units except 1976) until 1986. The amount of publication has doubled (around 20 units) per year since 1994. After that, the annual publications have elevated steadily and generated an immediate increase in the cumulative total publications. A remarkable increase in the number of publications was detected starting in 2017. In 2017, the number of publications exceeded 100 for the first time. The last decade's publication on fucoxanthin corresponded to 64.90% (1350 publications) of the total publications since 1928. In addition, the previous years (2017 to June 2021) recorded 35.96% of the total publications on fucoxanthin. These findings indicated that a strong interest in fucoxanthin research occurred recently.

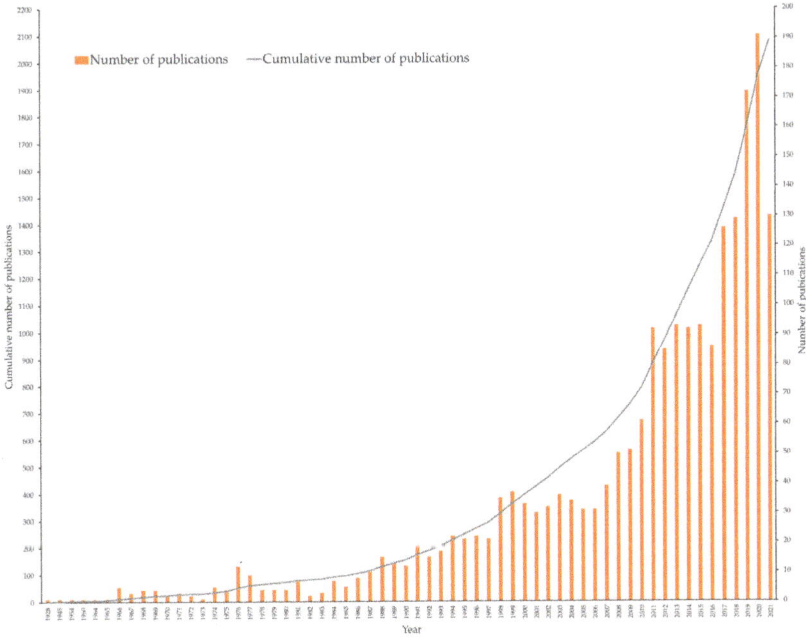

Figure 1. The annual and cumulative numbers of fucoxanthin publications from 1928 to June 2021.

The major publications on fucoxanthin were articles (82.55%) (Figure 2). Reviews were the second-largest amount of publications, which constituted 11.25% of the total publications. This is followed by book chapters (2.55%) and conference papers (2.31%). Other documents such as books, conference reviews, data papers, editorials, errata, letters, notes, and short surveys accounted for less than 2% of the total publications.

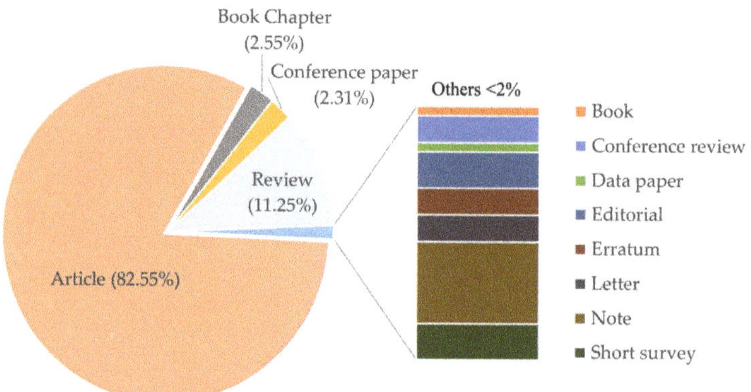

Figure 2. Distribution of publication types for fucoxanthin.

4. Global Production of Fucoxanthin Publications

Fucoxanthin studies have been conducted over the continentals in the world. The top ten countries have contributed to more than 70 publications from 1928 to June 2021 (Table 1), indicating these countries are the key players in studying fucoxanthin. From the time perspective, the United States predominates in producing fucoxanthin articles among the other countries from 1928 to 2000. Although more articles (n = 115) by the United States from 2001 to June 2021 were observed compared to the previous period, the United States could only be classified as a second or third place among the other countries in the last two periods. All these top ten countries demonstrated noticeably increased research performance on fucoxanthin from 1928 to June 2021. The most apparent upsurge of fucoxanthin publications was seen in 2011–June 2021 for the top ten countries except for Canada. For example, China showed the most remarkable productivity (almost 24-fold) on fucoxanthin articles; the number of articles escalated from nine (2001–2005) to 219 (2011–June 2021). Overall, Asian countries focused on fucoxanthin publications more than the other countries, as four Asian countries (Japan, China, South Korea, and India) occupied the top ten countries (1928–June 2021). The major three countries that produced the most fucoxanthin publications were Japan, the United States, and China. Japan leads the first ranking in the publications on fucoxanthin with 379 articles, suggesting that Japan plays a significant role in advancing fucoxanthin research.

Table 1. Top ten productive countries in producing fucoxanthin publications from 1928 to June 2021.

Country	TP	Number of Publications					
		1928–1970	1971–1980	1981–1990	1991–2000	2001–2010	2011–2021 *
Japan	379	0	4	13	37	98	227
United States	319	8	16	31	71	78	115
China	230	0	0	1	1	9	219
Germany	181	0	1	3	32	45	100
France	144	1	0	4	32	48	59
United Kingdom	132	5	6	5	28	26	62
South Korea	127	0	0	0	1	16	110
Spain	99	0	0	1	9	22	67
India	98	0	0	0	3	11	84
Canada	75	0	2	5	19	25	24

TP: Total publications; 2021 *: up to June 2021.

The top ten institutions contributed at least 33 fucoxanthin publications, respectively (Table 2). These top institutions were leaded by Hokkaido University, Japan (110 articles), CNRS Centre National de la Recherche Scientifique, France (61 articles), and Goethe-Universität Frankfurt am Main, Germany (49 articles). Among these ten institutions, five were European and five were Asian. Institutions from Japan occupied three of the top ten institutions: Hokkaido, Kyoto, and Kobe University. The contribution of these three major Japanese institutions resulted in Japan predominating in fucoxanthin research. Fewer fucoxanthin publications were produced from these top ten institutions in 1928–1990 except for Norges teknisk-naturvitenskapelige universitet, Norway. Most of the top ten institutions increased their publications at least one-fold from the 2001–2010 period to the 2011–June 2021 period. The most remarkable elevation of fucoxanthin publications was produced by the Chinese Academy of Sciences, China (around 18.5-fold).

Table 2. Top ten productive institutions in producing fucoxanthin publications from 1928 to June 2021.

Institution	Country	TP	Number of Publications					
			1928–1970	1971–1980	1981–1990	1991–2000	2001–2010	2011–2021 *
Hokkaido University	Japan	110	0	0	1	8	33	68
CNRS Centre National de la Recherche Scientifique	France	61	1	0	1	17	17	25
Goethe-Universität Frankfurt am Main	Germany	49	0	0	0	3	13	33
Norges teknisk-naturvitenskapelige universitet	Norway	44	0	7	12	12	8	5
Chinese Academy of Sciences	China	41	0	0	1	1	2	37
Sorbonne Universite	France	39	0	0	1	10	14	14
Kyoto University	Japan	34	0	0	3	8	9	14
Kobe University	Japan	34	0	0	0	1	7	26
Plymouth Marine Laboratory	United Kingdom	33	0	0	1	15	6	11
Pukyong National University	Korea	33	0	0	0	1	2	30

TP: Total publications; 2021 *: up to June 2021.

A distribution by communities of 31 countries with at least 20 articles on fucoxanthin was observed (Figure 3). These countries were distributed over four clusters. The first cluster comprised of eight European countries, along with Chile, Iran, Turkey, and New Zealand. The second cluster consisted of Mexico, Australia, Brazil, and four European countries. The third cluster was made of the United States and seven Asian countries, while Canada and three European countries constituted the last cluster. Based on Figure 3, the United Kingdom established a research collaboration with the highest number of countries (26 countries). The United States was the second country that formed the most partnerships with other countries (25 countries). Germany had the third highest number of country collaborations (23 countries). For the strength of research collaboration, the United States formed the most substantial collaboration network with other countries (total link strength: 214). This is followed by Germany (126), the United Kingdom (125), France (118), and Japan (103).

The top ten productive authors produced at least 18 publications (Table 3). Eight of the top ten productive authors were affiliated with Japanese institutions. No publication was generated from these top ten authors in the first period (1928–1970). The number of the publication produced by Liaaen-Jensen, Synnove (Germany) outweighed the other authors from the second period (1971–1980) to the fourth period (1991–2000). The last two periods were dominated by Miyashita, Kazuo (Japan) in terms of publications. Miyashita Kazuo was the most productive and impactful author on fucoxanthin research with 88 publications and a Scopus *h*-index of 50. The *h*-index was utilized to evaluate a researcher's contribution. The *h*-index not only indicates productivity, it also demonstrates the influence of a group's or scholar's published work [16]. The author recently studied the effect of fucoxanthin/fucoxanthinol on cancer, particularly pancreatic and colon cancer [17,18].

Masashi Hosokawa produced 68 publications and became the second most productive author on fucoxanthin research (Scopus *h*-index: 44). The author worked together with Kazuo Miyashita in examining the effect of fucoxanthin on dexamethasone-induced muscle atrophy and fat mass in mice [19]. The third most productive author was Claudia Buchel (41 publications, Scopus *h*-index: 34), who has investigated the structure and energy transfer of FCP complexes lately [20,21]. Synnove Liaaen-Jensen was among the leading researchers from Norway. The author demonstrated the iodine-catalyzed R/S isomerization of the allenic carotenoids fucoxanthin via the attack of I·on C7′ [22]. In addition, the author also studied the diphenyl diselenide-mediated photoisomerization of allenic carotenoids fucoxanthin in benzene solution to optimize synthetic yields of (6′S)-allenes [23]. A team led by Ryo Nagao studied the structure and function of a diatom photosystem I-light-harvesting supercomplex [24] and adaptation of light-harvesting and energy-transfer processes of diatoms [25,26].

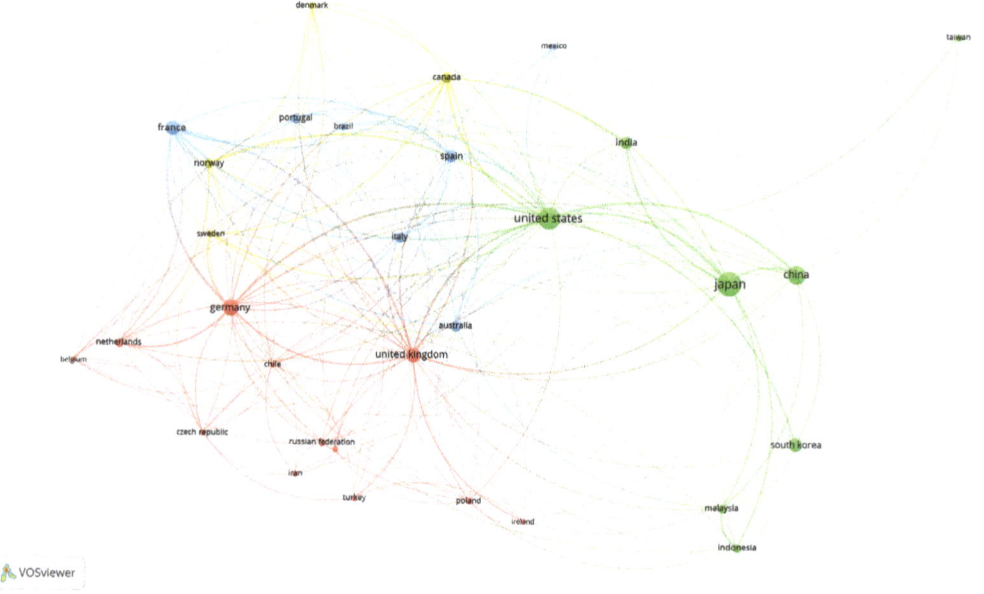

Figure 3. Visualization map of countries communities in fucoxathin research (minimum occurrence: 20).

Multidisciplinary collaboration networks among researchers are crucial in developing, evolving, and enriching the nature of fucoxanthin utilization research. The high productivity of these top ten authors could be attributed to the formation of effective research teams via co-authorship networks. The homophyllous network is a collaboration among researchers of the same gender, academic department, and research interest. This network is vital in establishing and sustaining the research capabilities of the members of the organization. On the other hand, heterophyllous collaboration allowed the cooperation among researchers of different gender, affiliations, and research interests. This cooperation could provide new insights or solutions to complex problems and create transformative research [27].

Table 3. Top ten prolific authors in producing fucoxanthin research from 1928 to June 2021.

Author	Country	TP	Scopus *h*-Index	Scopus ID	Number of Publications					
					1928–1970	1971–1980	1981–1990	1991–2000	2001–2010	2011–2021 *
Miyashita, Kazuo	Japan	88	50	• 55883552800	0	0	0	0	29	59
Hosokawa, Masashi	Japan	68	44	• 7202009871	0	0	0	1	25	42
Buchel, Claudia	Germany	41	34	• 7006466104	0	0	0	0	13	28
Maeda, Hayato	Japan	31	19	• 8396571500	0	0	0	0	11	20
Liaaen-Jensen, Synnove	Norway	29	40	• 35509195900	0	6	11	10	2	0
Nagao, Ryo	Japan	28	21	• 22941377100	0	0	0	0	2	26
Akimoto, Seiji	Japan	23	27	• 7102347852	0	0	0	0	0	23
Shen, Jian Ren	Japan	22	50	• 56374284200	0	0	0	0	0	22
Maoka, Takashi	Japan	19	37	• 7004037866	0	0	0	1	5	13
Hashimoto, Hideki	Japan	18	44	• 35253778600	0	0	0	0	2	16

TP: Total publications; 2021 *: up to June 2021.

5. Subject Category, Sources and Citations

At least 2.0% of total publications were assigned to each of the top ten subject categories for fucoxanthin research (Table 4). The global fucoxanthin publication landscape covered 27 subject categories. A total of 27.0% publications was classified under the agricultural and biological sciences category. This category's highest percentage of publications signifies the solid scientific interests in the fundamental knowledge of agricultural utilization [28], synthesis, and accumulation [29,30] of fucoxanthin. The rapid development of biochemistry [21,31] and molecular techniques such as genetic transformation [32,33], sequencing [34], and transcription [35] led to the dominance of biochemistry, genetics, and molecular biology category (18.1% publications). The high number of publications in the pharmacology, toxicology, and pharmaceutics (8.3% publications) category could be due to the increase of health-consciousness and the benefits of fucoxanthin [36,37]. However, a low percentage of articles in the economic and social science categories (data not shown) implies that fucoxanthin research is still in its infancy. Thus, it is crucial to focus on the economic viability and social acceptance of fucoxanthin in the future.

A total of 689 scientific journals published fucoxanthin articles from 1928 to June 2021. The top ten journals on fucoxanthin research belonged to seven different publishers (Table 5). Springer occupied three of the top ten journals. The rest of the journals were owned by Multidisciplinary Digital Publishing Institute (MDPI), Wiley Periodical LLC, Elsevier B.V., Inter-Research Science, American Chemical Society, and Oxford University Press. According to the InCites Journal Citation Reports, the top ten journals showed at least an impact factor of 2.400. The journal with the highest impact factor belonged to the Journal of Agricultural and Food Chemistry (5.279), while the Journal of Plankton Research was the lowest impact factor journal (2.455) among the top ten journals.

Table 4. Top ten prolific subject categories of fucoxanthin research from 1928 to June 2021.

Subject Category	TP (%)
Agricultural and Biological Sciences	27.0
Biochemistry, Genetics and Molecular Biology	18.1
Pharmacology, Toxicology and Pharmaceutics	8.3
Chemistry	7.9
Environmental Science	7.8
Earth and Planetary Sciences	7.5
Medicine	5.8
Chemical Engineering	3.8
Immunology and Microbiology	2.5
Engineering	2.3

TP: Total publications.

Table 5. Top ten core journals for research on fucoxanthin from 1928 to June 2021.

Journal	Publisher	IF (2020)	TP	TP (%)	Cummulative TP (%)
Marine Drugs	MDPI	5.118	123	5.91	5.91
Journal of Phycology	Wiley Periodical LLC	2.923	67	3.22	9.13
Journal of Applied Phycology	Springer	3.215	55	2.64	11.77
Photosynthesis Research	Springer	3.573	39	1.88	13.65
Biochimica Et Biophysica Acta (BBA)—Bioenergetics	Elsevier B.V.	3.991	38	1.83	15.48
Algal Research	Elsevier B.V.	4.401	32	1.54	17.02
Marine Ecology Progress Series	Inter-Research Science	2.824	30	1.44	18.46
Journal of Agricultural and Food Chemistry	American Chemical Society (ACS)	5.279	30	1.44	19.90
Journal of Plankton Research	Oxford University Press (OUP)	2.455	24	1.15	21.05
Marine Biology	Springer	2.573	24	1.15	22.20

MDPI: Multidisciplinary Digital Publishing Institute; IF: Impact factor; TP: Total publications.

Articles from these ten top journals constituted 22.20% of the total fucoxanthin publications (Table 5). The most productive journal was *Marine Drugs*, with 123 articles covering 5.91% of the total publications. This is followed by the *Journal of Phycology* (67 articles, 3.22%), *Journal of Applied Phycology* (55 articles, 2.64%), *Photosynthesis Research* (39 articles, 1.88%) and *Biochimica Et Biophysica Acta* (BBA)—*Bioenergetics* (38 articles, 1.83%).

The top ten articles in Table 6 demonstrated at least 178 citations. The data of highly cited publications are impactful, as these mirror the scientific advancement, offer revolutionary insights, and provide a significant perspective on scientific advancement [38]. Each of these articles was owned by one prestigious journal except *Marine Drugs* and the *Journal of Agricultural and Food Chemistry*. *Marine Drugs* possessed two top fucoxanthin publications with a high total citation: Peng et al. (2011) and Xia et al. (2013), while two highly cited fucoxanthin research articles were published in *Journal of Agricultural and Food Chemistry*: Sachindra et al. (2007) and Maeda et al. (2007). Furthermore, the top ten most cited publications were observed in only two of the top ten journals, namely *Marine Drugs* and *Journal of Agricultural and Food Chemistry*. The research article produced by Maeda et al. (2005) in *Biochemical and Biophysical Research Communications* was the publication that received the highest total citation on fucoxanthin research (411). The article demonstrated that fucoxanthin upregulates the expression of mitochondrial uncoupling protein 1 that may attribute to the reduction of white adipose tissue (WAT) weight [39]. The review article summarizes the topical state of understanding and provides an overview of the basic knowledge of fucoxanthin. The second highly cited publication was a review article (356). This review discussed the metabolism, safety, and bioactivities of fucoxanthin [2]. The other research article discovered that fucoxanthin was the major active compound in Japanese edible macroalgae, *Hijikia fusiformis*, which might be responsible for the high antioxidant activity [40]. This article was ranked third in the top ten most-cited publications (348). Most of the top ten cited fucoxanthin articles described and discussed the bioactivities of fucoxanthin (i.e., antiobesity, antioxidant, antiproliferation, anti-inflammatory) [2,6,39–44]. This could be inferred as the interest of researchers was focused on the bioactivities of fucoxanthin lately.

Table 6. Top ten most cited fucoxanthin publications from 1928 to June 2021.

Title	Authors	Year	Journal	Total Citation
Fucoxanthin from edible seaweed, *Undaria pinnatifida*, shows antiobesity effect through UCP1 expression in white adipose tissues	Maeda et al.	2005	*Biochemical and Biophysical Research Communications*	411
Fucoxanthin, a marine carotenoid present in brown seaweeds and diatoms: Metabolism and bioactivities relevant to human health	Peng et al.	2011	*Marine Drugs*	356
Fucoxanthin as the major antioxidant in *Hijikia fusiformis*, a common edible seaweed	Yan et al.	1999	*Bioscience, Biotechnology and Biochemistry*	348
Radical scavenging and singlet oxygen quenching activity of marine carotenoid fucoxanthin and its metabolites	Sachindra et al.	2007	*Journal of Agricultural and Food Chemistry*	308
Fucoxanthin induces apoptosis and enhances the antiproliferative effect of the PPARγ ligand, troglitazone, on colon cancer cells	Hosokawa et al.	2004	*Biochimica et Biophysica Acta (BBA)—General Subjects*	261
Dietary combination of fucoxanthin and fish oil attenuates the weight gain of white adipose tissue and decreases blood glucose in obese/diabetic KK-A^y mice	Maeda et al.	2007	*Journal of Agricultural and Food Chemistry*	226
A potential commercial source of fucoxanthin extracted from the microalga *Phaeodactylum tricornutum*	Kim et al.	2012	*Applied Biochemistry and Biotechnology*	218
Evaluation of anti-inflammatory effect of fucoxanthin isolated from brown algae in lipopolysaccharide-stimulated RAW 264.7 macrophages	Heo et al.	2010	*Food and Chemical Toxicology*	212
Fucoxanthin inhibits the inflammatory response by suppressing the activation of NF-κB and MAPKs in lipopolysaccharide-induced RAW 264.7 macrophages	Kim et al.	2010	*European Journal of Pharmacology*	204
Production, characterization, and antioxidant activity of fucoxanthin from the marine diatom *Odontella aurita*	Xia et al.	2013	*Marine Drugs*	178

6. Research Concepts

The keyword co-occurrence analysis provides an overview of fucoxanthin research concepts. Keywords serve as the fundamental of bibliographic research of academic literature [45]. A total of 40 relevant keywords were classified into four different clusters using a minimum occurrence of the 15-fold keyword (Figure 4). The first cluster (in red) consists of 18 keywords that revolved around "antioxidant activity", "neuroprotection", and "cancer inhibition". This cluster highlighted the bioactivities of fucoxanthin via in vitro study and in vivo study. Ten keywords such as "photosynthesis", "photosystem", "chlorophyll-binding proteins", and "energy transfer" were seen in the second cluster (in green). This cluster represents the research of photosynthesis that involved fucoxanthin. The third cluster (in blue) encompassed "chemistry", "biosynthesis", "extraction", "purification", and "biotechnology" keywords. Thus, this cluster depicted the optimization of processes in obtaining fucoxanthin as valuable compounds. The fourth cluster (in yellow) revolved around the major keywords such as "phytoplankton", "community structure", "pigment", and "environmental monitoring", which denoted the role of fucoxanthin in the environmental study of phytoplankton.

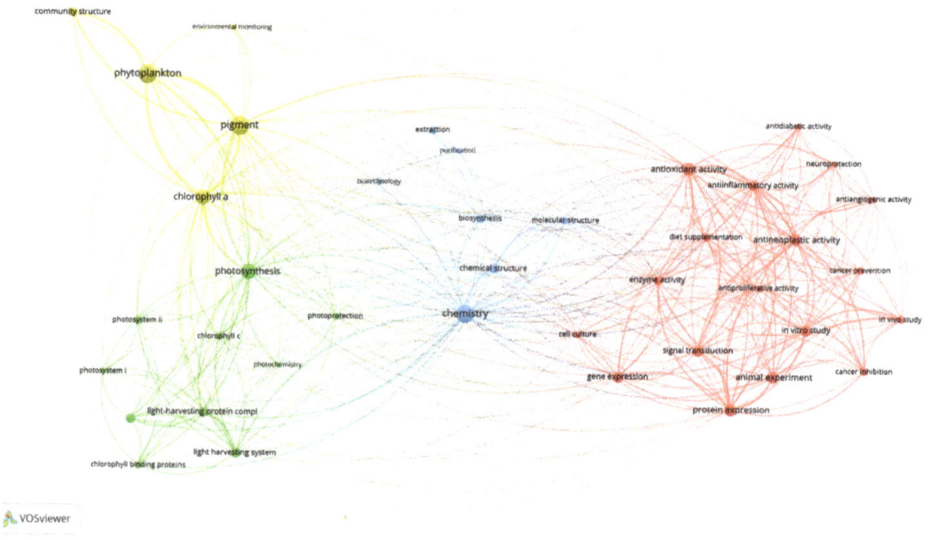

Figure 4. Visualization of keyword co-occurrence network analysis (minimum occurrences: 15).

6.1. Research Trends of Fucoxanthin Research from 1928 to June 2021

The four clusters (bioactivities, photosynthesis, optimization of process, environment) were utilized to analyze the fucoxanthin research trends from 1928 to June 2021. In the first period (1928–1970), the researchers focused on the optimization of process cluster (88.88%) (Figure 5). An equal percentage of fucoxanthin articles (5.56%) was observed in photosynthesis and environment clusters, respectively. No bioactivities of fucoxanthin publication were produced. In the second period (1971–1980), the percentage of articles in the optimization of process cluster decreased by 22.92% compared to the previous period. Both photosynthesis and environment clusters demonstrated the same percentage of publications, 17.02% (increased by 11.46%). The bioactivities cluster remained 0%. There was a further reduction (2.95%) of publications in the optimization of process cluster in the third period (1981–1990) compared to the previous period. In the same period, the bioactivities, photosynthesis, and environment clusters increased the percentage of publications, 1.37, 2.16, and 1.42%, respectively.

In the fourth period (1991–2000), bioactivities and environment clusters experienced an increase of publications percentage, 4.26, and 12.66%, respectively, compared to the previous period. On the other hand, the optimization of process and photosynthesis clusters exhibited a fall of publications percentage, 4.13, and 12.69%, respectively. In the fifth period (2001–2010), the only optimization of production cluster experienced a reduction of articles percentage to 38.13% compared to the earlier period. The remaining clusters displayed an increase in articles percentage. The biggest upsurge in terms of publications percentage (16.77%) was detected in the bioactivities cluster among the other clusters. The photosynthesis cluster showed 8.8% of publications, while 30.67% of publications were identified in the environment cluster. In the sixth period (2011–June 2021), the bioactivities cluster (39.97%) demonstrated a similar publications percentage with the optimization of process cluster (40.79%). The photosynthesis cluster displayed 8.03% of publications, whereas 11.22% of publications were seen in the environment cluster. Throughout the six periods (1928–June 2021), the interest in fucoxanthin research shifted from the optimization of the processing cluster to the bioactivities cluster. This could be because brown macroalgae and microalgae are considered a potential natural source of bioactive compounds (i.e., fucoxanthin), and these compounds provide various health benefits [46].

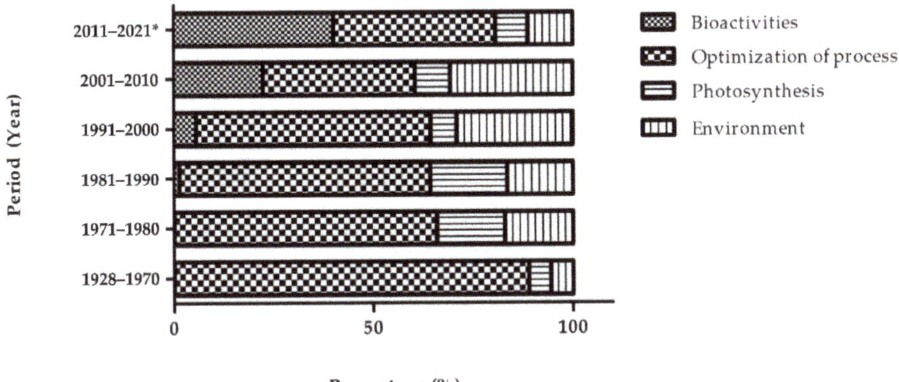

Figure 5. Research trends of fucoxanthin research from 1928 to June 2021. The asterisk after 2021 denotes that the data were extracted up to June 2021.

6.1.1. Bioactivities of Fucoxanthin

The interest in natural sources of bioactive compounds is growing immensely due to the numerous benefits to humans. According to the Scopus database, the first article on the bioactivities of fucoxanthin was published in 1990. In the first article on the bioactivities of fucoxanthin, Okuzumi et al. (1990) demonstrated the anti-tumor activity of fucoxanthin (isolated from brown algae, *Hijikia fusiforme*) on human neuroblastoma cells, GOTO cells. The authors showed that fucoxanthin caused the arrest in the G_0–G_1 phase of the cell cycle. In addition, fucoxanthin decreased the expression of the N-myc gene, and the authors suggested this mechanism could prevent the proliferation of cancer cells [47]. In the next several years, Okuzumi et al. (1993) showed that fucoxanthin inhibited duodenal carcinogenesis induced by N-ethyl-N'-nitro-N-nitrosoguanidine in mice [48]. Fucoxanthin was documented as a chemopreventive agent in the first bioactivities of fucoxanthin review based on Scopus search results [49].

The bioactivities of fucoxanthin have been extensively studied. The bioactivities of fucoxanthin in humans have been summarized in many review articles [2,4,5,8,50–53]. For instance, Peng et al. (2011) [2] reviewed fucoxanthin's metabolism, safety, and bioactivities. The bioactivities include antioxidant, anti-inflammatory, anti-obese, anti-diabetic, antimalarial, anticancer, and anti-angiogenic activity and hepatoprotective, skin protective, cerebrovascular protective, bone protective, and ocular protective effects [2]. Furthermore, Thiyagarasaiyar et al. (2020) [5] discussed the cosmeceutical potentials of fucoxanthin as skin whitening, anti-aging, anticancer, antioxidant, anti-inflammation, and antimicrobial. In addition, Sathasivam and Ki (2018) [54] summarized the publications on anti-angiogenic, anticancer, antidiabetic, anti-obesity, and antioxidant activity and then on the neuroprotective, cardioprotective, and osteo-protective effect of fucoxanthin. Despite these bioactivities, the mechanism of these bioactivities also has been reviewed. For example, Rengarajan et al. (2013) [55] summarized the mechanisms of anticancer effects of fucoxanthin. These mechanisms were anti-proliferation, induction of apoptosis, cell cycle arrest, and anti-angiogenesis. Moreover, the mechanisms of action of fucoxanthin on different types of cancers were also elucidated [56]. Additionally, the modulation of inflammatory and oxidative stress pathways using fucoxanthin was described [57]. Another review article summarized the possible underlying mechanism of fucoxanthin on lipid metabolism, adiposity, and related conditions [58].

The combined effect of fucoxanthin and the other compound(s) were also studied. Maeda et al. (2007) [6] demonstrated that feeding both fucoxanthin and fish oil to KK-A^y mice significantly reduced the WAT weight of mice compared with the mice fed fucoxanthin alone. Moreover, the reduction of serum levels of triacylglycerols, glucose, and leptin in diet-induced obese rats was observed using a combination of fucoxanthin and conjugated linoleic acid [59]. A combination of fucoxanthin and vitamin C has been shown to increase human lymphocytes' antioxidant and anti-inflammatory effects [60]. Another study demonstrated the combined effects of low-molecular-weight fucoidan and fucoxanthin in a mouse model of type II diabetes. The authors reported that the combination effectively decreased the urinary sugar, glucose, and lipid metabolism in the WAT of the mice than fucoidan or fucoxanthin alone [61]. The effects of fucoidan and fucoxanthin in combination were investigated in aging mice and hyperuricemic rats. The combination of these compounds improved the cardiac status of aging mice via decreased cardiac hypertrophy, cardiac fibrosis, reactive oxygen species level, and shortened QT interval in the mice [62]. For hyperuricemic rats, the combination inhibited xanthine oxidase activity in the liver and controlled the expression of uric acid-related transporters [63]. Furthermore, a combination of fucoxanthin and rosmarinic acid could offer photo-protective effects through the downregulation of NRLP3-inflammasome and increasing the Nrf2 signaling pathway in UVB-irradiated HaCaT keratinocytes [64]. Recently, a combination of fucoxanthin, myoinositol, D-chiro-inositol, and hydroxytyrosol has been reported to decrease systolic blood pressure and improve the vascular reactivity in a pregnant mouse model of hypertension [65].

Apart from this, the bioactivities of fucoxanthin were also studied in the other organisms. Fucoxanthin enhanced the phagocytic activities three times and increased the number of ovulated eggs of sea urchins by 3.25 times compared to the control group [66]. The planktonic larvae stage is crucial in the life cycle of coral. The percentage of a larval metamorphosis of coral *Pseudosiderastrea tayamai* was further enhanced by 60.3% in the presence of fucoxanthin compared to control group [67]. The effect of fucoxanthin was also examined in fly and worm. The median and maximum lifespan of *Drosophila melanogaster* (fly) was extended at least 33% and 12%, respectively using fucoxanthin. The decreased flies' fecundity, increased spontaneous locomotor activity, and resistance to oxidative stress were observed when feeding the flies with fucoxanthin. For *Caenorhabditis elegans* (worm), the feeding of fucoxanthin increased the mean and maximum lifespan by 14% and 24%, respectively [68].

6.1.2. Photosynthesis

Diatoms and brown algae utilized unique light-harvesting antennas, FCPs, to perform photosynthesis [69]. The pigment compositions and protein organization of FCPs are mostly distinct from those of light-harvesting complexes (Lhcs) in land plants and green algae [70,71]. At least 30 genes related to light harvesting have been identified based on the genome of two diatoms, *Thalassiosira pseudonana* [72] and *Phaeodactylum tricornutum* [73]. The products encoded by these genes were categorized into three groups, Lhcf, Lhcr, and Lhcx proteins. Proteins under the Lhcf family are the main light-harvesting proteins [74]. PSI antennae are composed of proteins from the Lhcr family [3], while Lhcx proteins are involved in non-photochemical quenching mechanisms to protect the photosystem [75]. A high similarity of Lhcx proteins to LI818 or LhcSR proteins of green alga was observed [76].

The FCP trimer, also known as FCPa, is the basic structure of different FCP proteins in the native thylakoid membrane of pennate and diatoms [77]. Diverse populations of FCP trimeric complexes, which differed in polypeptide composition and pigmentation, were identified through the sub-fractionation of FCP complexes of *P. tricornutum* [78]. Four different trimeric subtypes of FCPa in *Cyclotella meneghiniana* were revealed, FCPa1–4. Lhcf4/Lhcf6 proteins were found mainly in the FCPa2 trimer, whereas Lhcf1 was reported to be the major subunit in FCPa1, FCPa3, and FCPa4 [79]. Buchel (2003) [80] discovered that two FCP fractions differed in the polypeptide composition and oligomeric state from *C. meneghiniana*. The first fraction consisted of trimers with mainly 18 kDa polypeptides (FCPa), while the higher oligomers assembled from different trimers (19 kDa subunits) constituted the second fraction (FCPb). Two oligomeric subtypes, FCPb1 and FCPb2, with Lhcf3 being the main subunit in both antenna complexes in *C. meneghiniana* were revealed [79]. Different FCP complexes were observed using anion chromatography and divided into FCP complexes related to PSI, PSII core complexes, and peripheral FCP complexes. Various Lhcf proteins were detected in FCP complexes associated with PSI and PSII core complexes, whereas peripheral FCP complexes mainly contained Lhcf8 and Lhcf9. Subunits of the PSI core complex composed of Lhcr proteins and Lhcx proteins were the protein subunits that were identified in the PSII core complex [81].

The idea of the FCP trimer as the basic unit of photosynthesis antenna proteins in fucoxanthin-containing algae was contradicted by the findings based on cryo-electron miscopy [82–85] and X-ray crystallography [86]. Wang et al. (2019) [86] unraveled the X-ray crystal structure of an FCP of *P. tricornutum*, which had two monomers held together to form the dimeric structure of FCP within the PSII core. In addition, the cryo-electron microscopy data of the PSII–antenna supercomplex of *Chaetoceros gracilis* revealed a tetrameric organization of FCP proteins associated with the PSII [82]. Furthermore, 24 FCPs surrounding the PSI core of *C. gracilis* were in monomeric form based on cryo-electron microscopy [83].

Important characteristics of pigment organization of isolated FCP and the role of fucoxanthin molecules in excitation energy transfer have been unraveled using steady-state and ultrafast spectroscopic methods [87,88]. Efficient energy transfer was observed from fucoxanthin and chlorophyll *c* (Chl *c*) to Chl *a* based on spectroscopic studies [87,89,90]. At least three forms of fucoxanthin molecules differ in their photophysical and dipolar properties, Fx_{red}, Fx_{green}, and Fx_{blue} [91,92] and were confirmed using resonance Raman spectroscopy [93]. The Fx_{red} form transfers energy more efficiently, while the Fx_{blue} form demonstrated less efficiency in transferring excitation energy [92]. The time of energy transfer from fucoxanthin to Chl (around 300 fs) was shorter than the transfer from Chl *c* to Chl *a* (around 500 fs–6 ps), indicating that the fastest energy transfer was between fucoxanthin and Chl *a* [94]. Most of the pump-probe studies examined the dynamical energy transfer process in FCP of fucoxanthin-containing algae. For example, Papagianakis (2005) [87] characterized the energy transfer network in FCP. The energy transfer efficiency from Chl *c* to Chl *a* is 100%, whereas unequal efficiency was observed for fucoxanthins in the FCP. Furthermore, findings based on polarized transient absorption indicated that three fucoxanthin molecules in FCPa transferred their excitation energy directly to Chl *a*. The remaining fucoxanthin molecule was not associated with Chl *a* molecules and might transfer its excitation energy through another fucoxanthin molecule to Chl *a* [89]. In addition, a detailed model was proposed in describing the energy transfer in FCPa upon excitation at two different wavelengths [95]. Another study demonstrated the highly efficient energy transfer from Fx to Chl-a through the S_1/ICT state using the pump-probe method [96]. Apart from pump-probe techniques, two-dimensional electronic spectroscopy (2DES) has offered insights into energy change transfer dynamics, exciton diffusion, and molecular system relations [97–99]. The advances in knowledge of the mechanisms and dynamics of energy transfer in the FCPs of diatoms that have been accomplished using two-dimensional electronic spectroscopy (2DES) were reviewed [100].

In addition to FCPs' function as a light-harvesting complex, pigments in the FCPs are also involved in photoprotection. Non-photochemical quenching (NPQ) is the protection mechanism that most algal groups utilize to dissipate excess absorption energy as heat via molecular vibrations. One xanthophyll pigment, diadinoxanthin (Dd), was observed in a peripheral location of FCP [86]. This pigment is converted to diatoxanthin (Dt) under high light intensities [101]. The conversion showed a close relationship with the build-up of the NPQ mechanism [102]. Previously, the amount of diatoxanthin (Dt) was established to influence the NPQ mechanism in vivo, whereas the reduction of fluorescence yield of FCPa complexes in vitro was caused by Dt [103]. In addition, the acidification of thylakoid lumen regulated the ratio between Dd and Dt, which could affect the activities of the epoxidase and de-epoxidase in the NPQ mechanism [104]. Gundermann and Claudia (2012) [105] examined the factors determining the NPQ process in diatoms. The components involved in the NPQ mechanism in C. *meneghiniana* were reported to be heterogeneous and genuinely different from the NPQ type in *P. tricornutum* [106]. Apart from xanthophyll pigments

(Dd and Dt), Lhcx proteins play an essential role in the NPQ mechanism [76,107]. The amount of xanthophyll cycle pigments in various FCP preparations showed a relationship with the existence of Lhcx1 protein [108]. This protein was shown to have short-term photoprotection [107], which induced the conformational changes of the FCPs and reduced fluorescence yield [109]. The molecular structure, the arrangement of the different Lhc proteins in the complexes, the energy transfer abilities, and the photoprotection of other Lhc systems of Chl *c* containing organisms were reviewed recently [110].

6.1.3. Optimization of Process

The optimization of the process cluster focused mainly on the biosynthesis, biotechnology, extraction, and purification of fucoxanthin. Algal cultivation and fucoxanthin production are the major components of the biosynthesis section. Previous studies demonstrated that the fucoxanthin content of microalgae is higher than that of brown macroalgae [111,112]. The fucoxanthin extracted from fresh brown macroalgae ranged from 0.02 to 0.87 mg/g fresh weight, while the dried form of microalgae showed 1.81–15.33 mg/g dried weight (DW) fucoxanthin [111]. In diatoms, the fucoxanthin content ranged from 0.224% to 2.167% (around 0.224–21.67 mg/g) of dry weight [111,113]. Currently, *P. tricornutum* is the major natural fucoxanthin source in microalgae due to its substantial fucoxanthin [111,114]. Nevertheless, the lower dry well weight (g/L) of this microalgae [114] prompted the searching for alternative fucoxanthin sources via screening. The screening of high-performance microalgae strain is crucial, as it could determine the success of producing the desired amount of fucoxanthin prior to algal cultivation. Previous studies attempted to screen for potential microalgae with a high production of fucoxanthin [115,116]. For instance, Guo et al. (2016) [115] screened 13 diatoms strains to identify a promising strain with desired fucoxanthin production. Among these 13 diatoms strains, the highest fucoxanthin content was *Odontella aurita* (1.50% or 15.00 mg/g DW). A maximum biomass and fucoxanthin concentration of 6.36 g/L and 18.47 mg/g DW were reported for *O. aurita*, respectively [117]. Another study screened *Isochrysis* strains for their potential for concurrent DHA and fucoxanthin production [116]. *Isochyrsis* CCMP1324 demonstrated the comparable biomass concentration (2.72 g/L), DHA content (16.10 mg/g) and fucoxanthin content (14.50 mg/g). The accurate identification of microalgae is essential in the screening process to ensure repeatability, reproducibility, and quality assurance. A comprehensive identification method could assign a precise identity to the microalgae [118]. Successful algal cultivation is affected by the culture parameters. The past studies optimized several culture parameters to obtain the maximum amount of fucoxanthin [114,116]. For instance, McClure et al. (2018) [114] optimized the culture parameters such as light intensity, medium composition, and carbon dioxide addition on the fucoxanthin production of *P. tricornutum*. The authors obtained the maximum concentration of fucoxanthin (59.20 mg/g), which is nearly four times higher than that found by Kim et al. (2012) [111]. These parameters are required to optimize in order to develop a sustainable, feasible, and economically viable cultivation modus operandi for the microalgae [119].

Genetic transformation is one of the critical components of genetic engineering. Several genetic transformation protocols have been established for fucoxanthin-containing microalgae [120–123]. In addition, successful genetic engineering relies on a suitable promoter. Several past researchers searched for the promising promoter in enhancing the gene expression of fucoxanthin-containing algae [122,124]. For example, four different promoters were examined for the genetic transformation of brown algae [122]. The authors concluded that the FCP promoter of *P. tricornutum* was the most suitable promoter for the brown algae, as this promoter induced both integrated and transient expression in the algae. Erdene-Ochir et al. (2016) [124] discovered a potential endogenous promoter of *P. tricornutum*, which is a glutamine synthetase promoter. This promoter induced the gene expression constitutively, and it was at least four times higher than the FCP promoter at the stationary phase.

Furthermore, four additional novel promoters were found in *P. tricornutum* under varied nitrate availability [125]. Moreover, five putative endogenous gene promoters highly expressed in *P. tricornutum* were isolated [126]. The activity of the Vacuolar ATPase (V-ATPase) gene promoter was higher than the other promoters and could drive the gene expression under both light and dark conditions at the stationary phase. The overview of the fucoxanthin synthesis pathway is crucial in genetic engineering. Several review articles have discussed the biosynthetic pathway of fucoxanthin [127–129]. Previous studies adopted the genetic engineering of the carotenoid gene(s) to improve the fucoxanthin content [130,131]. For instance, the introduction of an additional endogenous 1-deoxy-D-xylulose 5-phosphate synthase and phytoene synthase (Psy) gene separately into the genome of *P. tricornutum* resulted in an elevation of carotenoids amounts such as fucoxanthin [130]. In addition, the overexpressing of the phytoene synthase gene (isolated from *P. tricornutum*) in *P. tricornutum* also enhanced the fucoxanthin content by around 1.45-fold compared to the wild-type diatom [131]. Recently, Manfellotto et al. (2020) [33] produced *P. tricornutum* transformants with the overexpression of violaxanthin de-epoxidase, Vde-related, and zeaxanthin epoxidase 3. These transformants demonstrated an increased accumulation of fucoxanthin content up to four-fold compared to the wild type.

The production of high-quality fucoxanthin depends on an effective extraction and purification method. Various protocols have been utilized to recover and purify fucoxanthin from algae. These included centrifugal partition chromatography [132], column chromatography [133], microwave irradiation [113], pressurized liquid extraction [134], supercritical carbon dioxide extraction [135], ultrasound-assisted extraction [136], and traditional solvent extraction followed by chromatographic methods [137,138]. Different extraction methods for fucoxanthin have been compared, and the extracted amount was variable [139]. In addition, other parameters such as solvent types, solvent volume, temperature, etc., are crucial in determining the concentration and purity of fucoxanthin. Therefore, optimum conditions of these parameters are necessary for obtaining the highest wield possible [140]. Several review articles have discussed in depth the extraction and purification of fucoxanthin. For example, carotenoids (i.e., fucoxanthin) extracted from algae using different innovative methods were summarized [141]. In addition, Lourenco-Lopes et al. (2020) discussed the available extraction, quantification, and purification methods with the purpose of recovering the highest ratio of fucoxanthin [142].

6.1.4. Environment

Specific pigment fractions have been utilized to identify the algal types in the phytoplankton community. Fucoxanthin is a specific pigment in Bacillariophyceae and Chrysophyceae, whereas Pyrrophyta consists of peridinin as their unique pigment. Chlorophyll *b* and/or lutein are exclusive to Chlorophyta and Euglenophyta, while the specific pigment of Cyanophyta is myxoxanthophyll [143,144]. Pigment analysis using chromatographic methods provided reliable results of different major algal types in the phytoplankton community than microscopy enumeration [143]. Therefore, pigment analysis has been applied in the environmental study of algae. For example, Barlow (1995) [145] investigated the phytoplankton community in the western Alboran Sea based on pigment biomarkers. Recently, Wang et al. (2020) [146] reported the seasonal differences of the phytoneuston community structure in Daya Bay based on algal pigment analysis. The unique pigment of algae could also be employed to monitor the phytoplankton changes during the bloom of phytoplankton and determine the causative species for the bloom [147]. Moreover, the freshwater phytoplankton dynamics affected by seasonally variable freshwater inputs were investigated using pigment biomarkers [148]. Another study employed phytoplankton pigment profiling as a potential bioindicator of stratification conditions in lakes [149]. Apart from these, the significance of algal pigments was highlighted for paleolimnological research [150,151]. On the other hand, the algal pigment was also utilized to study the diet of zooplankton [152] and the impact of grazing by zooplankton on phytoplankton in the environment [153].

7. Commercial Products and Potential Applications of Fucoxanthin

Until now, only two algae extracts containing fucoxanthin as commercial products are available in the market (Figure 6). Xanthigen is an antiobesity commercial product, which combined 0.4% fucoxanthin from brown macroalgae *Undaria pinnativida* and 35% punicic acid from pomegranate seed oil (https://nektium.com/branded-ingredient/xanthigen/, accessed on 30 July 2021). The antiobesity activity of this product was examined in a clinical trial [154]. The findings indicated the reduction of body and liver fat content and improved liver function in the subjects after orally administered Xanthigen for 16 weeks. The suppression of adipocyte differentiation and lipid accumulation by Xanthigen was via multiple mechanisms [155]. In addition, Xanthigen was demonstrated to be safe for consumption [156]. The other commercial product is FucoVital (https://www.algatech.com/algatech-product/fucovital/, accessed on 30 July 2021), which consists of 3% fucoxanthin, omega-3s, and other beneficial fatty acids (extracted from *P. tricornutum*). FucoVital is the first fucoxanthin food ingredient product to improve liver health that was approved by the United States Food and Drug Administration (NDI 1048, 2017).

Figure 6. Commercial products and potential applications of fucoxanthin.

Several studies demonstrated the potential in commercializing fucoxanthin. For renewable energy, brown macroalgae (*Sargassum wightii*) extract (contained fucoxanthin) was shown to be suitable as a sensitizer in a solar cell, as it is a low-cost and environmentally friendly alternative to the ruthenium metal complexes [157]. In addition, high open-circuit photovoltage was observed in the bio-photovoltaic devices fabricated with FCP complexes and titanium dioxide nanostructures [158]. These findings indicated the possibility of fucoxanthin to be exploited in solar cells at the commercial level.

On the other hand, the cosmetic properties of fucoxanthin have been illustrated in several studies. For instance, a topical formulation containing fucoxanthin has been developed that could prevent exacerbations related to skin inflammatory pathologies and protect skin against UV radiation [159]. In addition, solid lipid nanoparticle formulation loaded with fucoxanthin demonstrated the UV-blocking potential [160]. Furthermore, Kang et al. (2020) reported that fucoxanthin concentrate extracted from *P. tricornutum* could be utilized as an active ingredient in wrinkle care cosmetics [161].

The benefits of fucoxanthin as a feed to broiler chicken were highlighted in several studies. Sasaki et al. (2010) [162] demonstrated feeding fucoxanthin to the broiler chicken enhanced both the plasma antioxidative status and meat color of the chicken. Moreover, fucoxanthin also decreased the number of harmful microorganisms and regulated the antioxidant metabolism of the chicken meat [163]. The addition of 10–15% of brown macroalgae into the basal diet of laying hens augmented the carotenoid content of the yolks by 7.5–10-fold [164]. Apart from these, the incorporation of edible brown macroalgae (rich in fucoxanthin) into the semolina (wheat)-based pasta enhanced the nutritional value of the pasta [165]. Furthermore, Mok et al. (2016) developed whole milk and skimmed milk fortified with fucoxanthin [166]. These fucoxanthin incorporated products showed superior plasma absorption and organ tissue accumulation rates for fucoxanthin [167]. Overall, these studies revealed the possibility of fucoxanthin to be widely exploited at the commercial level in addition to the health supplements (Xanthigen and FucoVital).

8. Challenges and Possible Solutions

Recently, a broad spectrum of health benefits of fucoxanthin has been well illustrated. Thus, the demand for fucoxanthin is increasing rapidly. To fulfill the demand, the commercial production of fucoxanthin is a necessity. However, there are several challenges in fucoxanthin production. The amount of fucoxanthin produced using chemical synthesis is insufficient to meet the demand of the fucoxanthin market [168]. Hence, the fucoxanthin source has been shifted to natural sources such as brown macroalgae [40] and microalgae [111]. Microalgae are preferred as fucoxanthin sources due to their superior fucoxanthin content [111]. Commercial viability in fucoxanthin production was only examined in a few microalgae strains [169]. A comprehensive screening of high fucoxanthin-producing microalgae should be adopted globally via international collaboration. A further selection is needed to ensure the consistent production of biomass and fucoxanthin under variable conditions when cultivated in either the laboratory or outdoors [170]. Commercial fucoxanthin production relies on cultivation, extraction, and purification parameters. The optimization of these parameters is needed to enhance fucoxanthin production. With the optimized parameters, a standardized fucoxanthin production protocol could be developed to achieve a cost-effective production of fucoxanthin. Although biosynthesis pathways of fucoxanthin have been reviewed extensively [127–129], some pathways remain obscure due to the lack of experimental validation of these enzymes, intermediate products, and pathways [171]. Utilizing "omics" techniques could offer detailed insight into the ambiguous metabolic pathways and regulatory mechanisms in producing fucoxanthin [172].

Quality control is important in ensuring the safety, efficacy, and quality of a particular product. Heavy metal, pesticides, and other chemical contaminations might be present in the microalgae samples. Thus, monitoring the level of these contaminations according to the World Health Organization and the United States Food and Drug Administration is essential. The poor stability, low solubility, and weak bioaccessibility of fucoxanthin pose a challenge when incorporating fucoxanthin into food or supplements. The usage of nano/micro-encapsulation of fucoxanthin improved its stability and bioavailability [173,174]. Furthermore, most of the studies examined fucoxanthin effects only for a short duration [154]. Some bioactive compounds, such as alginates, have short period effects [175]. The effect of fucoxanthin might be overestimated regarding long-term effects. Thus, more intensive and longer studies, particularly human trials, are required to develop and verify fucoxanthin's "true" effects.

9. Conclusions

The research trends of fucoxanthin from 1928 to June 2021 were analyzed in the present study. A significant increase in fucoxanthin publications, especially in recent years, was observed. This trend corroborates with the increased market demand for fucoxanthin over recent years (Global Fucoxanthin Market 2020). According to the bibliometric analysis, the three most productive countries were Japan (379 articles), the United States (319 articles), and China (230 articles). The top productive institutions were led by Hokkaido University, Japan (110 articles), followed by the CNRS Centre National de la Recherche Scientifique, France (61 articles), and Goethe-Universität Frankfurt am Main, Germany (49 articles). The analysis revealed that the United States and Germany were significantly intertwined within the global research network. Miyashita, Kazuo from Japan was the most productive and impactful author on fucoxanthin research with 88 publications and a Scopus *h*-index of 50. Most of the fucoxanthin articles were published in the *Marine Drugs* journal (123 articles) and the agricultural and biological sciences category (27.0%). Based on the keyword analysis, the research concepts of fucoxanthin could be divided into four clusters: bioactivities of fucoxanthin, photosynthesis, optimization of process, and environment. The bioactivities of fucoxanthin were identified as the potential future direction as the number of these studies increased tremendously over the years. The current study revealed the research trends of fucoxanthin and offered valuable information (i.e., most productive author, institutions, country) to establish effective global collaborations in studying fucoxanthin. In addition, the current findings could be utilized as fundamental in constructing impactful fucoxanthin research to expand the scientific knowledge on this unique carotenoid.

Author Contributions: Conceptualization, Y.S.K., H.T.T. and F.M.Y.; methodology, Y.S.K.; software, Y.S.K. and H.T.T.; validation, Y.S.K. and F.M.Y.; formal Analysis, Y.S.K., H.T.T., M.F.N., N.A.S., A.R.O. and F.M.Y.; investigation, Y.S.K.; resources, F.M.Y.; data curation, Y.S.K. and H.T.T.; writing—original draft preparation, Y.S.K., H.T.T., N.A.I.N.M. and N.A.S.; writing—review and editing, Y.S.K., H.T.T., N.A.I.N.M., M.F.N., N.A.S., A.R.O. and F.M.Y.; visualization, Y.S.K.; supervision, N.A.S., A.R.O. and F.M.Y.; project administration, F.M.Y.; funding acquisition, F.M.Y. All authors have read and agreed to the published version of the manuscript.

Funding: This study was supported by the Ministry of Higher Education Malaysia under the HiCOE research grant (grant no. 6369100), and SATREPS-COSMOS (JPMJSA 1509) Japan-Malaysia collaborative project.

Data Availability Statement: Not applicable.

Acknowledgments: Technical assistance from staff and students of the Laboratory of Aquatic Animal Health and Therapeutics, Institute of Bioscience, Universiti Putra Malaysia is duly acknowledged.

Conflicts of Interest: The authors declare no conflict of interest.

References

1. Liaaen-Jensen, S. Marine Carotenoids. In *Chemical and Biochemical Perspective*; Academic Press: New York, NY, USA, 1978; pp. 1–73. [CrossRef]
2. Peng, J.; Yuan, J.-P.; Wu, C.-F.; Wang, J.-H. Fucoxanthin, a Marine Carotenoid Present in Brown Seaweeds and Diatoms: Metabolism and Bioactivities Relevant to Human Health. *Mar. Drugs* **2011**, *9*, 1806–1828. [CrossRef] [PubMed]
3. Veith, T.; Brauns, J.; Weisheit, W.; Mittag, M.; Büchel, C. Identification of a specific fucoxanthin-chlorophyll protein in the light harvesting complex of photosystem I in the diatom *Cyclotella meneghiniana*. *Biochim. Biophys. Acta Bioenerg.* **2009**, *1787*, 905–912. [CrossRef]
4. Krinsky, N.I.; Johnson, E.J. Carotenoid actions and their relation to health and disease. *Mol. Asp. Med.* **2005**, *26*, 459–516. [CrossRef] [PubMed]
5. Thiyagarasaiyar, K.; Goh, B.-H.; Jeon, Y.-J.; Yow, Y.-Y. Algae Metabolites in Cosmeceutical: An Overview of Current Applications and Challenges. *Mar. Drugs* **2020**, *18*, 323. [CrossRef] [PubMed]
6. Maeda, H.; Hosokawa, M.; Sashima, T.; Miyashita, K. Dietary Combination of Fucoxanthin and Fish Oil Attenuates the Weight Gain of White Adipose Tissue and Decreases Blood Glucose in Obese/Diabetic KK-AyMice. *J. Agric. Food Chem.* **2007**, *55*, 7701–7706. [CrossRef] [PubMed]
7. Hitoe, S.; Shimoda, H. Seaweed Fucoxanthin Supplementation Improves Obesity Parameters in Mild Obese Japanese Subjects. *Funct. Foods Heal. Dis.* **2017**, *7*, 246. [CrossRef]
8. Muthuirulappan, S.; Francis, S.P. Anti-Cancer Mechanism and Possibility of Nano-Suspension Formulation for a Marine Algae Product Fucoxanthin. *Asian Pac. J. Cancer Prev.* **2013**, *14*, 2213–2216. [CrossRef] [PubMed]

9. Woo, M.-N.; Jeon, S.-M.; Kim, H.-J.; Lee, M.-K.; Shin, S.-K.; Shin, Y.C.; Park, Y.-B.; Choi, M.-S. Fucoxanthin supplementation improves plasma and hepatic lipid metabolism and blood glucose concentration in high-fat fed C57BL/6N mice. *Chem. Interact.* **2010**, *186*, 316–322. [CrossRef] [PubMed]

10. Oliveira, C.Y.B.; Oliveira, C.D.L.; Müller, M.N.; Santos, E.P.; Dantas, D.M.M.; Gálvez, A.O. A Scientometric Overview of Global Dinoflagellate Research. *Publications* **2020**, *8*, 50. [CrossRef]

11. Garrido-Cardenas, J.A.; Manzano-Agugliaro, F.; Acien-Fernandez, F.G.; Molina-Grima, E. Microalgae research worldwide. *Algal Res.* **2018**, *35*, 50–60. [CrossRef]

12. Silva, S.C.; Ferreira, I.C.F.R.; Dias, M.M.; Barreiro, M.F. Microalgae-Derived Pigments: A 10-Year Bibliometric Review and Industry and Market Trend Analysis. *Molecules* **2020**, *25*, 3406. [CrossRef] [PubMed]

13. Li, Z.; Zhu, L. The scientometric analysis of the research on microalgae-based wastewater treatment. *Environ. Sci. Pollut. Res.* **2021**, *28*, 25339–25348. [CrossRef] [PubMed]

14. Van Eck, N.J.; Waltman, L. Software survey: VOSviewer, a computer program for bibliometric mapping. *Scientometrics* **2009**, *84*, 523–538. [CrossRef] [PubMed]

15. Aria, M.; Cuccurullo, C. Bibliometrix: An R-tool for comprehensive science mapping analysis. *J. Inf.* **2017**, *11*, 959–975. [CrossRef]

16. Tang, Y.; Xin, H.; Yang, F.; Long, X. A historical review and bibliometric analysis of nanoparticles toxicity on algae. *J. Nanoparticle Res.* **2018**, *20*, 92. [CrossRef]

17. Terasaki, M.; Inoue, T.; Murase, W.; Kubota, A.; Kojima, H.; Kojoma, M.; Ohta, T.; Maeda, H.; Miyashita, K.; Mutoh, M.; et al. Fucoxanthinol Induces Apoptosis in a Pancreatic Intraepithelial Neoplasia Cell Line. *Cancer Genom. Proteom.* **2021**, *18*, 133–146. [CrossRef] [PubMed]

18. Terasaki, M.; Hamoya, T.; Kubota, A.; Kojima, H.; Tanaka, T.; Maeda, H.; Miyashita, K.; Mutoh, M. Fucoxanthin Prevents Colorectal Cancer Development in Dextran Sodium Sulfate-treated ApcMin/+ Mice. *Anticancer. Res.* **2021**, *41*, 1299–1305. [CrossRef]

19. Yoshikawa, M.; Hosokawa, M.; Miyashita, K.; Nishino, H.; Hashimoto, T. Effects of Fucoxanthin on the Inhibition of Dexamethasone-Induced Skeletal Muscle Loss in Mice. *Nutrients* **2021**, *13*, 1079. [CrossRef] [PubMed]

20. Agostini, A.; Büchel, C.; Di Valentin, M.; Carbonera, D. A distinctive pathway for triplet-triplet energy transfer photoprotection in fucoxanthin chlorophyll-binding proteins from *Cyclotella meneghiniana*. *Biochim. Biophys. Acta Bioenerg.* **2020**, *1862*, 148310. [CrossRef] [PubMed]

21. Gelzinis, A.; Augulis, R.; Büchel, C.; Robert, B.; Valkunas, L. Confronting FCP structure with ultrafast spectroscopy data: Evidence for structural variations. *Phys. Chem. Chem. Phys.* **2020**, *23*, 806–821. [CrossRef] [PubMed]

22. He, Z.; Gao, G.; Hand, E.S.; Kispert, L.D.; Strand, A.; Liaaen-Jensen, S. Iodine-Catalyzed R/S Isomerization of Allenic Carotenoids. *J. Phys. Chem. A* **2002**, *106*, 2520–2525. [CrossRef]

23. Refvem, T.; Strand, A.; Kjeldstad, B.; Haugan, J.A.; Liaaen-Jensen, S. Stereoisomerization of Allenic Carotenoids-Kinetic, Thermodynamic and Mechanistic Aspects. *Acta Chem. Scand.* **1999**, *53*, 114–123. [CrossRef]

24. Nagao, R.; Kato, K.; Ifuku, K.; Suzuki, T.; Kumazawa, M.; Uchiyama, I.; Kashino, Y.; Dohmae, N.; Akimoto, S.; Shen, J.-R.; et al. Structural basis for assembly and function of a diatom photosystem I-light-harvesting supercomplex. *Nat. Commun.* **2020**, *11*, 2481. [CrossRef] [PubMed]

25. Oka, K.; Ueno, Y.; Yokono, M.; Shen, J.-R.; Nagao, R.; Akimoto, S. Adaptation of light-harvesting and energy-transfer processes of a diatom *Phaeodactylum tricornutum* to different light qualities. *Photosynth. Res.* **2020**, *146*, 227–234. [CrossRef] [PubMed]

26. Akimoto, S.; Ueno, Y.; Yokono, M.; Shen, J.-R.; Nagao, R. Adaptation of light-harvesting and energy-transfer processes of a diatom *Chaetoceros gracilis* to different light qualities. *Photosynth. Res.* **2020**, *146*, 87–93. [CrossRef] [PubMed]

27. Fagan, J.; Eddens, K.S.; Dolly, J.; Vanderford, N.L.; Weiss, H.; Levens, J.S. Assessing Research Collaboration through Co-authorship Network Analysis. *J. Res. Adm.* **2018**, *49*, 76–99. [PubMed]

28. Shawky, S M.; Fathalla, S.I.; Orabi, S.H.; El-Mosalhi, H.H.; Abu-Alya, I.S. Effect of *Amphora coffeaeformis* and Star Anise as Dietary Supplementson the Immunity and Growth Performance of Broiler Chickens. *J. World's Poult. Res.* **2020**, *10*, 631–642. [CrossRef]

29. Hao, T.-B.; Yang, Y.-F.; Balamurugan, S.; Li, D.-W.; Yang, W.-D.; Li, H.-Y. Enrichment of f/2 medium hyperaccumulates biomass and bioactive compounds in the diatom *Phaeodactylum tricornutum*. *Algal Res.* **2020**, *47*, 101872. [CrossRef]

30. Conceição, D.; Lopes, R.G.; Derner, R.; Cella, H.; Carmo, A.P.B.D.; D'Oca, M.G.M.; Petersen, R.; Passos, M.F.; Vargas, J.V.C.; Galli-Terasawa, L.V.; et al. The effect of light intensity on the production and accumulation of pigments and fatty acids in *Phaeodactylum tricornutum*. *Environ. Boil. Fishes* **2020**, *32*, 1017–1025. [CrossRef]

31. Tanabe, M.; Ueno, Y.; Yokono, M.; Shen, J.-R.; Nagao, R.; Akimoto, S. Changes in excitation relaxation of diatoms in response to fluctuating light, probed by fluorescence spectroscopies. *Photosynth. Res.* **2020**, *146*, 143–150. [CrossRef] [PubMed]

32. Ifuku, K.; Yan, D.; Miyahara, M.; Inoue-Kashino, N.; Yamamoto, Y.Y.; Kashino, Y. A stable and efficient nuclear transformation system for the diatom *Chaetoceros gracilis*. *Photosynth. Res.* **2014**, *123*, 203–211. [CrossRef] [PubMed]

33. Manfellotto, F.; Stella, G.R.; Falciatore, A.; Brunet, C.; Ferrante, M.I. Engineering the Unicellular Alga *Phaeodactylum tricornutum* for Enhancing Carotenoid Production. *Antioxidants* **2020**, *9*, 757. [CrossRef] [PubMed]

34. Méndez-Leyva, A.B.; Guo, J.; Mudd, E.A.; Wong, J.; Schwartz, J.-M.; Day, A. The chloroplast genome of the marine microalga *Tisochrysis lutea*. *Mitochondrial DNA Part B* **2019**, *4*, 253–255. [CrossRef]

35. Park, S.; Jung, G.; Hwang, Y.-S.; Jin, E. Dynamic response of the transcriptome of a psychrophilic diatom, *Chaetoceros neogracile*, to high irradiance. *Planta* **2009**, *231*, 349–360. [CrossRef]

36. Foo, S.C.; Yusoff, F.M.; Imam, M.U.; Foo, J.B.; Ismail, N.; Azmi, N.H.; Tor, Y.S.; Khong, N.M.; Ismail, M. Increased fucoxanthin in *Chaetoceros calcitrans* extract exacerbates apoptosis in liver cancer cells via multiple targeted cellular pathways. *Biotechnol. Rep.* **2018**, *21*, e00296. [CrossRef]

37. Manmuan, S.; Manmuan, P. Fucoxanthin Enhances 5-FU Chemotherapeutic Efficacy in Colorectal Cancer Cells by Affecting MMP-9 Invasive Proteins. *J. Appl. Pharm. Sci.* **2019**, *9*, 7–14.

38. Smith, D.R. Citation Indexing and Highly Cited Articles in the Australian Veterinary Journal. *Aust. Vet. J.* **2008**, *86*, 337–339. [CrossRef] [PubMed]

39. Maeda, H.; Hosokawa, M.; Sashima, T.; Funayama, K.; Miyashita, K. Fucoxanthin from edible seaweed, *Undaria pinnatifida*, shows antiobesity effect through UCP1 expression in white adipose tissues. *Biochem. Biophys. Res. Commun.* **2005**, *332*, 392–397. [CrossRef] [PubMed]

40. Yan, X.; Chuda, Y.; Suzuki, M.; Nagata, T. Fucoxanthin as the Major Antioxidant in *Hijikia fusiformis*, a Common Edible Seaweed. *Biosci. Biotechnol. Biochem.* **1999**, *63*, 605–607. [CrossRef]

41. Sachindra, N.M.; Sato, E.; Maeda, H.; Hosokawa, M.; Niwano, Y.; Kohno, M.; Miyashita, K. Radical Scavenging and Singlet Oxygen Quenching Activity of Marine Carotenoid Fucoxanthin and Its Metabolites. *J. Agric. Food Chem.* **2007**, *55*, 8516–8522. [CrossRef] [PubMed]

42. Hosokawa, M.; Kudo, M.; Maeda, H.; Kohno, H.; Tanaka, T.; Miyashita, K. Fucoxanthin Induces Apoptosis and Enhances the Antiproliferative Effect of the PPARγ Ligand, Troglitazone, on Colon Cancer Cells. *Biochim. Biophys. Acta General Subj.* **2004**, *1675*, 113–119. [CrossRef] [PubMed]

43. Kim, K.N.; Heo, S.J.; Yoon, W.J.; Kang, S.M.; Ahn, G.; Yi, T.H.; Jeon, Y.J. Fucoxanthin Inhibits the Inflammatory Response by Suppressing the Activation of NF-KB and MAPKs in Lipopolysaccharide-Induced RAW 264.7 Macrophages. *Eur. J. Pharmacol.* **2010**, *649*, 369–375. [CrossRef] [PubMed]

44. Heo, S.-J.; Yoon, W.-J.; Kim, K.-N.; Ahn, G.-N.; Kang, S.-M.; Kang, D.-H.; Affan, A.; Oh, C.; Jung, W.-K.; Jeon, Y.-J. Evaluation of anti-inflammatory effect of fucoxanthin isolated from brown algae in lipopolysaccharide-stimulated RAW 264.7 macrophages. *Food Chem. Toxicol.* **2010**, *48*, 2045–2051. [CrossRef]

45. Jones, K.S.; Jackson, D.M. The use of automatically-obtained keyword classifications for information retrieval. *Inf. Storage Retr.* **1970**, *5*, 175–201. [CrossRef]

46. Plaza, M.; Santoyo, S.; Jaime, L.; Reina, G.G.-B.; Herrero, M.; Señoráns, F.J.; Ibáñez, E. Screening for bioactive compounds from algae. *J. Pharm. Biomed. Anal.* **2010**, *51*, 450–455. [CrossRef]

47. Okuzumi, J.; Nishino, H.; Murakoshi, M.; Iwashima, A.; Tanaka, Y.; Yamane, T.; Fujita, Y.; Takahashi, T. Inhibitory effects of fucoxanthin, a natural carotenoid, on N-myc expression and cell cycle progression in human malignant tumor cells. *Cancer Lett.* **1990**, *55*, 75–81. [CrossRef]

48. Okuzumi, J.; Takahashi, T.; Yamane, T.; Kitao, Y.; Inagake, M.; Ohya, K.; Nishino, H.; Tanaka, Y. Inhibitory Effects of Fucoxanthin, a Natural Carotenoid, on N-Ethyl-N′-Nitro-N-Nitrosoguanidine-Induced Mouse Duodenal Carcinogenesis. *Cancer Lett.* **1993**, *68*, 159–168. [CrossRef]

49. Greenwald, P.; Kelloff, G.; Burch-Whitman, C.; Kramer, B.S. Chemoprevention. *CA Cancer J. Clin.* **1995**, *45*, 31–49. [CrossRef] [PubMed]

50. Balboa, E.M.; Conde, E.; Moure, A.; Falqué, E.; Domínguez, H. In vitro antioxidant properties of crude extracts and compounds from brown algae. *Food Chem.* **2013**, *138*, 1764–1785. [CrossRef] [PubMed]

51. Park, E.-J.; Pezzuto, J.M. Antioxidant Marine Products in Cancer Chemoprevention. *Antioxid. Redox Signal.* **2013**, *19*, 115–138. [CrossRef] [PubMed]

52. Manivasagan, P.; Bharathiraja, S.; Moorthy, M.S.; Mondal, S.; Seo, H.; Lee, K.D.; Oh, J. Marine natural pigments as potential sources for therapeutic applications. *Crit. Rev. Biotechnol.* **2017**, *38*, 745–761. [CrossRef] [PubMed]

53. Ojulari, O.V.; Lee, S.G.; Nam, J.-O. Therapeutic Effect of Seaweed Derived Xanthophyl Carotenoid on Obesity Management; Overview of the Last Decade. *Int. J. Mol. Sci.* **2020**, *21*, 2502. [CrossRef] [PubMed]

54. Sathasivam, R.; Ki, J.-S. A Review of the Biological Activities of Microalgal Carotenoids and Their Potential Use in Healthcare and Cosmetic Industries. *Mar. Drugs* **2018**, *16*, 26. [CrossRef] [PubMed]

55. Rengarajan, T.; Rajendran, P.; Nandakumar, N.; Balasubramanian, M.P.; Nishigaki, I. Cancer Preventive Efficacy of Marine Carotenoid Fucoxanthin: Cell Cycle Arrest and Apoptosis. *Nutrients* **2013**, *5*, 4978–4989. [CrossRef]

56. Martin, L.J. Fucoxanthin and Its Metabolite Fucoxanthinol in Cancer Prevention and Treatment. *Mar. Drugs* **2015**, *13*, 4784–4798. [CrossRef] [PubMed]

57. Kaulmann, A.; Bohn, T. Carotenoids, Inflammation, and Oxidative Stress—Implications of Cellular Signaling Pathways and Relation to Chronic Disease Prevention. *Nutr. Res.* **2014**, *34*, 907–929. [CrossRef] [PubMed]

58. Muradian, K.; Vaiserman, A.; Min, K.-J.; Fraifeld, V.E. Fucoxanthin and lipid metabolism: A minireview. *Nutr. Metab. Cardiovasc. Dis.* **2015**, *25*, 891–897. [CrossRef]

59. Hu, X.; Li, Y.; Li, C.; Fu, Y.; Cai, F.; Chen, Q.; Li, D. Combination of fucoxanthin and conjugated linoleic acid attenuates body weight gain and improves lipid metabolism in high-fat diet-induced obese rats. *Arch. Biochem. Biophys.* **2012**, *519*, 59–65. [CrossRef] [PubMed]

60. Molina, N.; Morandi, A.C.; Bolin, A.; Otton, R. Comparative effect of fucoxanthin and vitamin C on oxidative and functional parameters of human lymphocytes. *Int. Immunopharmacol.* **2014**, *22*, 41–50. [CrossRef]

61. Lin, H.-T.V.; Tsou, Y.-C.; Chen, Y.-T.; Lu, W.-J.; Hwang, P.-A. Effects of Low-Molecular-Weight Fucoidan and High Stability Fucoxanthin on Glucose Homeostasis, Lipid Metabolism, and Liver Function in a Mouse Model of Type II Diabetes. *Mar. Drugs* **2017**, *15*, 113. [CrossRef] [PubMed]

62. Chang, P.; Li, K.; Lin, Y. Fucoidan–Fucoxanthin Ameliorated Cardiac Function via IRS1GRB2 SOS1, GSK3βCREB Pathways and Metabolic Pathways in Senescent Mice. *Mar. Drugs* **2019**, *17*, 69. [CrossRef] [PubMed]

63. Chau, Y.-T.; Chen, H.-Y.; Lin, P.-H.; Hsia, S.-M. Preventive Effects of Fucoidan and Fucoxanthin on Hyperuricemic Rats Induced by Potassium Oxonate. *Mar. Drugs* **2019**, *17*, 343. [CrossRef] [PubMed]

64. Rodriguez-Luna, A.; Ávila-Román, J.; Oliveira, H.; Motilva, V.; Talero, E. Fucoxanthin and Rosmarinic Acid Combination Has Anti-Inflammatory Effects through Regulation of NLRP3 Inflammasome in UVB-Exposed HaCaT Keratinocytes. *Mar. Drugs* **2019**, *17*, 451. [CrossRef] [PubMed]

65. Menichini, D.; Alrais, M.; Liu, C.; Xia, Y.; Blackwell, S.C.; Facchinetti, F.; Sibai, B.M.; Longo, M. Maternal Supplementation of Inositols, Fucoxanthin, and Hydroxytyrosol in Pregnant Murine Models of Hypertension. *Am. J. Hypertens.* **2020**, *33*, 652–659. [CrossRef]

66. Kawakamia, T.; Tsushimab, M.; Katabamia, Y.; Mine, M.; Ishida, A.; Matsuno, M. Effect of b, b-Carotene, b-Echinenone, Astaxanthin, Fucoxanthin, Vitamin A and Vitamin E on the Biological Defense of the Sea Urchin *Pseudocentrotus depressus*. *J. Exp. Mar. Biol. Ecol.* **1998**, *226*, 165–174. [CrossRef]

67. Kitamura, M.; Koyama, T.; Nakano, Y.; Uemura, D. Characterization of a natural inducer of coral larval metamorphosis. *J. Exp. Mar. Biol. Ecol.* **2007**, *340*, 96–102. [CrossRef]

68. Lashmanova, E.; Proshkina, E.; Zhikrivetskaya, S.; Shevchenko, O.; Marusich, E.; Leonov, S.; Melerzanov, A.; Zhavoronkov, A.; Moskalev, A. Fucoxanthin increases lifespan of *Drosophila melanogaster* and *Caenorhabditis elegans*. *Pharmacol. Res.* **2015**, *100*, 228–241. [CrossRef] [PubMed]

69. Green, B.R.; Durnford, D.G. The chlorophyll-carotenoid proteins of oxygenic photosynthesis. *Annu. Rev. Plant Biol.* **1996**, *47*, 685–714. [CrossRef] [PubMed]

70. Nagao, R.; Takahashi, S.; Suzuki, T.; Dohmae, N.; Nakazato, K.; Tomo, T. Comparison of oligomeric states and polypeptide compositions of fucoxanthin chlorophyll a/c-binding protein complexes among various diatom species. *Photosynth. Res.* **2013**, *117*, 281–288. [CrossRef] [PubMed]

71. Depauw, F.A.; Rogato, A.; Ribera d'Alcalá, M.; Falciatore, A. Exploring the Molecular Basis of Responses to Light in Marine Diatoms. *J. Exp. Bot.* **2012**, *63*, 1575–1591. [CrossRef] [PubMed]

72. Armbrust, E.V.; Berges, J.A.; Bowler, C.; Green, B.R.; Martinez, D.; Putnam, N.; Zhou, S.; Allen, A.; Apt, K.E.; Bechner, M.; et al. The Genome of the Diatom *Thalassiosira pseudonana*: Ecology, Evolution, and Metabolism. *Science* **2004**, *306*, 79–86. [CrossRef] [PubMed]

73. Bowler, C.; Allen, A.; Badger, J.H.; Grimwood, J.; Jabbari, K.; Kuo, A.; Maheswari, U.; Martens, C.; Maumus, F.; Otillar, R.P.; et al. The *Phaeodactylum* genome reveals the evolutionary history of diatom genomes. *Nature* **2008**, *456*, 239–244. [CrossRef] [PubMed]

74. Lepetit, B.; Volke, D.; Gilbert, M.; Wilhelm, C.; Goss, R. Evidence for the Existence of One Antenna-Associated, Lipid-Dissolved and Two Protein-Bound Pools of Diadinoxanthin Cycle Pigments in Diatoms. *Plant Physiol.* **2010**, *154*, 1905–1920. [CrossRef] [PubMed]

75. Bailleul, B.; Rogato, A.; de Martino, A.; Coesel, S.; Cardol, P.; Bowler, C.; Falciatore, A.; Finazzi, G. An atypical member of the light-harvesting complex stress-related protein family modulates diatom responses to light. *Proc. Natl. Acad. Sci. USA* **2010**, *107*, 18214–18219. [CrossRef]

76. Zhu, S.-H.; Green, B.R. Photoprotection in the diatom *Thalassiosira pseudonana*: Role of LI818-like proteins in response to high light stress. *Biochim. Biophys. Acta Bioenerg.* **2010**, *1797*, 1449–1457. [CrossRef] [PubMed]

77. Lepetit, B.; Volke, D.; Szabo, M.; Hoffmann, R.; Garab, G.; Wilhelm, C.; Goss, R. Spectroscopic and Molecular Characterization of the Oligomeric Antenna of the Diatom *Phaeodactylum tricornutum*. *Biochemistry* **2007**, *46*, 9813–9822. [CrossRef] [PubMed]

78. Gundermann, K.; Schmidt, M.; Weisheit, W.; Mittag, M.; Büchel, C. Identification of several sub populations in the pool of light harvesting proteins in the pennate diatom *Phaeodactylum tricornutum*. *Biochim. Biophys. Acta Bioenerg.* **2013**, *1827*, 303–310. [CrossRef] [PubMed]

79. Gundermann, K.; Wagner, V.; Mittag, M.; Büchel, C. Fucoxanthin-Chlorophyll Protein Complexes of the Centric Diatom *Cyclotella meneghiniana* Differ in Lhcx1 and Lhcx6_1 Content. *Plant Physiol.* **2019**, *179*, 1779–1795. [CrossRef] [PubMed]

80. Büchel, C. Fucoxanthin-Chlorophyll Proteins in Diatoms: 18 and 19 kDa Subunits Assemble into Different Oligomeric States. *Biochemistry* **2003**, *42*, 13027–13034. [CrossRef]

81. Kansy, M.; Volke, D.; Sturm, L.; Wilhelm, C.; Hoffmann, R.; Goss, R. Pre-purification of diatom pigment protein complexes provides insight into the heterogeneity of FCP complexes. *BMC Plant Biol.* **2020**, *20*, 456. [CrossRef] [PubMed]

82. Wang, W.; Zhao, S.; Pi, X.; Kuang, T.; Sui, S.; Shen, J. Structural Features of the Diatom Photosystem II–light-harvesting Antenna Complex. *FEBS J.* **2020**, *287*, 2191–2200. [CrossRef] [PubMed]

83. Xu, C.; Pi, X.; Huang, Y.; Han, G.; Chen, X.; Qin, X.; Huang, G.; Zhao, S.; Yang, Y.; Kuang, T.; et al. Structural basis for energy transfer in a huge diatom PSI-FCPI supercomplex. *Nat. Commun.* **2020**, *11*, 5081. [CrossRef]

84. Nagao, R.; Kato, K.; Suzuki, T.; Ifuku, K.; Uchiyama, I.; Kashino, Y.; Dohmae, N.; Akimoto, S.; Shen, J.-R.; Miyazaki, N.; et al. Structural basis for energy harvesting and dissipation in a diatom PSII–FCPII supercomplex. *Nat. Plants* **2019**, *5*, 890–901. [CrossRef]

85. Pi, X.; Zhao, S.; Wang, W.; Liu, D.; Xu, C.; Han, G.; Kuang, T.; Sui, S.-F.; Shen, J.-R. The pigment-protein network of a diatom photosystem II–light-harvesting antenna supercomplex. *Science* **2019**, *365*, eaax4406. [CrossRef] [PubMed]

86. Wang, W.; Yu, L.-J.; Xu, C.; Tomizaki, T.; Zhao, S.; Umena, Y.; Chen, X.; Qin, X.; Xin, Y.; Suga, M.; et al. Structural basis for blue-green light harvesting and energy dissipation in diatoms. *Science* **2019**, *363*, eaav0365. [CrossRef]

87. Papagiannakis, E.; Van Stokkum, I.H.M.; Fey, H.; Büchel, C.; Van Grondelle, R. Spectroscopic Characterization of the Excitation Energy Transfer in the Fucoxanthin–Chlorophyll Protein of Diatoms. *Photosynth. Res.* **2005**, *86*, 241–250. [CrossRef] [PubMed]

88. Zigmantas, D.; Hiller, R.G.; Sharples, F.P.; Frank, H.A.; Sundström, V.; Polívka, T. Effect of a Conjugated Carbonyl Group on the Photophysical Properties of Carotenoids. *Phys. Chem. Chem. Phys.* **2004**, *6*, 3009–3016. [CrossRef]

89. Gildenhoff, N.; Herz, J.; Gundermann, K.; Büchel, C.; Wachtveitl, J. The excitation energy transfer in the trimeric fucoxanthin–chlorophyll protein from *Cyclotella meneghiniana* analyzed by polarized transient absorption spectroscopy. *Chem. Phys.* **2010**, *373*, 104–109. [CrossRef]

90. Ramanan, C.; Berera, R.; Gundermann, K.; van Stokkum, I.; Büchel, C.; van Grondelle, R. Exploring the Mechanism(s) of Energy Dissipation in the Light Harvesting Complex of the Photosynthetic Algae *Cyclotella meneghiniana*. *Biochim. Biophys. Acta Bioenerg.* **2014**, *1837*, 1507–1513. [CrossRef] [PubMed]

91. Premvardhan, L.; Sandberg, D.J.; Fey, H.; Birge, R.R.; Büchel, C.; van Grondelle, R. The Charge-Transfer Properties of the S2 State of Fucoxanthin in Solution and in Fucoxanthin Chlorophyll-a/c2 Protein (FCP) Based on Stark Spectroscopy and Molecular-Orbital Theory. *J. Phys. Chem. B* **2008**, *112*, 11838–11853. [CrossRef] [PubMed]

92. Szabó, M.; Premvardhan, L.; Lepetit, B.; Goss, R.; Wilhelm, C.; Garab, G. Functional heterogeneity of the fucoxanthins and fucoxanthin-chlorophyll proteins in diatom cells revealed by their electrochromic response and fluorescence and linear dichroism spectra. *Chem. Phys.* **2010**, *373*, 110–114. [CrossRef]

93. Premvardhan, L.; Bordes, L.; Beer, A.; Büchel, C.; Robert, B. Carotenoid Structures and Environments in Trimeric and Oligomeric Fucoxanthin Chlorophyll a/c2 Proteins from Resonance Raman Spectroscopy. *J. Phys. Chem. B* **2009**, *113*, 12565–12574. [CrossRef]

94. Akimoto, S.; Teshigahara, A.; Yokono, M.; Mimuro, M.; Nagao, R.; Tomo, T. Excitation Relaxation Dynamics and Energy Transfer in Fucoxanthin–chlorophyll a/c-Protein Complexes, Probed by Time-Resolved Fluorescence. *Biochim. Biophys. Acta Bioenerg.* **2014**, *1837*, 1514–1521. [CrossRef] [PubMed]

95. Gildenhoff, N.; Amarie, S.; Gundermann, K.; Beer, A.; Büchel, C.; Wachtveitl, J. Oligomerization and pigmentation dependent excitation energy transfer in fucoxanthin–chlorophyll proteins. *Biochim. Biophys. Acta Bioenerg.* **2010**, *1797*, 543–549. [CrossRef]

96. Kosumi, D.; Kita, M.; Fujii, R.; Sugisaki, M.; Oka, N.; Takaesu, Y.; Taira, T.; Iha, M.; Hashimoto, H. Excitation Energy-Transfer Dynamics of Brown Algal Photosynthetic Antennas. *J. Phys. Chem. Lett.* **2012**, *3*, 2659–2664. [CrossRef]

97. Romero, E.; Augulis, R.; Novoderezhkin, V.; Ferretti, M.; Thieme, J.; Zigmantas, D.; Van Grondelle, R. Quantum coherence in photosynthesis for efficient solar-energy conversion. *Nat. Phys.* **2014**, *10*, 676–682. [CrossRef]

98. Gelzinis, A.; Butkus, V.; Songaila, E.; Augulis, R.; Gall, A.; Büchel, C.; Robert, B.; Abramavicius, D.; Zigmantas, D.; Valkunas, L. Mapping energy transfer channels in fucoxanthin–chlorophyll protein complex. *Biochim. Biophys. Acta Bioenerg.* **2014**, *1847*, 241–247. [CrossRef] [PubMed]

99. Brixner, T.; Stenger, J.; Vaswani, H.M.; Cho, M.; Blankenship, R.E.; Fleming, G.R. Two-dimensional spectroscopy of electronic couplings in photosynthesis. *Nature* **2005**, *434*, 625–628. [CrossRef] [PubMed]

100. Lambrev, P.H.; Akhtar, P.; Tan, H.-S. Insights into the mechanisms and dynamics of energy transfer in plant light-harvesting complexes from two-dimensional electronic spectroscopy. *Biochim. Biophys. Acta Bioenerg.* **2019**, *1861*, 148050. [CrossRef] [PubMed]

101. Lavaud, J.; Rousseau, B.; Etienne, A.-L. Enrichment of the Light-Harvesting Complex in Diadinoxanthin and Implications for the Nonphotochemical Fluorescence Quenching in Diatoms. *Biochemistry* **2003**, *42*, 5802–5808. [CrossRef] [PubMed]

102. Lavaud, J.; Rousseau, B.; van Gorkom, H.J.; Etienne, A.-L. Influence of the Diadinoxanthin Pool Size on Photoprotection in the Marine Planktonic Diatom *Phaeodactylum tricornutum*. *Plant Physiol.* **2002**, *129*, 1398–1406. [CrossRef] [PubMed]

103. Gundermann, K.; Büchel, C. The fluorescence yield of the trimeric fucoxanthin–chlorophyll–protein FCPa in the diatom *Cyclotella meneghiniana* is dependent on the amount of bound diatoxanthin. *Photosynth. Res.* **2007**, *95*, 229–235. [CrossRef] [PubMed]

104. Goss, R.; Pinto, E.A.; Wilhelm, C.; Richter, M. The Importance of a Highly Active and ΔpH-Regulated Diatoxanthin Epoxidase for the Regulation of the PS II Antenna Function in Diadinoxanthin Cycle Containing Algae. *J. Plant Physiol.* **2006**, *163*, 1008–1021. [CrossRef]

105. Gundermann, K.; Büchel, C. Factors determining the fluorescence yield of fucoxanthin-chlorophyll complexes (FCP) involved in non-photochemical quenching in diatoms. *Biochim. Biophys. Acta Bioenerg.* **2012**, *1817*, 1044–1052. [CrossRef]

106. Grouneva, I.; Jakob, T.; Wilhelm, C.; Goss, R. A New Multicomponent NPQ Mechanism in the Diatom *Cyclotella meneghiniana*. *Plant Cell Physiol.* **2008**, *49*, 1217–1225. [CrossRef] [PubMed]

107. Taddei, L.; Chukhutsina, V.U.; Lepetit, B.; Stella, G.R.; Bassi, R.; Van Amerongen, H.; Bouly, J.-P.; Jaubert, M.; Finazzi, G.; Falciatore, A. Dynamic Changes between Two LHCX-Related Energy Quenching Sites Control Diatom Photoacclimation. *Plant Physiol.* **2018**, *177*, 953–965. [CrossRef] [PubMed]

108. Beer, A.; Gundermann, K.; Beckmann, J.; Büchel, C. Subunit Composition and Pigmentation of Fucoxanthin–Chlorophyll Proteins in Diatoms: Evidence for a Subunit Involved in Diadinoxanthin and Diatoxanthin Binding. *Biochemistry* **2006**, *45*, 13046–13053. [CrossRef] [PubMed]

109. Szabó, I.; Bergantino, E.; Giacometti, G.M. Light and oxygenic photosynthesis: Energy dissipation as a protection mechanism against photo-oxidation. *EMBO Rep.* **2005**, *6*, 629–634. [CrossRef]

110. Büchel, C. Light harvesting complexes in chlorophyll c-containing algae. *Biochim. Biophys. Acta Bioenerg.* **2019**, *1861*, 148027. [CrossRef] [PubMed]

111. Kim, S.M.; Jung, Y.-J.; Kwon, O.-N.; Cha, K.H.; Um, B.-H.; Chung, D.; Pan, C.-H. A Potential Commercial Source of Fucoxanthin Extracted from the Microalga *Phaeodactylum tricornutum*. *Appl. Biochem. Biotechnol.* **2012**, *166*, 1843–1855. [CrossRef] [PubMed]

112. Foo, S.C.; Yusoff, F.M.; Ismail, M.; Basri, M.; Yau, S.K.; Khong, N.M.; Chan, K.W.; Ebrahimi, M. Antioxidant capacities of fucoxanthin-producing algae as influenced by their carotenoid and phenolic contents. *J. Biotechnol.* **2017**, *241*, 175–183. [CrossRef] [PubMed]

113. Pasquet, V.; Chérouvrier, J.-R.; Farhat, F.; Thiéry, V.; Piot, J.M.; Bérard, J.-B.; Kaas, R.; Serive, B.; Patrice, T.; Cadoret, J.-P.; et al. Study on the microalgal pigments extraction process: Performance of microwave assisted extraction. *Process. Biochem.* **2011**, *46*, 59–67. [CrossRef]

114. McClure, D.D.; Luiz, A.; Gerber, B.; Barton, G.W.; Kavanagh, J.M. An investigation into the effect of culture conditions on fucoxanthin production using the marine microalgae *Phaeodactylum tricornutum*. *Algal Res.* **2018**, *29*, 41–48. [CrossRef]

115. Guo, B.; Liu, B.; Yang, B.; Sun, P.; Lu, X.; Liu, J.; Chen, F. Screening of Diatom Strains and Characterization of *Cyclotella cryptica* as A Potential Fucoxanthin Producer. *Mar. Drugs* **2016**, *14*, 125. [CrossRef]

116. Sun, Z.; Wang, X.; Liu, J. Screening of *Isochrysis* strains for simultaneous production of docosahexaenoic acid and fucoxanthin. *Algal Res.* **2019**, *41*, 101545. [CrossRef]

117. Xia, S.; Wang, K.; Wan, L.; Li, A.; Hu, Q.; Zhang, C. Production, Characterization, and Antioxidant Activity of Fucoxanthin from the Marine Diatom *Odontella aurita*. *Mar. Drugs* **2013**, *11*, 2667–2681. [CrossRef]

118. Khaw, Y.S.; Khong, N.M.H.; Shaharuddin, N.A.; Yusoff, F.M. A simple 18S rDNA approach for the identification of cultured eukaryotic microalgae with an emphasis on primers. *J. Microbiol. Methods* **2020**, *172*, 105890. [CrossRef] [PubMed]

119. Khan, M.I.; Shin, J.H.; Kim, J.D. The Promising Future of Microalgae: Current Status, Challenges, and Optimization of a Sustainable and Renewable Industry for Biofuels, Feed, and Other Products. *Microb. Cell Fact.* **2018**, *17*, 36. [CrossRef]

120. Falciatore, A.; Casotti, R.; Leblanc, C.; Abrescia, C.; Bowler, C. Transformation of Nonselectable Reporter Genes in Marine Diatoms. *Mar. Biotechnol.* **1999**, *1*, 239–251. [CrossRef] [PubMed]

121. Apt, K.E.; Grossman, A.R.; Kroth-Pancic, P.G. Stable nuclear transformation of the diatom *Phaeodactylum tricornutum*. *Mol. Genet. Genom.* **1996**, *252*, 572–579. [CrossRef]

122. Miyagawa, A.; Okami, T.; Kira, N.; Yamaguchi, H.; Ohnishi, K.; Adachi, M. Research Note: High Efficiency Transformation of the Diatom *Phaeodactylum tricornutum* with a Promoter from the Diatom *Cylindrotheca fusiformis*. *Phycol. Res.* **2009**, *57*, 142–146. [CrossRef]

123. Miyagawa-Yamaguchi, A.; Okami, T.; Kira, N.; Yamaguchi, H.; Ohnishi, K.; Adachi, M. Stable nuclear transformation of the diatom *Chaetoceros* sp. *Phycol. Res.* **2011**, *59*, 113–119. [CrossRef]

124. Erdene-ochir, E.; Shin, B.; Huda, N.; Hye, D.; Eun, K.; Lee, H. Cloning of a Novel Endogenous Promoter for Foreign Gene Expression in *Phaeodactylum tricornutum*. *Appl. Biol. Chem.* **2016**, *59*, 861–867. [CrossRef]

125. Adler-Agnon, Z.; Leu, S.; Zarka, A.; Boussiba, S.; Khozin-Goldberg, I. Novel promoters for constitutive and inducible expression of transgenes in the diatom *Phaeodactylum tricornutum* under varied nitrate availability. *Environ. Boil. Fishes* **2017**, *30*, 2763–2772. [CrossRef]

126. Watanabe, Y.; Kadono, T.; Kira, N.; Suzuki, K.; Iwata, O.; Ohnishi, K.; Yamaguchi, H.; Adachi, M. Development of endogenous promoters that drive high-level expression of introduced genes in the model diatom *Phaeodactylum tricornutum*. *Mar. Genom.* **2018**, *42*, 41–48. [CrossRef]

127. Mikami, K.; Hosokawa, M. Biosynthetic Pathway and Health Benefits of Fucoxanthin, an Algae-Specific Xanthophyll in Brown Seaweeds. *Int. J. Mol. Sci.* **2013**, *14*, 13763–13781. [CrossRef] [PubMed]

128. Iravani, N.; Hajiaghaee, R.; Zarekarizi, A. A Review on Biosynthesis, Health Benefits and Extraction Methods of Fucoxanthin, Particular Marine Carotenoids in Algae. *J. Med. Plants* **2018**, *17*, 6–30.

129. Zarekarizi, A.; Hoffmann, L.; Burritt, D. Approaches for the sustainable production of fucoxanthin, a xanthophyll with potential health benefits. *Environ. Boil. Fishes* **2018**, *31*, 281–299. [CrossRef]

130. Eilers, U.; Bikoulis, A.; Breitenbach, J.; Büchel, C.; Sandmann, G. Limitations in the biosynthesis of fucoxanthin as targets for genetic engineering in *Phaeodactylum tricornutum*. *Environ. Boil. Fishes* **2015**, *28*, 123–129. [CrossRef]

131. Kadono, T.; Kira, N.; Suzuki, K.; Iwata, O.; Ohama, T.; Okada, S.; Nishimura, T.; Akakabe, M.; Tsuda, M.; Adachi, M. Effect of an Introduced Phytoene Synthase Gene Expression on Carotenoid Biosynthesis in the Marine Diatom *Phaeodactylum tricornutum*. *Mar. Drugs* **2015**, *13*, 5334–5357. [CrossRef] [PubMed]

132. Kim, S.M.; Shang, Y.F.; Um, B.-H. A preparative method for isolation of fucoxanthin from *Eisenia bicyclis* by centrifugal partition chromatography. *Phytochem. Anal.* **2011**, *22*, 322–329. [CrossRef] [PubMed]

133. Tokarek, W.; Listwan, S.; Pagacz, J.; Leśniak, P.; Latowski, D. Column chromatography as a useful step in purification of diatom pigments. *Acta Biochim. Pol.* **2016**, *63*, 443–447. [CrossRef] [PubMed]

134. Shang, Y.F.; Kim, S.M.; Lee, W.J.; Um, B.-H. Pressurized liquid method for fucoxanthin extraction from *Eisenia bicyclis* (Kjellman) Setchell. *J. Biosci. Bioeng.* **2011**, *111*, 237–241. [CrossRef]

135. Quitain, A.T.; Kai, T.; Sasaki, M.; Goto, M. Supercritical Carbon Dioxide Extraction of Fucoxanthin from *Undaria pinnatifida*. *J. Agric. Food Chem.* **2013**, *61*, 5792–5797. [CrossRef] [PubMed]

136. Eom, S.J.; Kim, Y.E.; Kim, J.-E.; Park, J.; Kim, Y.H.; Song, K.-M.; Lee, N.H. Production of *Undaria pinnatifida* sporophyll extract using pilot-scale ultrasound-assisted extraction: Extract characteristics and antioxidant and anti-inflammatory activities. *Algal Res.* **2020**, *51*, 102039. [CrossRef]

137. Jaswir, I.; Noviendri, D.; Salleh, H.M.; Taher, M.; Miyashita, K. Isolation of Fucoxanthin and Fatty Acids Analysis of *Padina australis* and Cytotoxic Effect of Fucoxanthin on Human Lung Cancer (H1299) Cell Lines. *Afr. J. Biotechnol.* **2011**, *10*, 18855–18862.

138. Noviendri, D.; Salleh, H.M.; Taher, M.; Miyashita, K.; Ramli, N. Fucoxanthin Extraction and Fatty Acid Analysis of *Sargassum binderi* and *Sargassum duplicatum*. *J. Med. Plants Res.* **2011**, *5*, 2405–2412.

139. Heffernan, N.; Smyth, T.J.; FitzGerald, R.J.; Vila-Soler, A.; Mendiola, J.A.; Ibáñez, E.; Brunton, N. Comparison of extraction methods for selected carotenoids from macroalgae and the assessment of their seasonal/spatial variation. *Innov. Food Sci. Emerg. Technol.* **2016**, *37*, 221–228. [CrossRef]

140. Tan, H.T.; Khong, N.M.H.; Khaw, Y.S.; Ahmad, S.A.; Yusoff, F.M. Optimization of the Freezing-Thawing Method for Extracting Phycobiliproteins from *Arthrospira* sp. *Molecules* **2020**, *25*, 3894. [CrossRef]

141. Poojary, M.M.; Barba, F.J.; Aliakbarian, B.; Donsì, F.; Pataro, G.; Dias, D.A.; Juliano, P. Innovative Alternative Technologies to Extract Carotenoids from Microalgae and Seaweeds. *Mar. Drugs* **2016**, *14*, 214. [CrossRef] [PubMed]

142. Lourenço-Lopes, C.; Garcia-Oliveira, P.; Carpena, M.; Fraga-Corral, M.; Jimenez-Lopez, C.; Pereira, A.; Prieto, M.; Simal-Gandara, J. Scientific Approaches on Extraction, Purification and Stability for the Commercialization of Fucoxanthin Recovered from Brown Algae. *Foods* **2020**, *9*, 1113. [CrossRef] [PubMed]

143. Hallegraeff, G.M. Pigment Diversity in Freshwater Phytoplankton. II. Summer-succession in Three Dutch Lakes with Different Trophic Characteristics. *Int. Rev. der gesamten Hydrobiol. Hydrogr.* **1977**, *62*, 19–39. [CrossRef]

144. Jeffrey, S.W. Profiles of photosynthetic pigments in the ocean using thin-layer chromatography. *Mar. Biol.* **1974**, *26*, 101–110. [CrossRef]

145. Barlow, R.G.; Mantoura, R.F.C.; Peinert, R.D.; Miller, A.E.J.; Fileman, T.W. Distribution, sedimentation and fate of pigment biomarkers following thermal stratification in the western Alboran Sea. *Mar. Ecol. Prog. Ser.* **1995**, *125*, 279–291. [CrossRef]

146. Wang, Z.; Zhao, J.; Xiao, L. Response of Phytoneuston Community to Organic Nitrogen and Phosphorus Revealed by HPLC-Pigments Method. *J. Ocean Univ. China* **2020**, *19*, 853–864. [CrossRef]

147. Gieskes, W.W.; Kraay, G.W. Analysis of phytoplankton pigments by HPLC before, during and after mass occurrence of the microflagellate *Corymbellus aureus* during the spring bloom in the open northern North Sea in 1983. *Mar. Biol.* **1986**, *92*, 45–52. [CrossRef]

148. Terzic, S.; Ahel, M.; Cauwet, G.; Malej, A. Group-specific phytoplankton biomass/dissolved carbohydrate relationships in the Gulf of Trieste (Northern Adriatic). *Hydrobiologia* **1997**, *363*, 191–205. [CrossRef]

149. Van De Vyver, E.; Van Wichelen, J.; Vanormelingen, P.; Van Nieuwenhuyze, W.; Daveloose, I.; De Jong, R.; De Blok, R.; Urrutia, R.; Tytgat, B.; Verleyen, E.; et al. Variation in phytoplankton pigment composition in relation to mixing conditions in temperate South-Central Chilean lakes. *Limnologica* **2019**, *79*, 125715. [CrossRef]

150. Züllig, H. *Role of Carotenoids in Lake Sediments for Reconstructing Trophic History during the Late Quaternary*; Springer: Dordrecht, The Netherlands, 1990.

151. Jiménez, L.; Romero-Viana, L.; Conde-Porcuna, J.M.; Pérez-Martínez, C. Sedimentary photosynthetic pigments as indicators of climate and watershed perturbations in an alpine lake in southern Spain. *Limnetica* **2015**, *34*, 439–454. [CrossRef]

152. Paanakker, J.E.; Hallegraeff, G.M. A comparative study on the carotenoid pigmentation of the zooplankton of Lake Maarsseveen (Netherlands) and of Lac Pavin (Auvergne, France)—I. Chromatographic characterization of carotenoid pigments. *Comp. Biochem. Physiol. Part B Comp. Biochem.* **1978**, *60*, 51–58. [CrossRef]

153. Quiblier-Lloberas, C.; Bourdier, G.; Amblard, C.; Pepin, D. Impact of grazing on phytoplankton in Lake Pavin (France): Contribution of different zooplankton groups. *J. Plankton Res.* **1996**, *18*, 305–322. [CrossRef]

154. Abidov, M.; Ramazanov, Z.; Seifulla, R.; Grachev, S. The Effects of XanthigenTM in the Weight Management of Obese Premenopausal Women with Non-alcoholic Fatty Liver Disease and Normal Liver Fat. *Diabetes Obes. Metab.* **2010**, *12*, 72–81. [CrossRef]

155. Lai, C.S.; Tsai, M.L.; Badmaev, V.; Jimenez, M.; Ho, C.T.; Pan, M.H. Xanthigen Suppresses Preadipocyte Differentiation and Adipogenesis through Down-Regulation of PPARγ and C/EBPs and Modulation of SIRT-1, AMPK, and FoxO Pathways. *J. Agric. Food Chem.* **2012**, *60*, 1094–1101. [CrossRef] [PubMed]

156. López-Rios, L.; Vega, T.; Chirino, R.; Jung, J.C.; Davis, B.; Pérez-Machín, R.; Wiebe, J.C. Toxicological Assessment of Xanthigen® Nutraceutical Extract Combination: Mutagenicity, Genotoxicity and Oral Toxicity. *Toxicol. Rep.* **2018**, *5*, 1021–1031. [CrossRef]

157. Anand, M.; Suresh, S. Marine seaweed *Sargassum wightii* extract as a low-cost sensitizer for ZnO photoanode based dye-sensitized solar cell. *Adv. Nat. Sci. Nanosci. Nanotechnol.* **2015**, *6*, 035008. [CrossRef]

158. Zhang, T.; Liu, C.; Dong, W.; Wang, W.; Sun, Y.; Chen, X.; Yang, C.; Dai, N. Photoelectrochemical Complexes of Fucoxanthin-Chlorophyll Protein for Bio-Photovoltaic Conversion with a High Open-Circuit Photovoltage. *Chem. Asian J.* **2017**, *12*, 2996–2999. [CrossRef]

159. Rodriguez-Luna, A.; Ávila-Román, J.; González-Rodríguez, M.L.; Cózar, M.J.; Rabasco, A.M.; Motilva, V.; Talero, E. Fucoxanthin-Containing Cream Prevents Epidermal Hyperplasia and UVB-Induced Skin Erythema in Mice. *Mar. Drugs* **2018**, *16*, 378. [CrossRef]

160. Lee, Y.-J.; Nam, G.-W. Sunscreen Boosting Effect by Solid Lipid Nanoparticles-Loaded Fucoxanthin Formulation. *Cosmetics* **2020**, *7*, 14. [CrossRef]

161. Kang, S.Y.; Kang, H.; Lee, J.E.; Jo, C.S.; Moon, C.B.; Ha, J.; Hwang, J.S.; Choi, J. Antiaging Potential of Fucoxanthin Concentrate Derived from *Phaeodactylum tricornutum*. *J. Cosmet. Sci.* **2020**, *71*, 53–64.

162. Sasaki, K.; Ishihara, K.; Yamazaki, M.; Nakashima, K.; Abe, H.; Oyamada, C.; Motoyama, M.; Mitsumoto, M. Oral Administration of Fucoxanthin Increases Plasma Fucoxanthinol Concentration and Antioxidative Ability and Improves Meat Color in Broiler Chicks. *J. Poult. Sci.* **2010**, *47*, 316–320. [CrossRef]

163. Gumus, R.; Gelen, S.U.; Koseoglu, S.; Ozkanlar, S.; Ceylan, Z.; Imik, H. The Effects of Fucoxanthin Dietary Inclusion on the Growth Performance, Antioxidant Metabolism and Meat Quality of Broilers. *Brazilian J. Poult. Sci.* **2018**, *20*, 487–496. [CrossRef]

164. Jensen, A. The Effect of Seaweed Carotenoids on Egg Yolk Coloration. *Poult. Sci.* **1963**, *42*, 912–916. [CrossRef]

165. Prabhasankar, P.; Ganesan, P.; Bhaskar, N.; Hirose, A.; Stephen, N.; Gowda, L.R.; Hosokawa, M.; Miyashita, K. Edible Japanese seaweed, wakame (*Undaria pinnatifida*) as an ingredient in pasta: Chemical, functional and structural evaluation. *Food Chem.* **2009**, *115*, 501–508. [CrossRef]

166. Mok, I.-K.; Yoon, J.-R.; Pan, C.-H.; Kim, S.M. Development, Quantification, Method Validation, and Stability Study of a Novel Fucoxanthin-Fortified Milk. *J. Agric. Food Chem.* **2016**, *64*, 6196–6202. [CrossRef]

167. Mok, I.-K.; Lee, J.K.; Kim, J.H.; Pan, C.-H.; Kim, S.M. Fucoxanthin bioavailability from fucoxanthin-fortified milk: In vivo and in vitro study. *Food Chem.* **2018**, *258*, 79–86. [CrossRef]

168. Yamano, Y.; Tode, C.; Ito, M. Carotenoids and related polyenes. Part 3. First total synthesis of fucoxanthin and halocynthiaxanthin using oxo-metallic catalyst. *J. Chem. Soc. Perkin Trans. 1* **1995**, *15*, 1895–1904. [CrossRef]

169. Seth, K.; Kumar, A.; Rastogi, R.P.; Meena, M.; Vinayak, V. Harish Bioprospecting of fucoxanthin from diatoms—Challenges and perspectives. *Algal Res.* **2021**, *60*, 102475. [CrossRef]

170. Vella, F.M.; Sardo, A.; Gallo, C.; Landi, S.; Fontana, A.; D'Ippolito, G. Annual outdoor cultivation of the diatom *Thalassiosira weissflogii*: Productivity, limits and perspectives. *Algal Res.* **2019**, *42*, 101553. [CrossRef]

171. Arora, N.; Philippidis, G.P. Fucoxanthin Production from Diatoms: Current Advances and Challenges. In *Algae*; Springer: Singapore, 2021; pp. 227–242. [CrossRef]

172. García, J.L.; De Vicente, M.; Galán, B. Microalgae, old sustainable food and fashion nutraceuticals. *Microb. Biotechnol.* **2017**, *10*, 1017–1024. [CrossRef]

173. Ravi, H.; Kurrey, N.; Manabe, Y.; Sugawara, T.; Baskaran, V. Polymeric chitosan-glycolipid nanocarriers for an effective delivery of marine carotenoid fucoxanthin for induction of apoptosis in human colon cancer cells (Caco-2 cells). *Mater. Sci. Eng. C* **2018**, *91*, 785–795. [CrossRef]

174. Koo, S.Y.; Mok, I.-K.; Pan, C.-H.; Kim, S.M. Preparation of Fucoxanthin-Loaded Nanoparticles Composed of Casein and Chitosan with Improved Fucoxanthin Bioavailability. *J. Agric. Food Chem.* **2016**, *64*, 9428–9435. [CrossRef]

175. Lange, K.W.; Hauser, J.; Nakamura, Y.; Kanaya, S. Dietary seaweeds and obesity. *Food Sci. Hum. Wellness* **2015**, *4*, 87–96. [CrossRef]

2.3. Effect of AST on the Histopathology and Ultrastructural Pathology of Prostate Tissues

A histopathological examination, called the "golden standard," is one of the most important and reliable diagnostic methods for the evaluation of pathogenesis. The effect of AST on the histomorphology of ventral prostates in BPH model rats is shown in Figure 3A–F. The histological features found in the BPH model control rats were tall columnar epithelium, simple multifocal epithelial thickening with cellular crowding, and folding of the lining epithelium extending into lumina (Figure 3B). As compared with the BPH model control rats, these features were gradually ameliorated in the AST-treated rats (Figure 3C–E) in a dose-dependent manner, while similar morphological changes were observed in the EPR-treated rats (Figure 3F).

The epithelial thicknesses of ventral prostates in all rats were measured, and the average values of epithelial thickness in all groups are shown in Figure 3G. The average values of epithelial thickness in the AST-treated rats were significantly lower ($p < 0.01$) than in the BPH model control rats, shown in a dose-response manner; the average values of epithelial thickness in the EPR-treated rats also markedly decreased as compared with those in the BPH model control rats ($p < 0.01$).

Secretory granules in ventral prostates were observed using transmission electron microscopy (TEM), as shown in Figure 4. As compared with the BPH model control rats, there were fewer secretory granules observed in the AST-treated rats in a dose-dependent manner.

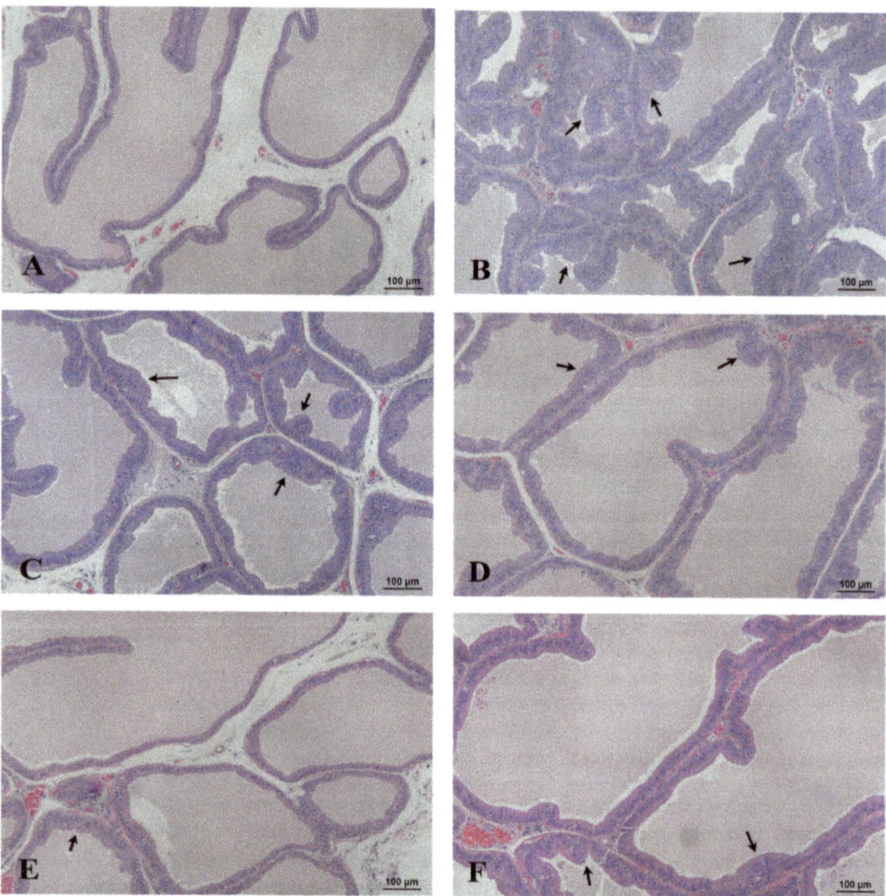

Figure 3. *Cont.*

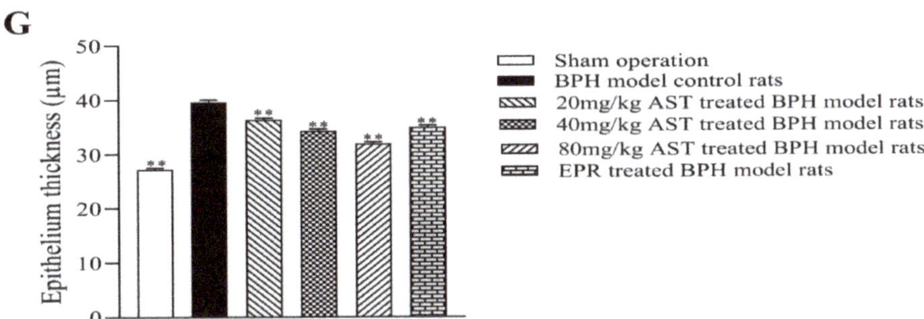

Figure 3. Effect of AST on histomorphology and epithelial thicknesses of ventral prostates in BPH rats. (**A**) No abnormal histological changes were observed in the sham operation; tall columnar epithelium, simple multifocal epithelial thickening with cellular crowding, and folding of the lining epithelium extending into lumina (arrows) were observed in: (**B**) the BPH model control rats; (**C**) 20 mg/kg, (**D**) 40 mg/kg, and (**E**) 80 mg/kg AST-treated BPH rats; (**F**) EPR-treated BPH rats; Hematoxylin and eosin (H&E) 100×; (**G**) epithelial thicknesses of prostates in the sham operation, BPH model control group, and AST or EPR treated groups. Data represent the mean ± SEM. (n = 12). ** Significantly different from BPH model control group, $p < 0.01$.

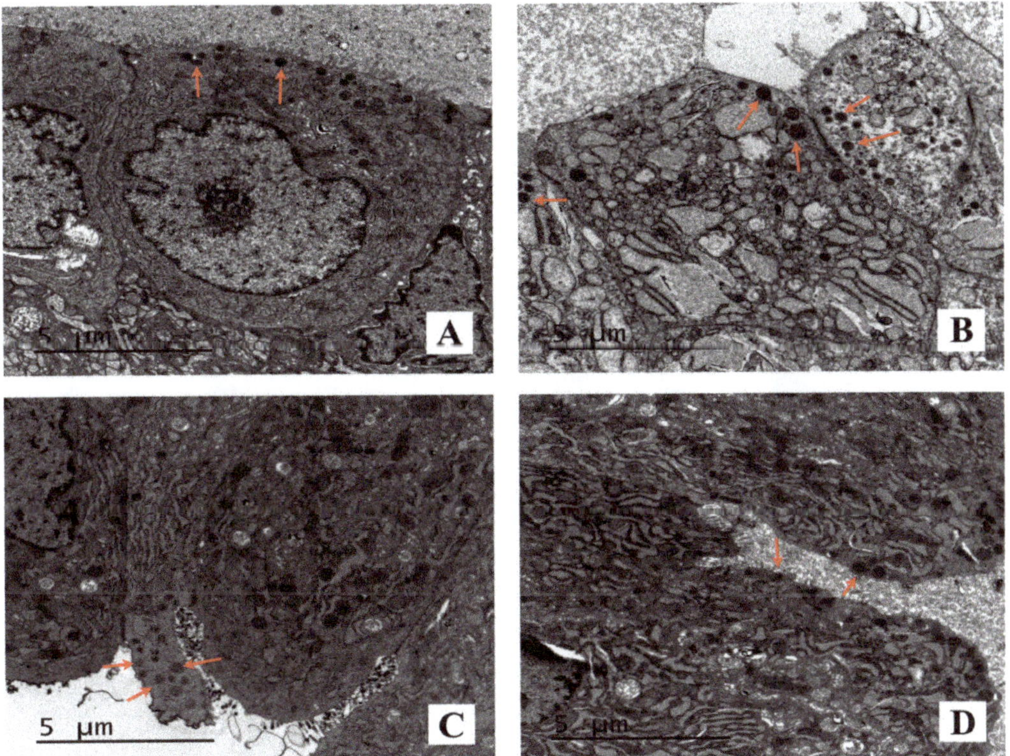

Figure 4. *Cont.*

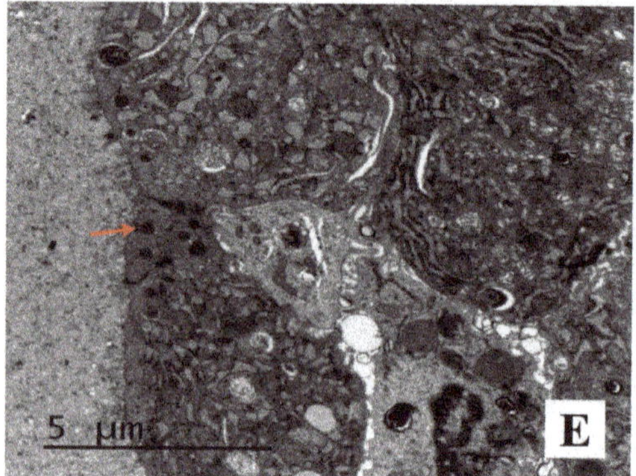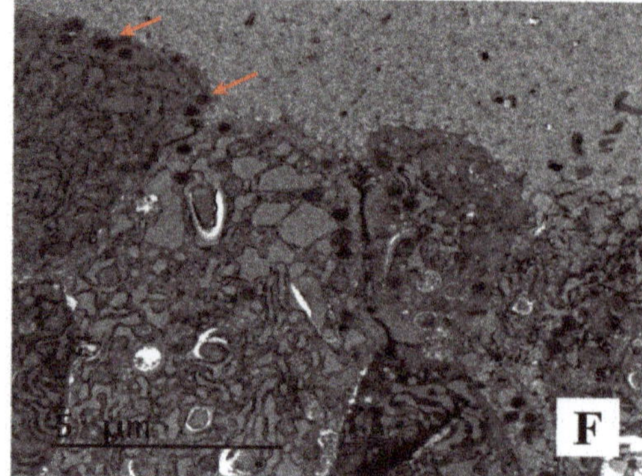

Figure 4. Images of ventral prostates under TEM, 4200×. Secretory granules (arrows) were observed in: (**A**) Sham operation; (**B**) BPH model control rats; (**C**) 20 mg/kg, (**D**) 40 mg/kg, and (**E**) 80 mg/kg AST-treated BPH rats; (**F**) EPR-treated BPH rats.

2.4. Effect of AST on the Superoxide Dismutase (SOD) Activity of Prostates

SOD is an important antioxidant enzyme in organisms. The effect of AST on the SOD activity of ventral prostates in the BPH model rats is shown in Figure 5. As compared with the BPH model control rats, the SOD activity of prostates in the AST-treated rats increased in a dose-dependent manner, while the SOD activity of prostates in the 40 mg/kg ($p < 0.05$) and 80 mg/kg ($p < 0.01$) AST-treated rats increased significantly.

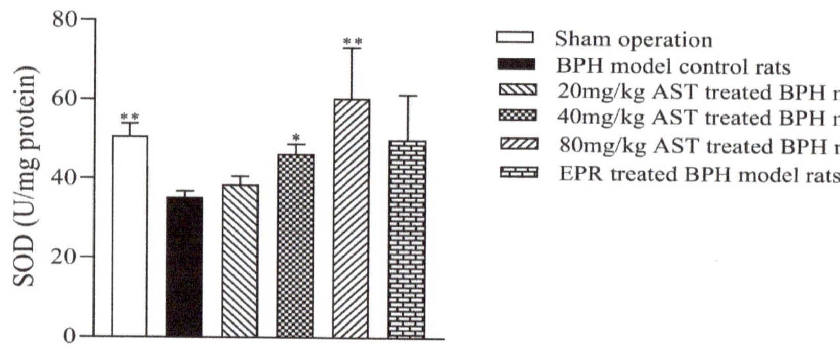

Figure 5. AST increased the SOD activity of ventral prostates. Data represent the mean ± SEM. ($n = 5$). * Significantly different from BPH model control group, $p < 0.05$, and ** significantly different from BPH model control group, $p < 0.01$.

2.5. Effect of AST on the Levels of T and DHT

T and DHT are the primary androgens in the prostate. The levels of T and DHT in ventral prostates in all groups are shown in Figure 6A,B. As compared with the BPH model control rats, the prostate T and DHT levels in the AST-treated rats decreased, while the T levels in the 40 mg/kg and 80 mg/kg AST-treated rats significantly decreased ($p < 0.05$); the DHT levels of prostates in the 40 mg/kg AST-treated rats ($p < 0.05$) also decreased significantly.

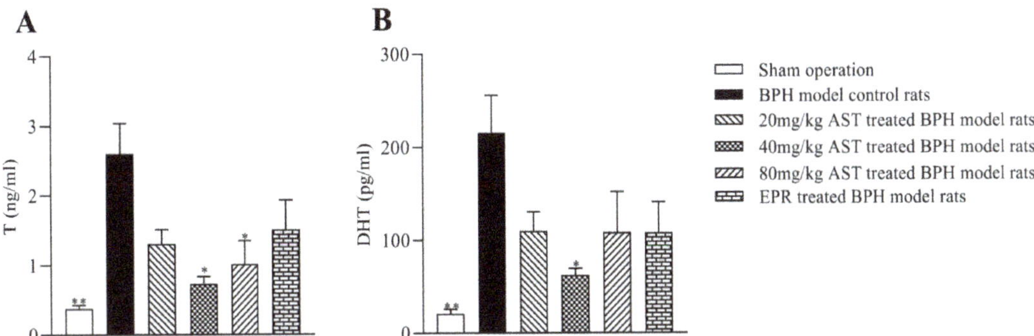

Figure 6. AST decreased levels of T (**A**) and DHT (**B**) in ventral prostates. Data represent the mean ± SEM. (n = 5). * Significantly different from BPH model control group, $p < 0.05$, and ** significantly different from BPH model control group, $p < 0.01$.

3. Discussion

To the best of our knowledge, this is the first study to evaluate the inhibitory effect of AST on BPH in vivo. BPH is characterized by an enlarged prostate. Castration was performed to eliminate the influence of endogenous androgen, and rats were given a 5 mg/kg BW/day dose of exogenous androgen to build the BPH model, considered to be relatively closer to the pathogenesis of clinical BPH, as previously reported [17,18]. In the present study, except for the sham-operated rats, BPH model rats were randomly divided into five groups: BPH model control rats, AST (20 mg/kg, 40 mg/kg, and 80 mg/kg)-treated BPH model control rats, and EPR-treated BPH model control rats. A decline in the prostate and ventral prostate weights and index of prostates and ventral prostates in the AST-treated groups were observed, and a significant decline was found in the 80 mg/kg AST-treated group as compared with the BPH model control rats. The prostate and ventral prostate weights and the prostate index values of prostates and ventral prostates in the 80 mg/kg AST-treated rats were also lower than those in the EPR-treated rats. However, the dorsal prostate weights and dorsal prostate index values of the AST-treated rats did not differ significantly from those of the BPH model control rats. Hence, the results suggest that AST might have a more significant inhibitory effect on ventral prostates than dorsal prostates in T-induced rats.

Histological observation of BPH is described as the proliferation of the stroma and epithelia. After treatment for four weeks, the histomorphology of prostates in the T-induced BPH rats was evaluated. The proliferation characteristics of epithelia such as tall columnar epithelia, simple multifocal epithelial thickening with cellular crowding, and folding of the lining epithelia extending into the lumina were markedly observed in the BPH model control group, which were gradually alleviated following AST treatment. The epithelial thicknesses of prostates in all rats were measured, and secretory granules in the epithelia were also observed under TEM. The epithelial thickness of prostates in the AST-treated rats was markedly lower than that in the BPH model control rats. It was also reported that the secretory granules that release their contents into the gland lumen increased in the T-induced BPH model as compared with the normal control [19,20]. In the present study, fewer secretory granules in the epithelia were observed in the AST-treated rats than in the BPH model control rats. These results indicate that AST might ameliorate the epithelial thicknesses of prostates and decrease the number of secretory granules in epithelia in T-induced BPH rats.

T and DHT are the principal androgens in the prostate, possibly related to the development of BPH [21]. The levels of T and DHT of prostates in the three AST-treated groups decreased as compared with the BPH model control group; a significant decline was found in the T level of prostates in the 40 mg/kg and 80 mg/kg AST-treated rats and the DHT level of prostates in the 40 mg/kg AST-treated rats, albeit not in a dose-response relationship. These results indicated that AST might cause the decrease in T and DHT levels in prostate tissues of T-induced BPH rats.

Oxidative stress, defined as the imbalance between the production and elimination of reactive oxygen species, can be alleviated by antioxidants and is one of several parameters considered to play a pivotal role in the development of BPH [22–24]. SOD, a group of antioxidant enzymes, plays an important role in oxidative stress in cells [25]. Research has suggested that erythrocyte SOD activity was significantly decreased in BPH patients as compared with age- and sex-matched healthy subjects [26]. Moreover, T injection in the BPH model has been shown to weaken cellular antioxidant mechanisms, including a decrease in SOD activity [27–29]. In this study, the SOD activity of prostates in the AST-treated rats increased, while the SOD activity of prostates in the 40 mg/kg and 80 mg/kg AST-treated rats was significantly higher than that in the BPH model control rats. These results demonstrated that AST might inhibit T-induced BPH in rats by regulating SOD activity. Further experiments on the mechanism by which AST regulates SOD activity in the present study will be carried out.

4. Materials and Methods

4.1. Materials

AST was purchased from Shanghai Aladdin Biochemical Technology Co., Ltd. (Shanghai, China) (lot no. F2030151). EPR was manufactured by Jiangsu Lianhuan Pharmaceutical Group Co., Ltd. (Jiangsu, China) (lot no. 20191002). Testosterone propionate was provided by Ningbo Second Hormone Factory (Ningbo, China) (lot no. 2003071). The SOD assay kit was obtained from Nanjing Jiancheng Bioengineering Institute (Nanjing, China) (lot no. 20210515). The T and DHT assay kits were purchased from Cusabio Biotech Co., Ltd. (Wuhan, China) (lot nos. C0307110304 and C0307100303).

4.2. Animals

The 72 male pathogen-free Sprague Dawley rats at 6–7 weeks of age were purchased from the Shanghai Sippr- BK Laboratory Animal Co., Ltd (Shanghai, China). Rats were housed 4 per cage with ad libitum access to water and feed in an environmentally controlled animal room (temperature 24 ± 2 °C, relative humidity 50 ± 10%, 12 h light-dark cycle) at the Shanghai Institute for Biomedical and Pharmaceutical Technologies. All experimental procedures were approved by the Animal Care and Use Committee of Shanghai Institute for Biomedical and Pharmaceutical Technologies (certificate no. 2020-29), and performed in conformity with the National Institutes of Health Guide for the Care and Use of Laboratory Animals.

4.3. Study Design and Treatment

After 5 days of acclimatization, the rats, except for twelve in the sham operation group, were castrated under pentobarbital sodium anesthesia before the experiment to prevent the influence of intrinsic testosterone. Except for the sham-operated rats which were just submitted to anesthesia and testicles exposure, castration was performed by removing testicles with epididymal fat via the scrotal route, as published previously [30].

Twelve normal rats, administrated olive oil orally and injected olive oil subcutaneously daily, served as the sham operation group. The other sixty castrated rats were randomly assigned into five groups (n = 12/group): (A) BPH model control group administrated olive oil orally and testosterone (5 mg/kg/day, s.c.), (B–D) AST groups treated with AST (20, 40 and 80 mg/kg /day, respectively) dissolved in olive oil orally and testosterone (5 mg/kg/day, s.c.), (E) EPR group (positive control group) given EPR (2 mg/kg/day) by oral gavage and testosterone (5 mg/kg/day, s.c.). EPR is used to treat benign prostatic hyperplasia to decrease prostate size and relieve the symptoms [31]. All rats in this experiment were treated once a day for four consecutive weeks. All animals' body weights were measured once a week. At the end of experiment, under pentobarbital sodium anesthesia, prostates of all the rats were removed and weighed. One part of each ventral prostate lobe was fixed in 10% neutral buffered formalin for histopathological examination, while the remainder of each prostate was flash frozen in liquid nitrogen, and then stored at $-80\ ^\circ$C for other analyses.

4.4. Prostate Index

The prostate index was calculated as the ratio of the prostate weight to the rat's body weight.

4.5. Histopathological Examination

Following dehydration, the fixed prostates were embedded in paraffin, cut into 4 μm sections, and then stained with hematoxylin and eosin (H&E). Coverslips were mounted on the microscope slides containing tissue sections using mounting medium. Then, morphological changes were investigated and epithelium thicknesses of prostates were measured under a light microscope.

4.6. TEM

The ventral prostates were washed with normal saline solution and cut into pieces of about 1 mm^3 on ice, and fixed with a 2.5% glutaraldehyde solution for 2 h at 4 $^\circ$C. Following three washes with 0.1 M PBS, the tissues were postfixed in 1% osmium tetroxide for 2 h. After dehydration in ascending concentrations of ethanol and acetone, the tissues were embedded in Epon 618. Thin 70 nm sections stained with lead citrate were examined under a transmission electron microscope.

4.7. Analysis of SOD Activity

The frozen prostate tissue was homogenized in cold normal saline solution. The homogenate was centrifuged at 1520× *g* for 10 min at 4 $^\circ$C. Then, the supernatants were collected, and the Bradford protein assay (Bradford assay) was used to quantify the protein amount. The SOD activity was assessed strictly according to the manufacturer's protocol, as previously reported [32–34].

4.8. Analysis of Levels of T and DHT

The frozen prostate tissue was homogenized in cold normal saline solution. After two freeze-thaw cycles, the homogenate was centrifuged at 5000× *g* for 5 min at 4 $^\circ$C. The supernatants were collected, while protein amounts were quantified by Bradford assay. Levels of T and DHT were assessed strictly according to the manufacturer's protocol, as previously reported [35–37].

4.9. Statistical Analysis

Data obtained from this experiment are presented as means ± standard error of the mean (SEM). When equal variance was assumed, the data were evaluated using one-way analysis of variance (ANOVA) followed by post hoc LSD for multiple comparisons. When equal variance was not assumed, the data were analyzed by the non-parametric Kruskal–Wallis test with the Mann–Whitney U test for multiple comparisons. A significant difference was defined as $p < 0.05$ or $p < 0.01$.

5. Conclusions

The results of this study indicated that AST might have an inhibitory effect on T-induced BPH, especially in the ventral prostate, possibly due to the regulation of SOD activity and levels of T and DHT. Hence, AST may be a candidate as a novel therapeutic agent to inhibit BPH.

Author Contributions: L.W. and Z.S. designed the study; L.W., Y.H., R.W., Q.P., D.L. and H.Y. performed experiments; L.W. analyzed the data and wrote the paper; L.W. and Z.S. revised the paper. All authors have read and agreed to the published version of the manuscript.

Funding: This work was supported by funding from the Shanghai Institute for Biomedical and Pharmaceutical Technologies (no. Q2019-7).

Acknowledgments: We thank Congcong Shao, Xin Su, and Rongfu Yang for assistance in the process of castration.

Conflicts of Interest: The authors have declared no conflict of interest.

Abbreviations

astaxanthin (AST); benign prostatic hyperplasia (BPH); dihydrotestosterone (DHT); episteride (EPR); hematoxylin and eosin (H&E); superoxide dismutase (SOD); testosterone (T); transmission electron microscopy (TEM).

References

1. Chughtai, B.; Forde, J.C.; Thomas, D.D.; Laor, L.; Hossack, T.; Woo, H.H.; Te, A.E.; Kaplan, S.A. Benign prostatic hyperplasia. *Nat. Rev. Dis. Primers* **2016**, *2*, 16031. [CrossRef] [PubMed]
2. Bae, W.J.; Park, H.J.; Koo, H.C.; Kim, D.R.; Ha, U.S.; Kim, K.S.; Kim, S.J.; Cho, H.J.; Hong, S.H.; Lee, J.Y.; et al. The effect of seoritae extract in men with mild to moderate lower urinary tract symptoms suggestive of benign prostatic hyperplasia. *Evid. Based Compl. Alt.* **2016**, *2016*, 1960926. [CrossRef]
3. Sun, Z.; Li, Y.; Liu, X.; Wu, J.; Zhou, L. *Prostate Pharmacology*; Shanghai Scientific and Technical Publishers: Shanghai, China, 2013; pp. 191–206. (In Chinese)
4. Thomas, D.; Chughtai, B.; Kini, M.; Te, A. Emerging drugs for the treatment of benign prostatic hyperplasia. *Expert Opin. Emerg. Dr.* **2017**, *22*, 201–212. [CrossRef]
5. Fakhri, S.; Abbaszadeh, F.; Dargahi, L.; Jorjani, M. Astaxanthin: A mechanistic review on its biological activities and health benefits. *Pharmacol. Res.* **2018**, *136*, 1–20. [CrossRef] [PubMed]
6. McCall, B.; McPartland, C.K.; Moore, R.; Frank-Kamenetskii, A.; Booth, B.W. Effects of astaxanthin on the proliferation and migration of breast cancer cells in vitro. *Antioxidants* **2018**, *7*, 135. [CrossRef]
7. Chang, M.X.; Xiong, F. Astaxanthin and its Effects in inflammatory responses and inflammation-associated diseases: Recent advances and future directions. *Molecules* **2020**, *25*, 5342. [CrossRef]
8. Kumar, A.; Dhaliwal, N.; Dhaliwal, J.; Dharavath, R.N.; Chopra, K. Astaxanthin attenuates oxidative stress and inflammatory responses in complete Freund-adjuvant-induced arthritis in rats. *Pharmacol. Rep.* **2020**, *72*, 104–114. [CrossRef]
9. Gao, F.; Wu, X.; Mao, X.; Niu, F.; Zhang, B.; Dong, J.; Liu, B. Astaxanthin provides neuroprotection in an experimental model of traumatic brain injury via the Nrf2/HO-1 pathway. *Am. J. Transl. Res.* **2021**, *13*, 1483–1493.
10. Ni, X.; Yu, H.; Wang, S.; Zhang, C.; Shen, S. Astaxanthin inhibits PC-3 xenograft prostate tumor growth in nude mice. *Mar. Drugs* **2017**, *15*, 66. [CrossRef] [PubMed]
11. Buesen, R.; Schulte, S.; Strauss, V.; Treumann, S.; Becker, M.; Groters, S.; Carvalho, S.; van Ravenzwaay, B. Safety assessment of [3S, 3'S]-astaxanthin–Subchronic toxicity study in rats. *Food Chem. Toxicol.* **2015**, *81*, 129–136. [CrossRef]
12. Creasy, D.; Bube, A.; de Rijk, E.; Kandori, H.; Kuwahara, M.; Masson, R.; Nolte, T.; Reams, R.; Regan, K.; Rehm, S.; et al. Proliferative and nonproliferative lesions of the rat and mouse male reproductive system. *Toxicol. Pathol.* **2012**, *40* (Suppl. S6), 40s–121s. [CrossRef]
13. Ginja, M.; Pires, M.J.; Gonzalo-Orden, J.M.; Seixas, F.; Correia-Cardoso, M. Anatomy and imaging of rat prostate: Practical monitoring in experimental cancer-induced protocols. *Diagnostics* **2019**, *9*, 68. [CrossRef]
14. Madersbacher, S.; Sampson, N.; Culig, Z. Pathophysiology of benign prostatic hyperplasia and benign prostatic enlargement: A mini-review. *Gerontology* **2019**, *65*, 458–464. [CrossRef]
15. Anderson, M.L. A preliminary investigation of the enzymatic inhibition of 5α-reductase and growth of prostatic carcinoma cell line LNCap-FGC by natural astaxanthin and saw palmetto lipid extract in vitro. *J. Herb. Pharmacother.* **2009**, *5*, 17–26. [CrossRef]
16. Sun, S.Q.; Zhao, Y.X.; Li, S.Y.; Qiang, J.W.; Ji, Y.Z. Anti-tumor effects of astaxanthin by inhibition of the expression of STAT3 in prostate cancer. *Mar. Drugs* **2020**, *18*, 415. [CrossRef] [PubMed]
17. Liu, J.; Fang, T.; Li, M.; Song, Y.; Li, J.; Xue, Z.; Li, J.; Bu, D.; Liu, W.; Zeng, Q.; et al. Pao pereira extract attenuates testosterone-induced benign prostatic hyperplasia in rats by inhibiting 5alpha-reductase. *Sci. Rep.* **2019**, *9*, 19703. [CrossRef]
18. Wang, C.T.; Wang, Y.Y.; Liu, W.S.; Cheng, C.M.; Chiu, K.H.; Liu, L.L.; Liu, X.Z. Rhodobacter sphaeroides extract lycogen™ attenuates testosterone-induced benign prostate hyperplasia in Rats. *Int. J. Mol. Sci.* **2018**, *19*, 1137. [CrossRef] [PubMed]
19. Wang, W. The ultrastructure of the rat ventral prostate. *Acta Acad. Med. Qingdao* **1986**, *22*, 22–28. (In Chinese)
20. He, D.; Mi, J.; Wang, X.; Zheng, L.; Shi, Z. The histological and ultrastructural observation of experimental prostatic hyperplasia dogs. *Chin. J. Vet. Med.* **2009**, *45*, 52–53. (In Chinese)
21. Wang, Y.R.; Xu, Y.; Jiang, Z.Z.; Zhang, L.Y.; Wang, T. Triptolide reduces prostate size and androgen level on testosterone-induced benign prostatic hyperplasia in Sprague Dawley rats. *Chin. J. Nat. Med.* **2017**, *15*, 341–346. [CrossRef]

22. Minciullo, P.L.; Inferrera, A.; Navarra, M.; Calapai, G.; Magno, C.; Gangemi, S. Oxidative stress in benign prostatic hyperplasia: A systematic review. *Urol. Int.* **2015**, *94*, 249–254. [CrossRef]

23. Zabaiou, N.; Mabed, D.; Lobaccaro, J.M.; Lahouel, M. Oxidative stress in benign prostate hyperplasia. *Andrologia* **2016**, *48*, 69–73. [CrossRef] [PubMed]

24. Roumeguere, T.; Sfeir, J.; El Rassy, E.; Albisinni, S.; Van Antwerpen, P.; Boudjeltia, K.Z.; Fares, N.; Kattan, J.; Aoun, F. Oxidative stress and prostatic diseases. *Mol. Clin. Oncol.* **2017**, *7*, 723–728. [CrossRef]

25. Ismy, J.; Sugandi, S.; Rachmadi, D.; Hardjowijoto, S.; Mustafa, A. The effect of exogenous superoxide dismutase (SOD) on caspase-3 activation and apoptosis induction in Pc-3 prostate cancer cells. *Res. Rep. Urol.* **2020**, *12*, 503–508. [CrossRef]

26. Aydin, A.; Arsova-Sarafinovska, Z.; Sayal, A.; Eken, A.; Erdem, O.; Erten, K.; Ozgok, Y.; Dimovski, A. Oxidative stress and antioxidant status in non-metastatic prostate cancer and benign prostatic hyperplasia. *Clin. Biochem.* **2006**, *39*, 176–179. [CrossRef] [PubMed]

27. El-Sherbiny, M.; El-Shafey, M.; El-Din El-Agawy, M.S.; Mohamed, A.S.; Eisa, N.H.; Elsherbiny, N.M. Diacerein ameliorates testosterone-induced benign prostatic hyperplasia in rats: Effect on oxidative stress, inflammation and apoptosis. *Int. Immunopharmacol.* **2021**, *100*, 108082. [CrossRef] [PubMed]

28. Shoieb, S.M.; Esmat, A.; Khalifa, A.E.; Abdel-Naim, A.B. Chrysin attenuates testosterone-induced benign prostate hyperplasia in rats. *Food Chem. Toxicol.* **2018**, *111*, 650–659. [CrossRef]

29. Abo-Youssef, A.M.; Afify, H.; Azouz, A.A.; Abdel-Rahman, H.M.; Abdel-Naim, A.B.; Allam, S. Febuxostat attenuates testosterone-induced benign prostatic hyperplasia in rats via inhibiting JAK/STAT axis. *Life Sci.* **2020**, *260*, 118414. [CrossRef] [PubMed]

30. Coppenolle, F.V.; Bourhis, X.L.; Carpentier, F.O.; Delaby, G.; Cousse, H.; Raynaud, J.-P.; Dupouy, J.-P.; Prevarskaya, N. Pharmacological effects of the lipidosterolic extract of serenoa repens (permixonT) on ratprostate hyperplasia induced by hyperprolactinemia: Comparison with finasteride. *Prostate* **2000**, *43*, 49–58. [CrossRef]

31. Sun, Z.Y.; Wu, H.Y.; Wang, M.Y.; Tu, Z.H. The mechanism of episteride against benign prostatic hyperplasia. *Eur. J. Pharmacol.* **1999**, *371*, 227–233. [CrossRef]

32. Mao, G.X.; Zheng, L.D.; Cao, Y.B.; Chen, Z.M.; Lv, Y.D.; Wang, Y.Z.; Hu, X.L.; Wang, G.F.; Yan, J. Antiaging effect of pine pollen in human diploid fibroblasts and in a mouse model induced by D-galactose. *Oxid. Med. Cell. Longev.* **2012**, *2012*, 750963. [CrossRef]

33. Liu, L.; Liu, Y.; Cui, J.; Liu, H.; Liu, Y.B.; Qiao, W.L.; Sun, H.; Yan, C.D. Oxidative stress induces gastric submucosal arteriolar dysfunction in the elderly. *World J. Gastroenterol.* **2013**, *19*, 9439–9446. [CrossRef]

34. Du, J.; Yin, G.; Hu, Y.; Shi, S.; Jiang, J.; Song, X.; Zhang, Z.; Wei, Z.; Tang, C.; Lyu, H. Coicis semen protects against focal cerebral ischemia-reperfusion injury by inhibiting oxidative stress and promoting angiogenesis via the TGFβ/ALK1/Smad1/5 signaling pathway. *Aging* **2020**, *13*, 877–893. [CrossRef] [PubMed]

35. Moustafa, A. Changes in nitric oxide, carbon monoxide, hydrogen sulfide and male reproductive hormones in response to chronic restraint stress in rats. *Free Radic. Biol. Med.* **2021**, *162*, 353–366. [CrossRef]

36. Liu, Q.; Yu, W.; Fan, S.; Zhuang, H.; Han, Y.; Zhang, H.; Yuan, Z.; Weng, Q. Seasonal expressions of androgen receptor, estrogen receptors, 5α-reductases and P450arom in the epididymis of the male muskrat (Ondatra zibethicus). *J. Steroid Biochem. Mol. Biol.* **2019**, *194*, 105433. [CrossRef] [PubMed]

37. Baig, M.S.; Kolasa-Wołosiuk, A.; Pilutin, A. Finasteride-induced inhibition of 5α-reductase type 2 could lead to kidney damage-animal, experimental study. *Int. J. Environ. Res. Public Health* **2019**, *16*, 1726. [CrossRef] [PubMed]

Article

Fucoxanthin Pretreatment Ameliorates Visible Light-Induced Phagocytosis Disruption of RPE Cells under a Lipid-Rich Environment via the Nrf2 Pathway

Yunjun Liu [1], Zixin Guo [1], Shengnan Wang [1], Yixiang Liu [1,2,*] and Ying Wei [3,*]

[1] College of Ocean Food and Biological Engineering, Jimei University, Xiamen 361021, China; lyj28528x@163.com (Y.L.); zzxxgguo@163.com (Z.G.); wsnynlfighting@163.com (S.W.)

[2] Collaborative Innovation Center of Provincial and Ministerial Co-Construction for Marine Food Deep Processing, Dalian Polytechnic University, Dalian 116034, China

[3] The Department of Food Engineering, China National Research Institute of Food & Fermentation Industries Corporation Limited, Beijing 100015, China

* Correspondence: lyxjmu@jmu.edu.cn (Y.L.); proudwy@126.com (Y.W.); Tel.: +86-0592-618-1915 (Y.L.); +86-0105-321-8288 (Y.W.)

Abstract: Fucoxanthin, a special xanthophyll derived from marine algae, has increasingly attracted attention due to its diverse biological functions. However, reports on its ocular benefits are still limited. In this work, the ameliorative effect of fucoxanthin on visible light and lipid peroxidation-induced phagocytosis disruption in retinal pigment epithelium (RPE) cells was investigated in vitro. Marked oxidative stress, inflammation, and phagocytosis disruption were evident in differentiated RPE cells following their exposure to visible light under a docosahexaenoic acid (DHA)-rich environment. Following pretreatment with fucoxanthin, however, the activated nuclear factor erythroid-derived-2-like 2 (Nrf2) signaling pathway was observed and, furthermore, when the fucoxanthin -pretreated RPE cells were irradiated with visible light, intracellular reactive oxygen species (ROS), malondialdehyde (MDA) levels and inflammation were obviously suppressed, while phagocytosis was significantly improved. However, following the addition of Nrf2 inhibitor ML385, the fucoxanthin exhibited no ameliorative effects on the oxidative stress, inflammation, and phagocytosis disruption in the RPE cells, thus indicating that the ameliorative effect of fucoxanthin on the phagocytosis of RPE cells is closely related to the Nrf2 signaling pathway. In conclusion, these results suggest that fucoxanthin supplementation might be beneficial to the prevention of visible light-induced retinal injury.

Keywords: fucoxanthin; RPE cells; phagocytosis; Nrf2; visible light; docosahexaenoic acid

Citation: Liu, Y.; Guo, Z.; Wang, S.; Liu, Y.; Wei, Y. Fucoxanthin Pretreatment Ameliorates Visible Light-Induced Phagocytosis Disruption of RPE Cells under a Lipid-Rich Environment via the Nrf2 Pathway. *Mar. Drugs* **2022**, *20*, 15. https://doi.org/10.3390/md20010015

Academic Editor: Hayato Maeda

Received: 26 November 2021
Accepted: 21 December 2021
Published: 23 December 2021

Publisher's Note: MDPI stays neutral with regard to jurisdictional claims in published maps and institutional affiliations.

1. Introduction

The proliferation of electronic devices (such as computers, widescreen phones, and televisions) and diverse lighting products has led to a dramatic increase in the incidences of light-induced photochemical eye damage and has become a major cause of visual health problems in modern society [1,2]. As the site of visual imaging and visible light focusing, retinal tissue is a vulnerable target for photochemical damage [3]. Not only does the retina metabolize and function under hyperoxic conditions, but the outer segments of its photoreceptors are also rich in photosensitizer molecules and polyunsaturated fatty acids (PUFAs) [4,5]. When the eyes are exposed to natural or artificial light sources for long periods of time, retinas become vulnerable to oxidative stress, thus greatly increasing the risk of retinopathy [6–8]. It is believed that excessive illumination can result in photoreceptor apoptosis, disturbance in the blood-retinal barrier, and inflammatory infiltration in the retina [9,10]. It has also been suggested that light-induced retinal injury can initiate age-related macular degeneration (AMD), which is a major cause of vision deterioration and blindness in the elderly [4,11]. Consequently, the prevention of retinal photo-oxidative damage through dietary nutritional supplementation has become a significant research focus for food scientists and nutritionists.

As one of the main targets of retinal photo-oxidative damage, and due to their important role in maintaining the physiological function of the retina, retinal pigment epithelium (RPE) cells and the potential to improve vision by maintaining their health through dietary supplementation have garnered increasing scientific attention. There is considerable evidence to confirm that RPE cells act as part of the outer blood-retinal barrier (BRB), controlling the exchange of nutrients and waste products between choroidal vessels and photoreceptor cells, and thus supporting the survival and normal functioning of photoreceptor cells [12]. In fact, the recycling of the photoreceptor outer segments (POSs) damaged by oxidation is completed via the phagocytosis of RPE cells [13]. Lipid peroxidation of PUFAs in the POSs can subject RPE cells to intense oxidative stress [4] and, therefore, the inhibition of photo-oxidative damage in RPE cells through dietary antioxidants is increasingly emphasized in research. In one study, berry-derived anthocyanins were reported to efficiently scavenge intracellular reactive oxide species (ROS) and down-regulate the expression of vascular endothelial growth factor (VEGF) in RPE cells under visible light exposure [14]. In another, epigallocatechin-3-gallate, a polyphenolic compound found in green tea, displayed a regulatory role in ultraviolet light irradiation-induced autophagy in RPE cells [15], while quercetin-3-O-α-L-arabinopyranoside reportedly exhibited an inhibitory effect on blue light-induced cell apoptosis and inflammation in RPE cells [16]. However, few studies have investigated dietary active ingredients to ameliorate visible light and lipid peroxidation-induced phagocytic disorder in RPE cells. In addition, while the RPE is a polarized monolayer of highly differentiated epithelial cells in vivo, current in vitro RPE culture models are unable to preserve many of their specific properties or to reproduce the functional features and gene expression patterns that RPE exhibits in vivo [17].

Nuclear factor erythroid-derived 2-like 2 (Nrf2), also known as nuclear factor erythroid 2-related factor 2, plays an important role as a main antioxidant pathway in a variety of diseases [18]. Under oxidative or other stress conditions, Nrf2 in the cytoplasm is translocated to the nucleus, thus regulating antioxidant response element (ARE)-mediated phase II detoxification and the expression of antioxidant proteins/enzymes, including glutamate-cysteine ligase (GCL), heme oxygenase-1 (HO-1), and NAD(P)H: quinone oxidoreductase (NQO1) [19,20]. There is growing evidence that the inhibition of Nrf2 signaling pathway activation further aggravates oxidative damage in cells [21]. It was recently reported that, after pigmented rabbits were exposed to visible light, the oral administration of dietary polyphenols could reduce light-induced retinal oxidative stress and further up-regulate the expression level of HO-1 mRNA [22]. Therefore, the regulation of Nrf2, as an upstream signaling molecule, would make it an attractive candidate gene as a regulatory target for improving phagocytosis in RPE cells during photo-oxidative stress.

Fucoxanthin, a special xanthophyll derived from edible brown seaweeds and some microalgae, has attracted attention due to its biological functions and unique structural properties, including epoxide, allenic, and acetyl groups [23,24]. In our earlier studies, fucoxanthin supplementation was found to provide comparatively superior performance to lutein in protecting the retina against visible light-induced damage, both in vitro and in vivo [23]. However, the mechanisms by which fucoxanthin ameliorates visible light-induced retinal damage have not yet been demonstrated. Accordingly, the aim of this study is to investigate the preventive effect of fucoxanthin on visible light-induced phagocytic dysfunction of RPE cells and the underlying mechanisms. In order to better mimic the tissue properties of pigment epithelium, differentiated RPE cell monolayers were employed in the construction of in vitro evaluation models. Additionally, as the most abundant polyunsaturated fatty acid (PUFA) in the POSs, docosahexaenoic acid (DHA) was used to create the lipid-rich environment to which the RPE cells are exposed in vivo. Under this in vitro model, the effects of the intensity and duration of visible light exposure on oxidative damage and phagocytic function of RPE cells were observed. Subsequently, the mechanism by which fucoxanthin improves the phagocytosis of RPE cells through the Nrf2 signaling pathway was elucidated. This study, thus, provides a theoretical basis for the use of marine fucoxanthin to prevent visual impairment caused by prolonged light exposure.

2. Results

2.1. Cytotoxicity of Fucoxanthin to RPE Cells in a Lipid-Rich Environment

The cytotoxicity of fucoxanthin and lutein to RPE cells in a high-lipid environment was evaluated, with the results shown in Figure 1. When the medium contained 25.0 μmol/L of DHA, a slight proliferative effect was observed on the RPE cells after 24 h incubation, with no statistically significant difference between the fucoxanthin and lutein. It was therefore clear that under the condition of 25.0 μmol/L DHA, no obvious ($p < 0.05$) neither fucoxanthin nor lutein exerted a significant inhibitory effect on RPE cells' proliferation when their dosages ≤50.0 μg/mL. Compared with the control group, there was also no significant difference ($p < 0.05$) in the lactic dehydrogenase (LDH) expression in all treatment groups. The LDH analysis further confirmed that 25.0 μmol/L DHA plus ≤50.0 μg/mL fucoxanthin or lutein exerted no cytotoxicity on the RPE cells.

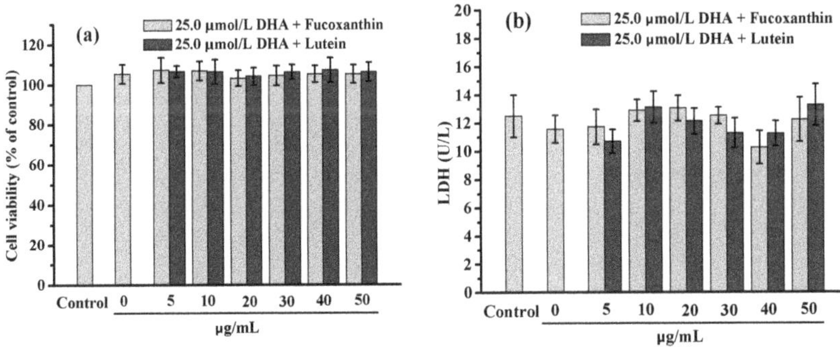

Figure 1. Cytotoxicity of fucoxanthin and lutein on RPE cells under a lipid-rich environment: (**a**) cell viability; (**b**) lactic dehydrogenase (LDH) levels. The DHA concentration was 25.0 μmol/L.

2.2. Effects of Visible Light Exposure on Oxidative Damage in Differentiated RPE Cells

The effects of different light exposure conditions on the oxidative damage to and phagocytic function of RPE cells were investigated, the results of which are presented in Figure 2. When the differentiated RPE cells were irradiated with 1500 lux visible light for 6–24 h, no obvious ($p < 0.05$) oxidative damage, inflammation, or phagocytic dysfunction were observed. However, when the light intensity was increased to 3500 or 5000 lux, visible light-induced injury was induced with prolonged light exposure. After 24 h light exposure under 3500 lux, the intracellular ROS and malondialdehyde (MDA) level had increased by approximately 1.46 and 1.05 times, respectively, while superoxide dismutase (SOD) activity had decreased by 29.85%. Compared with the control group, the inflammatory cytokines interleukin (IL) 6 (351.13 ± 25.44 pg/mL of control) and tumor necrosis factor-α (TNF-α) (114.13 ± 5.39 pg/mL of control) in the media were found to have increased to 632.36 ± 22.39 pg/mL and 432.36 ± 21.35 pg/mL, respectively, and a 37.64% loss was simultaneously observed in the phagocytic index. It is clear that under the more intense light exposure conditions and consequently more extensive oxidative damage, the phagocytic function of the RPE cells was further degraded. For example, when the RPE cells were exposed to light irradiation at 5000 lux for 24 h, their phagocytic index remained only 45.34 ± 5.68%. Considering the degree of oxidative damage and degradation of the phagocytic function of RPE cells, light exposure at 3500 lux for 24 h was employed in the subsequent experiments.

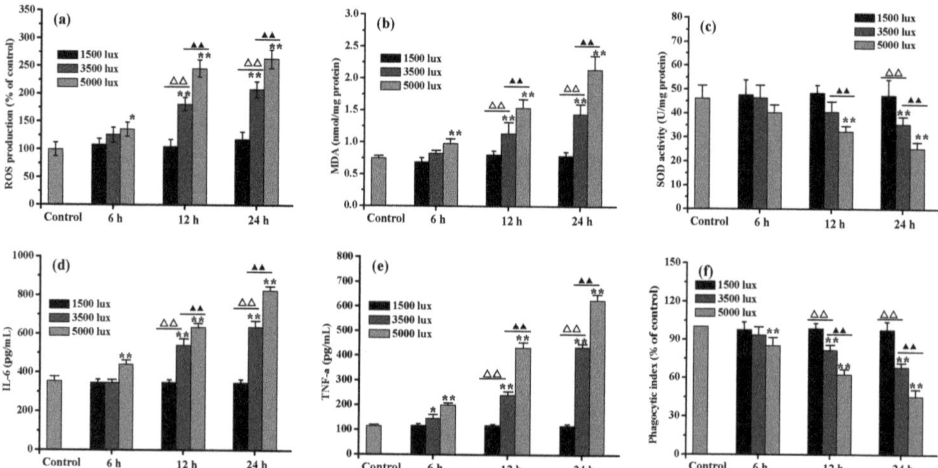

Figure 2. The effects of visible light exposure on oxidative damage, inflammation and phagocytosis in differentiated RPE cells: (**a**) intracellular ROS levels; (**b**) intracellular MDA levels; (**c**) SOD activity; (**d**) inflammatory factor IL-6; (**e**) inflammatory factor TNF-α; (**f**) phagocytic index. (* $p < 0.05$ and ** $p < 0.01$ vs. control; $^{\triangle\triangle}$ $p < 0.01$ means 1500 vs. 3500 lux; $^{\blacktriangle\blacktriangle}$ $p < 0.01$ means 3500 vs. 5000 lux).

2.3. Fucoxanthin Pretreatment Activated the Nrf2 Signal Pathway in RPE Cells

In this study, the results showed that adequate pretreatment time with fucoxanthin was beneficial in activating the antioxidant system of the RPE cells. As shown in Figure 3, after RPE cells were pretreated by fucoxanthin for 6–24 h, the expressions of nuclear Nrf2 (Nucl-Nrf2) protein and its regulated downstream antioxidant proteins or detoxification enzymes were investigated. When the RPE cells were co-incubated with 20.0 μmol/L fucoxanthin for 6 and 12 h (Figure 3a), it was found that, compared with the control group, Nucl-Nrf2 activity was increased by approximately 1.28 and 1.48 times, respectively. However, there was no further significant ($p < 0.05$) increase in Nucl-Nrf2 expression when the fucoxanthin pretreatment time was extended to 24 h. Interestingly, lutein, which is also a xanthophyll, performed worse ($p < 0.01$) than fucoxanthin in activating Nucl-Nrf2. After 12 h pretreatment, the Nucl-Nrf2 level in the lutein group had only increased 1.21 times. As shown in Figure 3b–e, similar phenomena were observed in the antioxidant proteins, including glutamate-cysteine ligase catalytic subunit (GCLC), glutathione peroxidase (GPx), thioredoxin reductase (TrxR), and HO-1, as well as the detoxification enzyme NQO1. The above results, thus, indicate that 12 h fucoxanthin pretreatment was sufficient to effectively activate the Nrf2 signaling pathway in the differentiated RPE cells.

2.4. Fucoxanthin Attenuated Visible Light-Induced Oxidative Stress and Phagocytosis Disorder in RPE Cells

Following 12 h pretreatment with fucoxanthin, the differentiated RPE cells were subjected to visible light exposure at 3500 lux for 24 h. As shown in Figure 4, when the concentration of fucoxanthin was 5.0 μmol/L, both the intracellular ROS and MDA levels began to be effectively ($p < 0.01$) inhibited, although no obvious ($p < 0.05$) ameliorative effects on inflammation or phagocytosis were evident. However, when the fucoxanthin concentrations were further increased to 10.0 or 20.0 μmol/L, the oxidative stress and inflammatory response in the RPE cells were significantly ($p < 0.01$) ameliorated. Under the condition of 20.0 μmol/L fucoxanthin, the intracellular ROS level decreased from 208.24 ± 8.56% (model group) to 118.56 ± 7.68%, and, compared with the model group, the levels of MDA, IL-6 and TNF-α were reduced by 44.87, 39.16, and 63.89%, respectively. Simultaneously, the phagocytic index of RPE cells recovered from 62.36 ± 4.15% to 89.56 ± 6.36%.

2.5. Fucoxanthin Protected against Phagocytosis Disorder of RPE Cells via the Nrf2-Mediated Pathway

In order to further elucidate the mechanism by which fucoxanthin provides protection against visible light-induced phagocytosis disorder of RPE cells, the Nrf2-mediated signaling pathway was investigated. As shown in Figure 5a, when the RPE cells were irradiated with visible light, the expression levels of Nucl-Nrf2 and its regulated NQO1 and HO-1 were slightly increased compared with those of the control group. Interestingly, when the RPE cells were administered with 20.0 μmol/L fucoxanthin (12 h pretreatment time plus 24 h light exposure time), the levels of Nucl-Nrf2, NQO1, and HO-1 were increased by approximately 1.64, 1.54, and 1.78, respectively, compared with those of the control group. Correspondingly, the intracellular ROS level was close to that of the control, while the TNF-α was reduced by 64.21%, and the phagocytic index recovered to approximately 86.04%. However, when the Nrf2 inhibitor ML385 was added at the same time as the fucoxanthin pretreatment of the RPE cells, the levels of Nucl-Nrf2, NQO1, and HO-1 after the light exposure were only 24.56, 26.21, and 21.16%, respectively, of those of the control group. Moreover, under the same ML385 pretreatment condition, the fucoxanthin did not exhibit ameliorative effects on the oxidative stress, inflammatory response, or phagocytosis disruption in the RPE cells. As shown in Figure 5b–d, there were obvious ($p < 0.05$) differences between the fucoxanthin +ML385 group and the model group at the intracellular ROS level, inflammatory factor TNF-α level, and the phagocytic index. Thus, it is clear that ML385 can effectively block activation of the Nrf2 signaling pathway, resulting in the inability of fucoxanthin to improve the phagocytosis of RPE cells.

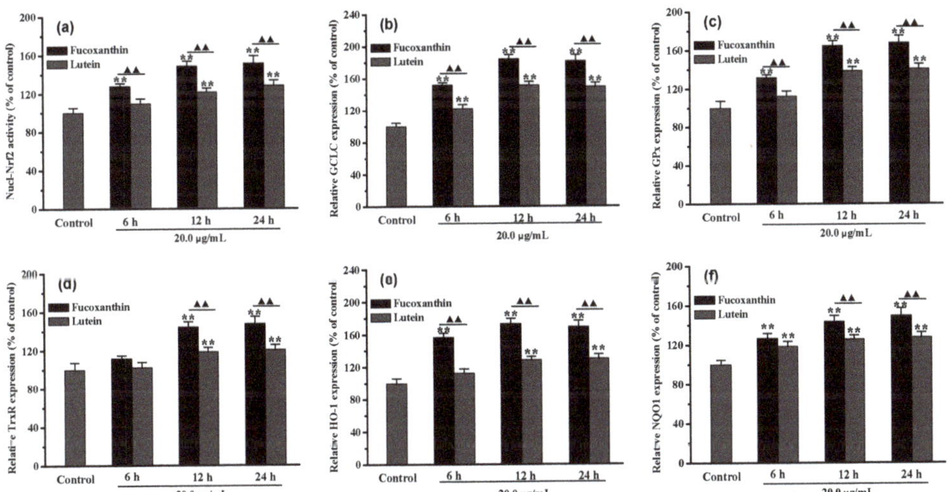

Figure 3. The effect of pretreatment time on Nrf2 signaling pathway activated by fucoxanthin in differentiated RPE cells. (**a**) Nucl-Nrf2 activity; (**b**) GCLC expression level; (**c**) GPx expression level; (**d**) TrxR expression level; (**e**) HO-1 expression level; (**f**) NQO1 expression level. (** $p < 0.01$ vs. control; ▲▲ $p < 0.01$ vs. lutein).

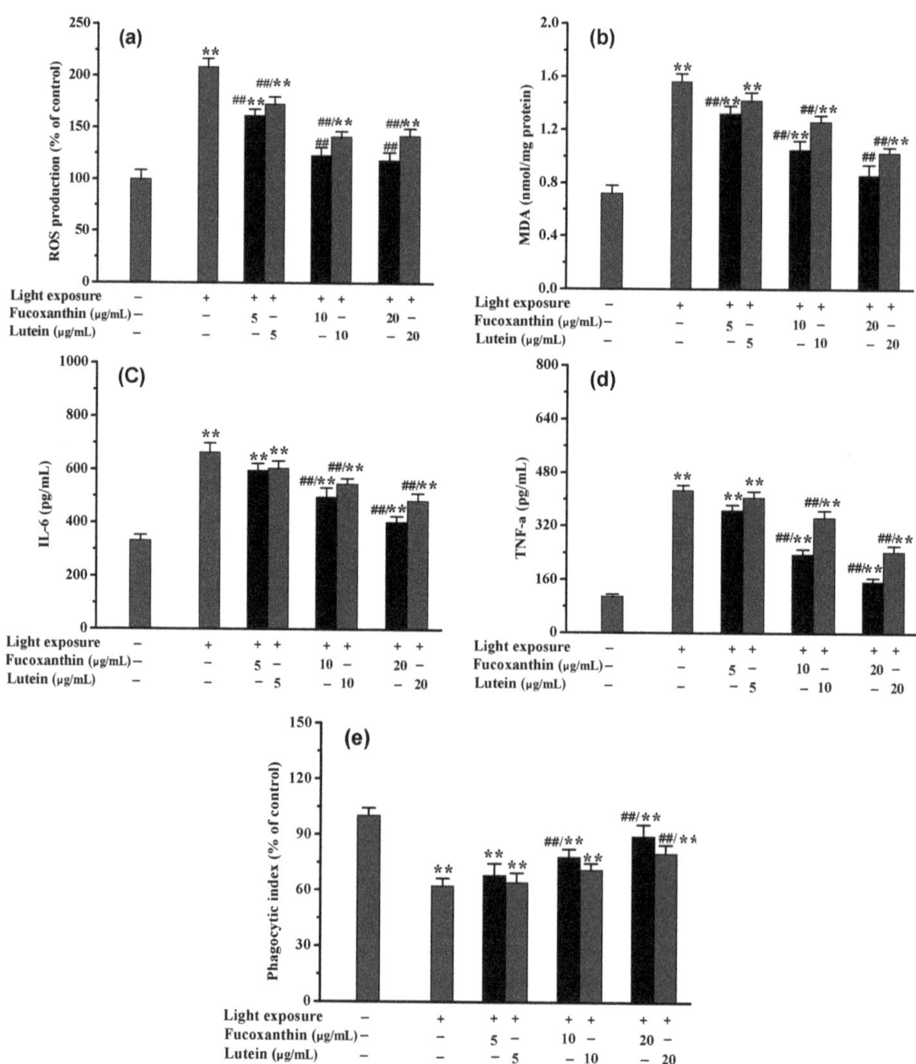

Figure 4. Inhibitory effects of fucoxanthin on oxidative damage and phagocytosis disorder in RPE cells induced by visible light: (**a**) intracellular ROS levels; (**b**) intracellular MDA levels; (**c**) inflammatory factor IL-6; (**d**) inflammatory factor TNF-α; (**e**) phagocytic indexes. (** $p < 0.01$ vs. control; ## $p < 0.01$ vs. light exposure).

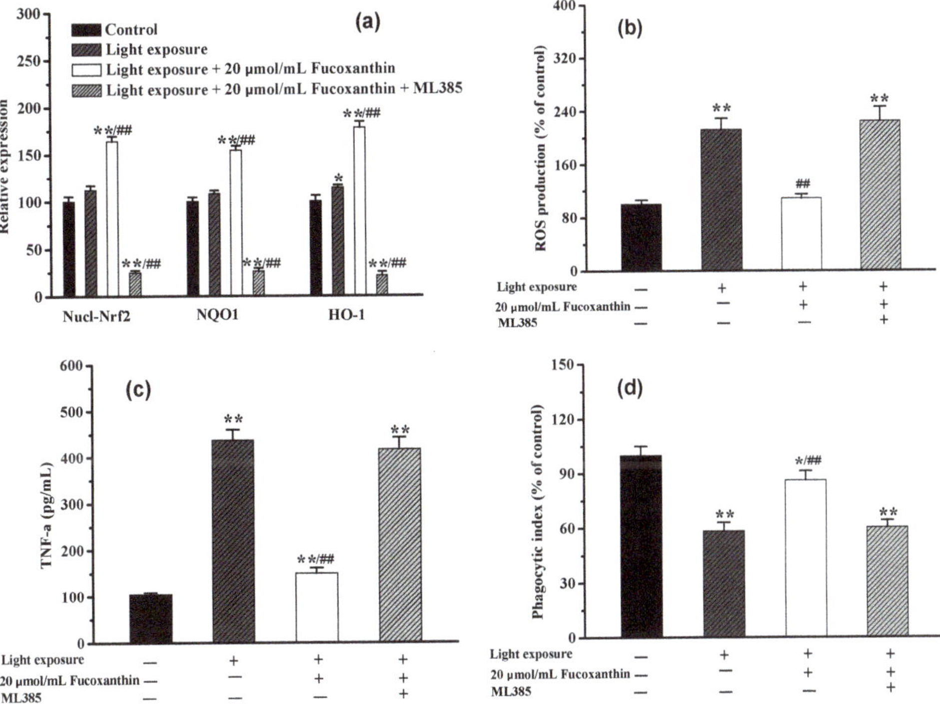

Figure 5. Ameliorative effects of fucoxanthin on phagocytosis disorder in RPE cells via the Nrf2 signal pathway: (**a**) the expressions of Nucl-Nrf2, NQO1, and HO-1 when RPE cells were treated with fucoxanthin or fucoxanthin +ML385; (**b**) ROS production when RPE cells were treated with fucoxanthin or fucoxanthin + ML385; (**c**) TNF-α levels when RPE cells were treated with fucoxanthin or fucoxanthin +ML385; (**d**) phagocytic indexes when RPE cells were treated with fucoxanthin or fucoxanthin +ML385. (* $p < 0.05$ and ** $p < 0.01$ vs. control; ## $p < 0.01$ vs. light exposure).

3. Discussion

This work used a polarized RPE cell culture model exhibiting similar morphological and physiological characteristics to the intact RPE monolayer, including the apical microvilli, well-defined tight junctions, membrane transport capability, and melanocytic pigmentation [25]. In our previous studies, when unpolarized RPE cells were exposed to 3500 lux visible light plus 25.0 μmol/L DHA for 12 h, the phagocytic index remained at only approximately 50% [1]. However, a phagocytic index with 81.49 ± 4.36% was observed when the differentiated RPE cell monolayer was exposed to the same visible light and lipid environment, suggesting that the differentiated RPE cells exhibited greater tolerance to visible light exposure. Therefore, it was hypothesized that, compared with the conventional in vitro cell model, differentiated RPE cells would more effectively reflect the true physiological conditions of retinal photodamage.

The RPE is considered a major target for retinal photodamage, which has long been attributed to the intracellular accumulation of lipofuscin particles during the phagocytizing of POSs [1,26]. Upon irradiation with visible light, N-retinylidene-N-retinylethanol-amine (A2E), one of the bis-retinoids derived from lipofuscin, produces singlet oxygen and other ROS, thus becoming a source of oxidative stress in RPE cells [27,28]. Due to their abundance of PUFAs, lipid peroxidation of POSs occurs before being swallowed in the presence of light exposure [29]. Given the close contact between RPE and POSs, this lipid peroxidation appears to be an important cause of visible light-induced damage to RPE cells. The results of this present study further confirmed that under the condition of visible light plus DHA exposure, obvious oxidative injury occurred in the RPE cells. Numerous studies have indicated that excessive light exposure can not only lead to oxidative stress by triggering ROS generation but also induce retinal cell dysfunction. When Wistar rats were illuminated by white light-emitting diode light, disruption to the outer blood-retinal barrier, which is constructed by RPE cells, was observed [30]. Furthermore, excessive light exposure could result in increased vascular endothelial growth factor (VEGF) secretion in cultured RPE cells, thus increasing the risk of neovascularization and edema [31]. In our previous studies, cellular senescence was also discovered when RPE cells were subjected to visible light exposure [1], while, in this work, it was further found that intense visible light radiation could lead to the degradation of phagocytic functioning in differentiated RPE cells. It is, thus, apparent that the protection of RPE cells against visible light-induced oxidative damage and phagocytosis disorder within a PUFA-rich environment should be an important pathway for vision-protecting dietary nutrients.

Antioxidant activity has been shown to be one of the important ways in which dietary nutrients perform their vision-protective functions. For example, dietary flavonoids (such as fisetin, luteolin, and quercetin) and vitamins C and E are capable of protecting RPE cells against hydrogen peroxide (H_2O_2)- or *t*-butyl hydroperoxide (*t*-BOOH)-induced death by scavenging intracellular ROS [32]. Curcumin pretreatment was shown to display a protective effect on light-induced retinal degeneration in a rat model through antioxidant pathways [33], while malvidin-3-galactoside/glucoside, the characteristic component of blueberry anthocyanins, could reduce oxidative stress in RPE cells by decreasing the levels of ROS and MDA [34]. It was also reported that madecassoside, a major bioactive triterpenoid saponin, could attenuate H_2O_2-induced ROS and MDA production in RPE cells [35]. As mentioned above, RPE cells are highly susceptible to cell damage induced by lipid peroxidation in the body; however, as shown via the results in this work, fucoxanthin can effectively inhibit oxidative stress and inflammation induced by visible light plus DHA in RPE cells.

There is increasing evidence that the activation of endogenous cellular antioxidant systems is an important way in which dietary functional ingredients exert their antioxidant physiological activities. As a nuclear transcription factor, Nrf2 controls the expression and coordinated induction of a battery of defensive genes encoding detoxifying enzymes and antioxidant proteins, which are critically important mechanisms for cellular protection and cell survival [36]. Resveratrol pretreatment reportedly significantly restored the SOD activity and upregulated the protein and mRNA expressions of Nrf2 and HO-1 in rat brain [37] and, as a carotenoid, lycopene pretreatment has also been reported to be effective in enhancing the expressions of Nrf2 and HO-1 in rat brain, thereby exerting a neurocytoprotective function [38]. Furthermore, when RPE cells were pretreated with hesperetin for between 1 and 6 h, the effect on Nucl-Nrf2 activation was discovered to be treatment time-dependent [20]. Based on our present results, when RPE cells were pretreated with fucoxanthin for 6–24 h, the expressions of Nucl-Nrf2 and its regulated downstream antioxidant proteins (such as GCLC, GPx, TrxR, HO-1, and NQO1) increased progressively. In addition, the addition of Nrf2 inhibitor ML385 during the whole experiment was found to effectively inhibit the expressions of Nucl-Nrf2, NQO1, and HO-1, and no protective effect from fucoxanthin on the RPE cells was observed. Thus, it was apparent that fucoxanthin performed an ameliorative effect on the visible light-induced phagocytic disorder of the RPE cells via the Nrf2 signaling pathway.

4. Materials and Methods

4.1. Materials and Chemical Reagents

The fucoxanthin and lutein standards were purchased from Dexter Biotechnology Co., Ltd. (Chengdu, China). Dimethyl sulfoxide, Dulbecco's modified Eagle's/Ham's F12 media, DHA, 3-(4,5-dimethylthiazol-2-yl)-2,5-diphenyl tetrazolium bromide (MTT), 2′,7′-dichlorofluorescin diacetate (DCFH-DA), blue fluorescent amine-modified microspheres (0.05 μm) and fetal bovine serum (FBS) were obtained from Sigma-Aldrich (MO, USA).

Commercial test kits, including LDH (CAS: A020-2-2), MDA (CAS: A003-4-1), SOD (CAS: A001-3-1), and HO-1 (CAS: H246-1), were purchased from Nanjing Jiancheng Bioengineering Institute (Nanjing, Jiangsu, China). Penicillin, streptomycin, and Hanks' balanced salt solution (HBSS) were obtained from Gibco Life Technologies (Grand Island, NY, USA). The IL-6 (CAS: SEKM-0007) and TNF-α (CAS: SEKM-0034)kits were purchased from Beijing Solarbio Science & Technology Co., Ltd. (Beijing, China). The nuclear extraction kit (CAS: ab113474), Nrf2 transcription factor assay kit (CAS: ab207223), GCLC (CAS: ab233632), NQO1 (CAS: ab184867) were purchased from Abcam Shanghai Trading Co., Ltd. (Shanghai, China). The antioxidant protein test kits, including GPx (CAS: BC1195) and TrxR (CAS: BC1155), were obtained from Beijing Solarbio Science & Technology Co., Ltd. (Beijing, China). The Nrf2 inhibitor ML385 was provided by Sigma-Aldrich (MO). All other reagents were analytical reagent-grade and purchased from the China National Pharmaceutical Industry Corporation Ltd. (Shanghai, China).

4.2. Cell Culture

The human RPE cell line, ARPE-19 (ATCC CRL–2302), was provided by the American Type Culture Collection (Mantissa, VA, USA). The cell cultures were maintained in Dulbecco's modified Eagle's/Ham's F12 media (Invitrogen, Carlsbad, CA, USA), to which 10% fetal bovine serum (Sigma-Aldrich, St. Louis, MO, USA), 100 U/mL penicillin, and 100 mg/mL streptomycin were added at 37 °C under a humidified 5% CO_2 atmosphere.

4.3. Cytotoxicity Evaluation

The MTT assay and LDH release assay were employed to measure cell viability and cell membrane integrity, respectively, according to a previous study [23]. Briefly, RPE cells were plated in 96-well plates at the concentration of 5×10^5 cells/mL for incubation 48 h. The cells were then treated with serum-free F12 medium containing 25.0 μmol/L DHA and different concentrations of fucoxanthin or lutein. After 24 h incubation, the cell supernatant was collected for LDH analysis. Subsequently, 150 μL of serum-free F12 medium containing 0.50 mg/mL MTT was added into each plate well and incubated for 4 h. The medium was then replaced with dimethyl sulfoxide (DMSO), and the absorbance was measured at 570 nm. The LDH activity in the cell supernatant was determined using a commercial LDH kit, according to the manufacturer's instructions.

4.4. Construction of Differentiated RPE Cell Monolayer

The differentiated RPE cell monolayer was cultured as described in previous studies [39,40]. RPE cells at passage 4 were seeded onto transwell inserts with polyester membranes (6.5 mm diameter, 0.4 μm pores) from Corning Inc. (Corning, NY, USA), at a seeding density of 1×10^4 cells per well (about 30,000 cells/cm^2). In the apical and basolateral chambers, 200.0 and 600.0 μL of serum-free F12 media were added, respectively. The cultures were supplied with 5% CO_2 in a humidified incubator (37 °C). Transepithelial electrical resistance (TER) was used to evaluate the differentiation degree of the RPE cell monolayer (Figure 6). It was deemed to be usable for subsequent light damage experiments when the differentiated RPE cell monolayer was cultured for 4 weeks, and the net TERs were ≥25.0 Ohm·cm^2 [39].

4.5. Visible Light-Induced RPE Cell Injury In Vitro

Visible light- and lipid-induced RPE cell damage was performed based on our previous reports [41], with some modifications. Subsequent to the formation of the differentiated RPE cell monolayer, serum-free F12 medium containing 25.0 μmol/L of DHA was added into the apical chamber, whereafter the RPE cell monolayer was subjected to white light irradiation. The light intensities were set as 1500, 3500, or 5000 lx, with light exposure of 6, 12, or 24 h, after which periods oxidative damages to the RPE cells were assessed.

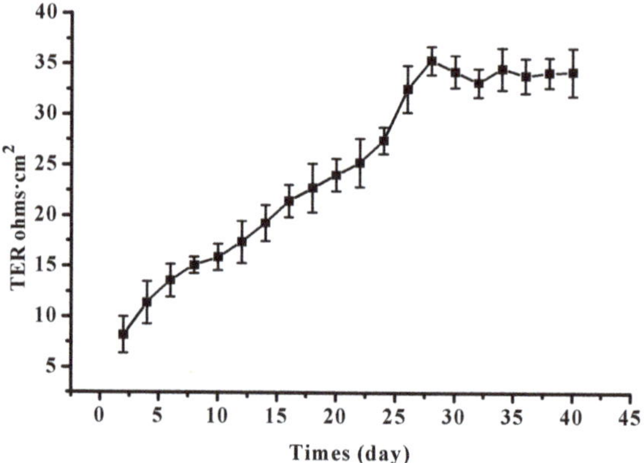

Figure 6. TER changes of the retinal pigment epithelium (RPE) cell monolayer during the cultivating processes.

4.6. Fucoxanthin Pretreatment of RPE Cells

In order to effectively improve the resistance of RPE cells to oxidative stress, the effect of fucoxanthin pretreatment time on the Nrf2-regulated antioxidant system was observed. Fucoxanthin or lutein was dissolved with a small amount of DMSO and added to the serum-free F12 medium at a final concentration of 20.0 μmol/L. Subsequently, the medium (200.0 μL) was added into the apical basolateral chamber. After incubation for 6, 12, or 24 h, the expressions of Nrf2 protein, as well as its regulated downstream antioxidant proteins, were analyzed.

4.7. Protective Effect of Fucoxanthin against Visible Light-Induced Injury of RPE Cells

After 24 h pretreatment with fucoxanthin, the medium of the apical chamber was replaced by a serum-free F12 medium containing 20.0 μmol/L fucoxanthin and 25.0 μmol/L DHA. Thereafter, the cell monolayer was subjected to light irradiation for 24 h at the light intensity of 3500 lx. The protective effects of FUCO on visible light-induced phagocytic disorder and oxidative damage in the RPE cells were subsequently observed. In order to verify whether fucoxanthin had enhanced the ability of RPE cells to resist photo-oxidative stress through the Nrf2 pathway, the cells were pretreated with the Nrf2 inhibitor ML385 (10.0 μmol/L) for 24 h, and the inhibitor was also added during the light exposure [21].

4.8. Detection of ROS and MDA Content

The intracellular ROS and MDA levels were measured as described in previous studies [23,42], with some modifications. The medium in the apical chamber was replaced by new medium containing 25.0 μmol of DCFH-DA. After 1 h incubation at 37 °C, the supernatant containing DCFH-DA was removed, and the cell monolayer was washed with Hanks' balanced salt solution (HBSS). The cell monolayer was then completely digested by RIPA lysis buffer, whereafter the cell lysate was added into a transparent black 96-well plate for fluorescence analysis. The fluorescence intensity was recorded using a microplate reader (Molecular Devices, San Jose, CA, USA) at 485 nm excitation and 530 nm emission. The MDA level in the cell lysate was measured according to the commercial kit's instruction, which was based on the thiobarbituric acid reactive substance (TBARS) assay.

4.9. Detection of Intracellular SOD, HO-1, GCLC, GPx, NQO1, and TrxR Activity

After the cell monolayer was washed with HBSS and fully lysed by RIPA lysis buffer, the expressions of antioxidative enzymes, including SOD, HO-1, GCLC, GPx, NQO1, and TrxR, in the cell lysate were determined, according to the manufacturer's protocols.

4.10. Detection of Inflammatory Cytokines

The inflammatory cytokines, including IL-6 and TNF-α, were measured using commercial ELISA kits following the manufacturer's instructions.

4.11. Measurement of Nucl-Nrf2

The expression of Nucl-Nrf2 in the RPE cell monolayer was measured as described in a previous study, with some modifications [43]. Briefly, the differentiated RPE cell monolayer was treated with trypsin buffer at 37 °C for 10.0 min, whereafter the cells were collected and the nuclear proteins extracted using the nuclear extraction kit, according to the kit manufacturer's instructions. The protein concentration of each sample was normalized to total protein content in the cell pellet using the Bradford assay. Next, 20.0 µg nuclear protein was added into the 96-well plate based on the instructions of the Nrf2 transcription factor assay kit. The optical density was recorded using a microplate reader (Molecular Devices, San Jose, CA, USA) at 450 nm.

4.12. Investigation of Phagocytosis of RPE Cells

The phagocytosis of the RPE cells was determined according to our previous studies [1,23], with some modifications. Following light irradiation, the supernatant in the apical chamber was replaced by a serum-free medium containing fluorescent microspheres (1×10^7/mL). After 24 h incubation, the uningested microspheres were washed away with fresh serum-free medium, and, after being washed twice using HBSS, the cells were lysed by RIPA lysis buffer. Thereafter, the cell lysates were transferred to a black, clear-bottomed 96-well plate, and the fluorescence intensity was recorded using a microplate reader (Molecular Devices, San Jose, CA, USA) at excitation/emission 360/420 nm. The phagocytic index was expressed as fluorescence intensity and presented as a percentage relative to the control.

4.13. Statistical Analyses

All data were expressed as mean ± SD of at least three individual experiments. Statistical analyses were performed using SPSS Statistical 24 Software. Comparisons between groups were performed using one-way analysis of variance (ANOVA) with Duncan's range tests. A normality test showed that all raw data displayed a normal distribution, while a variance test indicated that all groups exhibited equal variance. $p < 0.05$ (two-sided) was regarded as significant (*, $p < 0.05$; **, $p < 0.01$).

5. Conclusions

The results of this present study confirm that when differentiated RPE cells were exposed to both visible light (3500 lux or 5000 lux) and a PUFA-rich environment (25.0 µmol/L DHA) for 12–24 h, elevated oxidative stress and degenerative phagocytosis occur. However, fucoxanthin pretreatment at 20.0 µg/mL was found to effectively suppress the oxidative damage, inflammatory response, and phagocytosis disorder in RPE cells. Furthermore, the enhanced expression of Nucl-Nrf2 and the restored activity of detoxification enzymes such as HO-1 and NQO1 were simultaneously observed. Therefore, it is the conclusion of this work that the ameliorative effect of fucoxanthin on the phagocytosis of RPE cells should be closely related to the Nrf2 signaling pathway. However, recent studies have indicated that ARPE19 cells are different from natural RPE cells in terms of phenotype, including tight junction protein expression and pigment deposition. Therefore, in vivo animal experiments should be considered in future studies.

Author Contributions: Conceptualization and validation, Y.L. (Yixiang Liu) and Y.W.; methodology and writing—review and editing, Y.L. (Yixiang Liu); formal analysis, Z.G.; investigation, S.W.; resources, Y.L. (Yixiang Liu); data curation, Y.W.; data curation and writing—original draft preparation, Y.L. (Yunjun Liu); supervision and projectadministration, Y.W.; funding acquisition, Y.L. (Yixiang Liu) All authors have read and agreed to the published version of the manuscript.

Funding: This research was funded by Fujian Science Foundation for Distinguished Young Scholars (2020J06024) and Jimei University Young Talent Program (ZR2020001).

Conflicts of Interest: The authors declare no conflict of interest.

References

1. Liu, Y.; Liu, M.; Chen, Q.; Liu, G.M.; Cao, M.J.; Sun, L.; Lu, Z.; Guo, C. Blueberry polyphenols ameliorate visible light and lipid-induced injury of retinal pigment epithelial cells. *J. Agric. Food Chem.* **2018**, *66*, 12730–12740. [CrossRef] [PubMed]
2. Rong, R.; Yang, R.; Li, H.; You, M.; Liang, Z.; Zeng, Z.; Zhou, R.; Xia, C.; Ji, D. The roles of mitochondrial dynamics and NLRP3 inflammasomes in the pathogenesis of retinal light damage. *Ann. N. Y. Acad. Sci.* **2021**. [CrossRef] [PubMed]
3. Hunter, J.J.; Morgan, J.I.; Merigan, W.H.; Sliney, D.H.; Sparrow, J.R.; Williams, D.R. The susceptibility of the retina to photochemical damage from visible light. *Prog. Retin. Eye Res.* **2012**, *31*, 28–42. [CrossRef] [PubMed]
4. Baksheeva, V.E.; Tiulina, V.V.; Tikhomirova, N.K.; Gancharova, O.S.; Komarov, S.V.; Philippov, P.P.; Zamyatnin, A.A.; Senin, I.I.; Zernii, E.Y. Suppression of light-induced oxidative stress in the retina by mitochondria-targeted antioxidant. *Antioxidants* **2019**, *8*, 3. [CrossRef]
5. Beatty, S.; Koh, H.-H.; Phil, M.; Henson, D.; Boulton, M. The role of oxidative stress in the pathogenesis of age-related macular degeneration. *Surv. Ophthalmol.* **2000**, *45*, 115–134. [CrossRef]
6. Zernii, E.Y.; Baksheeva, V.E.; Iomdina, E.N.; Averina, O.A.; Permyakov, S.E.; Philippov, P.; Zamyatnin, A.; Senin, I. Rabbit models of ocular diseases: New relevance for classical approaches. *CNS Neurol. Disord. Drug Targets* **2016**, *15*, 267–291. [CrossRef]
7. Van Norren, D.; Vos, J.J. Light damage to the retina: An historical approach. *Eye* **2016**, *30*, 169–172. [CrossRef] [PubMed]
8. Organisciak, D.T.; Vaughan, D.K. Retinal light damage: Mechanisms and protection. *Prog. Retin. Eye Res.* **2010**, *29*, 113–134. [CrossRef]
9. Novikova, Y.P.; Gancharova, O.S.; Eichler, O.; Philippov, P.; Grigoryan, E. Preventive and therapeutic effects of SkQ1-containing Visomitin eye drops against light-induced retinal degeneration. *Biochemistry* **2014**, *79*, 1101–1110. [CrossRef] [PubMed]
10. Zernii, E.Y.; Nazipova, A.A.; Gancharova, O.S.; Kazakov, A.S.; Serebryakova, M.V.; Zinchenko, D.V.; Tikhomirova, N.K.; Senin, I.I.; Philippov, P.P.; Permyakov, E.A. Light-induced disulfide dimerization of recoverin under ex vivo and in vivo conditions. *Free Radic. Biol. Med.* **2015**, *83*, 283–295. [CrossRef]
11. Marquioni-Ramella, M.D.; Suburo, A.M. Photo-damage, photo-protection and age-related macular degeneration. *Photochem. Photobiol. Sci.* **2015**, *14*, 1560–1577. [CrossRef]
12. Liu, Y.; Song, X.; Zhang, D.; Zhou, F.; Wang, D.; Wei, Y.; Gao, F.; Xie, L.; Jia, G.; Wu, W.; et al. Blueberry anthocyanins: Protection against ageing and light-induced damage in retinal pigment epithelial cells. *Brit. J. Nutr.* **2012**, *108*, 16–27. [CrossRef] [PubMed]
13. Miceli, M.V.; Liles, M.R.; Newsome, D.A. Evaluation of oxidative processes in human pigment epithelial cells associated with retinal outer segment phagocytosis. *Exp. Cell Res.* **1994**, *214*, 242–249. [CrossRef]
14. Wang, Y.; Zhang, D.; Liu, Y.; Wang, D.; Liu, J.; Ji, B.P. The protective effects of berry-derived anthocyanins against visible light-induced damage in human retinal pigment epithelial cells. *J. Sci. Food Agric.* **2015**, *95*, 936–944. [CrossRef] [PubMed]
15. Li, C.P.; Yao, J.; Tao, Z.F.; Li, X.M.; Jiang, Q.; Yan, B. Epigallocatechin-gallate (EGCG) regulates autophagy in human retinal pigment epithelial cells: A potential role for reducing UVB light-induced retinal damage. *Biochem. Biophys. Res. Commun.* **2013**, *438*, 739–745. [CrossRef] [PubMed]
16. Kim, J.; Jin, H.L.; Jang, D.S.; Jeong, K.W.; Choung, S.Y. Quercetin-3-O-α-l-arabinopyranoside protects against retinal cell death via blue light-induced damage in human RPE cells and Balb-c mice. *Food Funct.* **2018**, *9*, 2171–2183. [CrossRef]
17. Hazim, R.A.; Karumbayaram, S.; Jiang, M.; Dimashkie, A.; Lopes, V.S.; Li, D.; Burgess, B.L.; Vijayaraj, P.; Alva-Ornelas, J.A.; Zack, J.A.; et al. Differentiation of RPE cells from integration-free iPS cells and their cell biological characterization. *Stem Cell Res. Ther.* **2017**, *8*, 217. [CrossRef]
18. Zhang, H.; Davies, K.J.A.; Forman, H.J. Oxidative stress response and Nrf2 signaling in aging. *Free Radic. Biol. Med.* **2015**, *88*, 314–336. [CrossRef] [PubMed]
19. Deng, Y.; Zhu, J.; Mi, C.; Xu, B.; Jiao, C.; Li, Y.; Xu, D.; Liu, W.; Xu, Z. Melatonin antagonizes Mn-induced oxidative injury through the activation of keap1-Nrf2-ARE signaling pathway in the striatum of mice. *Neurotoxic. Res.* **2015**, *27*, 156–171. [CrossRef]
20. Zhu, C.; Dong, Y.; Liu, H.; Ren, H.; Cui, Z. Hesperetin protects against H_2O_2-triggered oxidative damage via upregulation of the Keap1-Nrf2/HO-1 signal pathway in ARPE-19 cells. *Biomed. Pharmacother.* **2017**, *88*, 124–133. [CrossRef]
21. Jin, W.; Zhu, X.; Yao, F.; Xu, X.; Chen, X.; Luo, Z.; Zhao, D.; Li, X.; Leng, X.; Sun, L. Cytoprotective effect of Fufang Lurong Jiangu capsule against hydrogen peroxide-induced oxidative stress in bone marrow stromal cell-derived osteoblasts through the Nrf2/HO-1 signaling pathway. *Biomed. Pharmacother.* **2020**, *121*, 109676. [CrossRef]
22. Wang, Y.; Huo, Y.; Zhao, L.; Lu, F.; Wang, O.; Yang, X.; Ji, B.; Zhou, F. Cyanidin-3-glucoside and its phenolic acid metabolites attenuate visible light-induced retinal degeneration in vivo via activation of Nrf2/HO-1 pathway and NF-B suppression. *Mol. Nutr. Food Res.* **2016**, *60*, 1564–1577. [CrossRef]
23. Liu, Y.; Liu, M.; Zhang, X.; Chen, Q.; Chen, H.; Sun, L.; Liu, G. Protective effect of fucoxanthin isolated from *Laminaria japonica* against visible light-induced retinal damage both in vitro and in vivo. *J. Agric. Food Chem.* **2016**, *64*, 416–424. [CrossRef] [PubMed]

24. Li, D.; Zhang, Q.; Huang, L.; Chen, Z.; Zou, C.; Ma, Y.; Cao, M.J.; Liu, G.M.; Liu, Y.; Wang, Y. Fabricating hydrophilic particles with oleic acid and bovine serum albumin to improve the dispersibility and bioaccessibility of fucoxanthin in water. *Food Hydrocoll.* **2021**, *118*, 106752. [CrossRef]

25. Sonoda, S.; Spee, C.; Barron, E.; Ryan, S.J.; Kannan, R.; Hinton, D.R. A protocol for the culture and differentiation of highly polarized human retinal pigment epithelial cells. *Nat. Protoc.* **2009**, *4*, 662–673. [CrossRef]

26. Boulton, M.; Różanowska, M.; Różanowski, B. Retinal photodamage. *J. Photochem. Photobiol. B Biol.* **2001**, *64*, 144–161. [CrossRef]

27. Wang, Y.; Kim, H.J.; Sparrow, J.R. Quercetin and cyanidin-3-glucoside protect against photooxidation and photodegradation of A2E in retinal pigment epithelial cells. *Exp. Eye Res.* **2017**, *160*, 45–55. [CrossRef]

28. Rozanowska, M.B. Light-induced damage to the retina: Current understanding of the mechanisms and unresolved questions: A symposium-in-print. *Photochem. Photobiol.* **2012**, *88*, 1303–1308. [CrossRef] [PubMed]

29. Winkler, B.S. An hypothesis to account for the renewal of outer segments in rod and cone photoreceptor cells: Renewal as a surrogate antioxidant. *Investig. Ophthalmol. Vis. Sci.* **2008**, *49*, 3259–3261. [CrossRef]

30. Jaadane, I.; Villalpando Rodriguez, G.E.; Boulenguez, P.; Chahory, S.; Carré, S.; Savoldelli, M.; Jonet, L.; Behar-Cohen, F.; Martinsons, C.; Torriglia, A. Effects of white light-emitting diode (LED) exposure on retinal pigment epithelium in vivo. *J. Cell. Mol. Med.* **2017**, *21*, 3453–3466. [CrossRef]

31. Klettner, A.; Kampers, M.; Töbelmann, D.; Roider, J.; Dittmar, M. The influence of melatonin and light on VEGF secretion in primary RPE cells. *Biomolecules* **2021**, *11*, 114. [CrossRef]

32. Hanneken, A.; Lin, F.F.; Johnson, J.; Maher, P. Flavonoids protect human retinal pigment epithelial cells from oxidative-stress–induced death. *Investig. Ophthalmol. Vis. Sci.* **2006**, *47*, 3164–3177. [CrossRef]

33. Mandal, M.N.A.; Patlolla, J.M.R.; Zheng, L.; Agbaga, M.-P.; Tran, J.T.A.; Wicker, L.; Kasus-Jacobi, A.; Elliott, M.H.; Rao, C.V.; Anderson, R.E. Curcumin protects retinal cells from light-and oxidant stress-induced cell death. *Free Radic. Biol. Med.* **2009**, *46*, 672–679. [CrossRef]

34. Huang, W.Y.; Wu, H.; Li, D.J.; Song, J.F.; Xiao, Y.D.; Liu, C.Q.; Zhou, J.Z.; Sui, Z.Q. Protective effects of blueberry anthocyanins against H_2O_2-induced oxidative injuries in human retinal pigment epithelial cells. *J. Agric. Food Chem.* **2018**, *66*, 1638–1648. [CrossRef] [PubMed]

35. Zhou, J.; Chen, F.; Yan, A.; Xia, X. Madecassoside protects retinal pigment epithelial cells against hydrogen peroxide-induced oxidative stress and apoptosis through the activation of Nrf2/HO-1 pathway. *Biosci. Rep.* **2020**, *40*, BSR20194347. [CrossRef] [PubMed]

36. Kaspar, J.W.; Niture, S.K.; Jaiswal, A.K. Nrf2: INrf2 (Keap1) signaling in oxidative stress. *Free Radic. Biol. Med.* **2009**, *47*, 1304–1309. [CrossRef]

37. Ren, J.; Fan, C.; Chen, N.; Huang, J.; Yang, Q. Resveratrol pretreatment attenuates cerebral ischemic injury by upregulating expression of transcription factor Nrf2 and HO-1 in rats. *Neurochem. Res.* **2011**, *36*, 2352–2362. [CrossRef] [PubMed]

38. Lei, X.; Lei, L.; Zhang, Z.; Cheng, Y. Neuroprotective effects of lycopene pretreatment on transient global cerebral ischemia-reperfusion in rats: The role of the Nrf2/HO-1 signaling pathway. *Mol. Med. Rep.* **2016**, *13*, 412–418. [CrossRef]

39. Liu, Y.; Liu, G.M.; Cao, M.J.; Chen, Q.; Sun, L.; Ji, B. Potential retinal benefits of dietary polyphenols based on their permeability across the Blood–Retinal Barrier. *J. Agric. Food Chem.* **2017**, *65*, 3179–3189. [CrossRef]

40. Hazim, R.A.; Volland, S.; Yen, A.; Burgess, B.L.; Williams, D.S. Rapid differentiation of the human RPE cell line, ARPE-19, induced by nicotinamide. *Exp. Eye Res.* **2019**, *179*, 18–24. [CrossRef]

41. Liu, Y.; Zhang, D.; Wu, Y.; Ji, B. Docosahexaenoic acid aggravates photooxidative damage in retinal pigment epithelial cells via lipid peroxidation. *J. Photochem. Photobiol. B Biol.* **2014**, *140*, 85–93. [CrossRef] [PubMed]

42. Cui, B.; Zhang, S.; Wang, Y.; Guo, Y. Farrerol attenuates β-amyloid-induced oxidative stress and inflammation through Nrf2/Keap1 pathway in a microglia cell line. *Biomed. Pharmacother.* **2019**, *109*, 112–119. [CrossRef] [PubMed]

43. Clementi, M.E.; Sampaolese, B.; Sciandra, F.; Tringali, G. Punicalagin protects human retinal pigment epithelium cells from ultraviolet radiation-induced oxidative damage by activating Nrf2/HO-1 signaling pathway and reducing apoptosis. *Antioxidants* **2020**, *9*, 473. [CrossRef] [PubMed]

Article

Improved Productivity of Astaxanthin from Photosensitive *Haematococcus pluvialis* Using Phototaxis Technology

Kang Hyun Lee [1,†], Youngsang Chun [2,†], Ja Hyun Lee [3], Chulhwan Park [4,*], Hah Young Yoo [1,*] and Ho Seok Kwak [2,*]

[1] Department of Biotechnology, Sangmyung University, 20, Hongjimun-2Gil, Jongno-gu, Seoul 03016, Korea; oys7158@naver.com

[2] Department of Bio-Convergence Engineering, Dongyang Mirae University, 445-8, Gyeongin-ro, Guro-gu, Seoul 02841, Korea; youngsangchun@gmail.com

[3] Department of Convergence Bio-Chemical Engineering, Soonchunhyang University, 22, Soonchunhyang-ro, Asan-si 31538, Korea; jhlee84@sch.ac.kr

[4] Department of Chemical Engineering, Kwangwoon University, 20, Kwangwoon-ro, Nowon-gu, Seoul 01897, Korea

* Correspondence: chpark@kw.ac.kr (C.P.); y2h2000@smu.ac.kr (H.Y.Y.); hskwak@dongyang.ac.kr (H.S.K.)

† These authors contributed equally to this work.

Citation: Lee, K.H.; Chun, Y.; Lee, J.H.; Park, C.; Yoo, H.Y.; Kwak, H.S. Improved Productivity of Astaxanthin from Photosensitive *Haematococcus pluvialis* Using Phototaxis Technology. *Mar. Drugs* **2022**, *20*, 220. https://doi.org/10.3390/md20040220

Academic Editors: Masashi Hosokawa and Hayato Maeda

Received: 8 February 2022
Accepted: 19 March 2022
Published: 22 March 2022

Publisher's Note: MDPI stays neutral with regard to jurisdictional claims in published maps and institutional affiliations.

Abstract: *Haematococcus pluvialis* is a microalgae actively studied for the production of natural astaxanthin, which is a powerful antioxidant for human application. However, it is economically disadvantageous for commercialization owing to the low productivity of astaxanthin. This study reports an effective screening strategy using the negative phototaxis of the *H. pluvialis* to attain the mutants having high astaxanthin production. A polydimethylsiloxane (PDMS)-based microfluidic device irradiated with a specific light was developed to efficiently figure out the phototactic response of *H. pluvialis*. The partial photosynthesis deficient (PP) mutant (negative control) showed a 0.78-fold decreased cellular response to blue light compared to the wild type, demonstrating the positive relationship between the photosynthetic efficiency and the phototaxis. Based on this relationship, the *Haematococcus* mutants showing photosensitivity to blue light were selected from the 10,000 random mutant libraries. The M1 strain attained from the phototaxis-based screening showed 1.17-fold improved growth rate and 1.26-fold increases in astaxanthin production (55.12 ± 4.12 mg g^{-1}) in the 100 L photo-bioreactor compared to the wild type. This study provides an effective selection tool for industrial application of the *H. pluvialis* with improved astaxanthin productivity.

Keywords: *Haematococcus pluvialis*; astaxanthin; photosynthetic efficiency; microfluidics; photobioreactor

1. Introduction

The ketocarotenoid astaxanthin (3,3′-dihydroxy-β-carotene-4,4′-dione) straddles the cell membrane bilayer and acts as a super antioxidant that significantly reduces free radicals and oxidative stress [1–3]. Astaxanthin has been widely applied as a promising functional substance in the functional foods, pharmaceutical, cosmetics, and aquaculture industries due to its potential to prevent Alzheimer's disease, atherosclerotic, and cardiovascular diseases [4–7]. The global astaxanthin market was estimated at USD 288.7 million in 2017 and forecasted to grow to USD 426.9 million in 2022 [8]. Astaxanthin can be produced synthetically or obtained from natural sources [9]. However, synthetic astaxanthin is not approved for human consumption directly by the Food and Drug Administration (FDA) as a food or dietary supplement due to several issues, such as food safety, potential toxicity in the final product, and sustainability [10]. Compared with synthetic astaxanthin, natural astaxanthin has various advantages such as high safety and antioxidant activity, environmental friendliness, sustainability, and renewability [11]. For these reasons, the demand for natural astaxanthin has increased.

Traditionally, natural astaxanthin has been produced from a variety of seafood, including shrimp, lobster, fish eggs, and microalgae [12–14]. Microalgae has advantages in astaxanthin production compared to other seafood in terms of productivity, production area, safety control, and stability [15]. Among the various microalgae, *Haematococcus pluvialis* is considered to be the most abundant source of natural astaxanthin (3–5% of dry biomass) [16]. In addition, natural astaxanthin obtained from *H. pluvialis* was confirmed as a food ingredient for humans by the European Commission (EC) and the European Food Safety Authority (EFSA) [17]. However, there are limitations to the utilization of natural astaxanthin on an industrial scale. Various studies have been actively conducted to improve astaxanthin productivity from *H. pluvialis* [16,18,19]. Astaxanthin productivity can be improved by maximizing the cell density in the green phase and by process optimization such as culture parameters and the design of photobioreactors in the red phase during *H. pluvialis* culture [16,18,19].

In order to improve astaxanthin productivity from *H. pluvialis*, various approaches have been used such as development of strains, optimization of culture conditions, protoplast fusion with other microalgae, and addition of nanoparticles [20–23]. Among these approaches, the development and selection of strains is the most important strategy in industrial production [24]. Random mutagenesis by ultraviolet (UV) irradiation is a common and cost-effective technique for strain development [25]. Various studies have been conducted to develop genetically modified mutants with high growth rate and astaxanthin productivity through random mutagenesis [26–28]. However, one of the disadvantages of random mutagenesis is that it is difficult to select strains with improved specific efficacy among numerous variants, and the selection process requires a lot of time and labor. Microfluidic-based technologies with specific light can rapidly screen strains at the single-cell level, taking into account the size, shape, and characteristics of microalgae [29]. In a previous study [30], strains with improved photosynthetic productivity were selected using a microfluidic device fabricated by applying the phototaxis technique.

In this study, the *H. pluvialis* strain was developed with improved astaxanthin productivity by using the phototaxis technology and scaling it up using the photo-bioreactor for industrial application (Figure 1). At the micro scale, the phototactic response of *H. pluvialis* and the correlation between photosynthesis and phototaxis were investigated to select mutated *H. pluvialis* with high astaxanthin productivity. Mutants with high negative phototaxis selected by the microfluidic device were scaled up to a 100 L flat panel photo-bioreactor, and astaxanthin production was determined. This multi scale approach can be easily applied to various microalgae and provides an efficient selection method to develop the microalgal mutants with improved productivity of diverse metabolites such as the astaxanthin. Currently, most studies using the phototaxis properties of microalgae are related to photosynthetic efficiency. To the best of the authors' knowledge, this is the first study utilizing phototaxis properties to screen *H. pluvialis* for improved astaxanthin productivity.

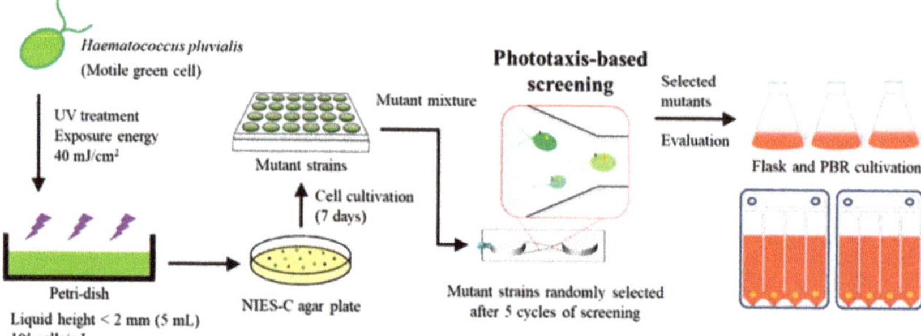

Figure 1. Schematic diagram of a phototaxis-based screening technique for the isolation of photosensitive strains using microfluidic devices. The screening was performed on a mixture of 10,000 mutants generated from the wild-type strain by UV random mutagenesis.

2. Results and Discussion

2.1. Investigation of Phototactic Properties of H. pluvialis

Phototaxis is a reaction in which the direction of movement changes depending on the intensity of light [31]. Positive phototaxis indicates the characteristic of cells moving toward the light source at low-light intensity, whereas negative phototaxis indicates that the cell moves away from the light source at high-light intensity [32]. It is reported that the control of negative phototaxis is easier and more consistent than the positive phototaxis of cells [33]. To investigate the optical properties of the *H. pluvialis*, the cells, which were photo-autotrophically pre-cultured for 10 days, were diluted to a concentration of 10,000 cells mL^{-1}, and then 50 µL aliquot (about 3000 cells) of the dilute solution was put into the inlet of the microfluidic device (Figure 2). Then, various light emitting diode (LED) lights (dark, blue, green, red, and white) were irradiated to the inlet part of the microfluidic device for 30 min to observe the negative phototaxis of the cells.

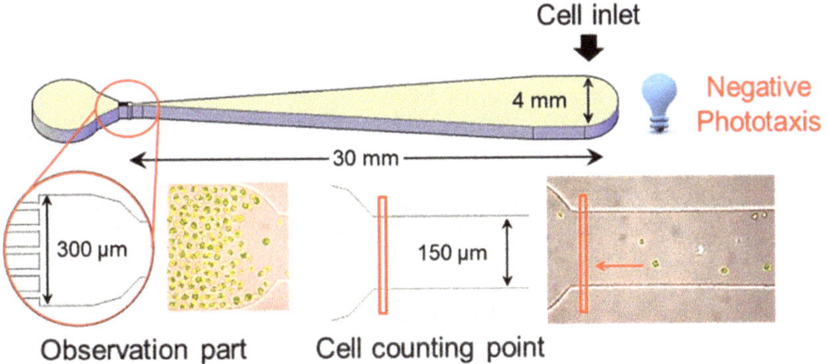

Figure 2. Microfluidic device design for detection of phototaxis of the *H. pluvialis*. Cells grown exponentially were loaded into an inlet chamber in the microfluidic device and exposed to blue LED (70 µmol photons m^{-2} s^{-1}). The phototactic movements of cells were monitored under an inverted microscope and analyzed using video analyzing software.

Figure 3A represents the negative phototaxis of the *H. pluvialis* to five different light wavelengths and *H. pluvialis* shows the most sensitive response to the blue light wavelength (470 nm). About 69.33% of the cells injected moved to the opposite chamber (observation part) for 30 min by the blue light. In the green light (540 nm), the phototactic response was reduced by 25% compared to the result in the blue light. In the red light (700 nm), there was no reaction to the light during 30 min as in the dark condition (control). These results show a trend similar to that of the model microalgae, *Chlamydomonas reinhardtii*, which has been reported to exhibit phototaxis by detecting the direction, intensity, and wavelength of light because it has an eyespot in the thylakoid membrane [30,34]. Tamaki et al. [35] reported that carotenoids, such as astaxanthin, fucoxanthin, and diadinoxanthin are major constituents of eyespot and are essential for the phototactic response. In addition, it has been reported that blue light significantly improves astaxanthin productivity of *H. pluvialis* [36]. Therefore, it can be inferred that *H. pluvialis*, which has sensitive phototaxis properties, contains a large amount of astaxanthin. Figure 3B is a result showing the light responses to different light intensities, and shows that the negative phototaxis of the *H. pluvialis* increased up to the 70 µmol photons m^{-2} s^{-1}. However, there was little change at the intensity above 70 µmol photons m^{-2} s^{-1}. Therefore, 70 µmol photons m^{-2} s^{-1} was set as the standard for measuring the phototactic behavior in this study.

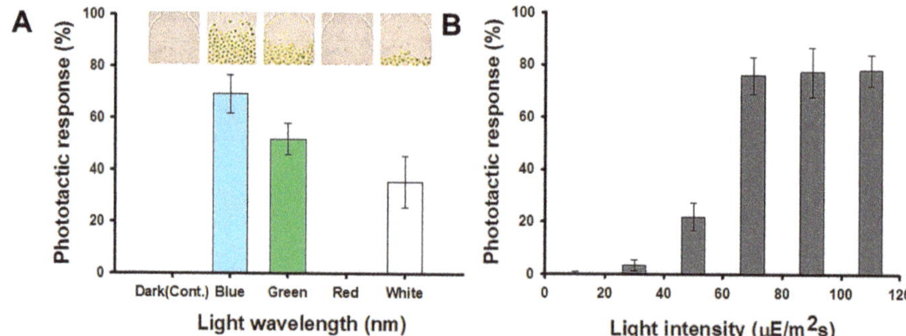

Figure 3. Light conditions for detecting phototaxis of *H. pluvialis*: (**A**) The phototactic responses of the wild type at five different light wavelength conditions; dark (control), blue (470 nm), green (540 nm), red (700 nm), white (all). Photo image showing the observation part of the microfluidic device after 30 min. (**B**) The phototactic responses of wild type at different light intensities; 10, 30, 50, 70, 90, 110 μmol photons m^{-2} s^{-1}.

2.2. Correlation between Photosynthesis and Phototaxis

To confirm the correlation between photosynthesis and phototaxis of the *H. pluvialis*, the partial photosynthesis deficient (PP) mutant was used as a negative control to compare with wild-type strain. The chlorophyll contents and the photosystem II (PSII) operating efficiency (Y(II)), indicating the photosynthetic efficiency of PP mutant, showed 6.7% and 64% less than that of wild type, respectively (Figure 4A,B). Hong et al. [26], reported that the total cell numbers of PP mutant in the green stage decreased by 32.2% compared to the wild-type strain because of partial photosynthesis deficient. These results are consistent with Wan et al. [37], who determined that *H. pluvialis* is suitable for mass culture on an industrial scale because it is easy to acquire large quantities of cells due to its high photosynthetic efficiency. Figure 4C,D represent the phototactic activities of the wild type and PP mutant to the blue light for 30 min. In the wild-type strain, 78.8% of the cells injected into the inlet of the microfluidic device reached the observation chamber 30 mm away from the inlet. The PP mutant showed 61.5% of the cells responded to the blue light, and it is 0.78-fold decrease compared to the wild-type strain. In addition, the cells that reached the opposite chamber took 12.4 min on average for the wild-type strain, but 19.1 min for the PP mutant. This means that the wild-type strain is highly reactivity to blue light, thereby increasing the cell migration rate. Figure 4D indicates the accumulative histogram of arriving cell numbers for 30 min. The average time required to reach 80% of the cells responding to the blue light was 13.3 min for the wild-type strain and more than 20 min for the PP mutant. These results indicate that the high Y(II) in the *H. pluvialis* is correlated with sensitive reactivity to the blue light. This positive correlation between phototaxis and photosynthetic efficiency was also proved in model microalgae, *C. reinhardtii* [30].

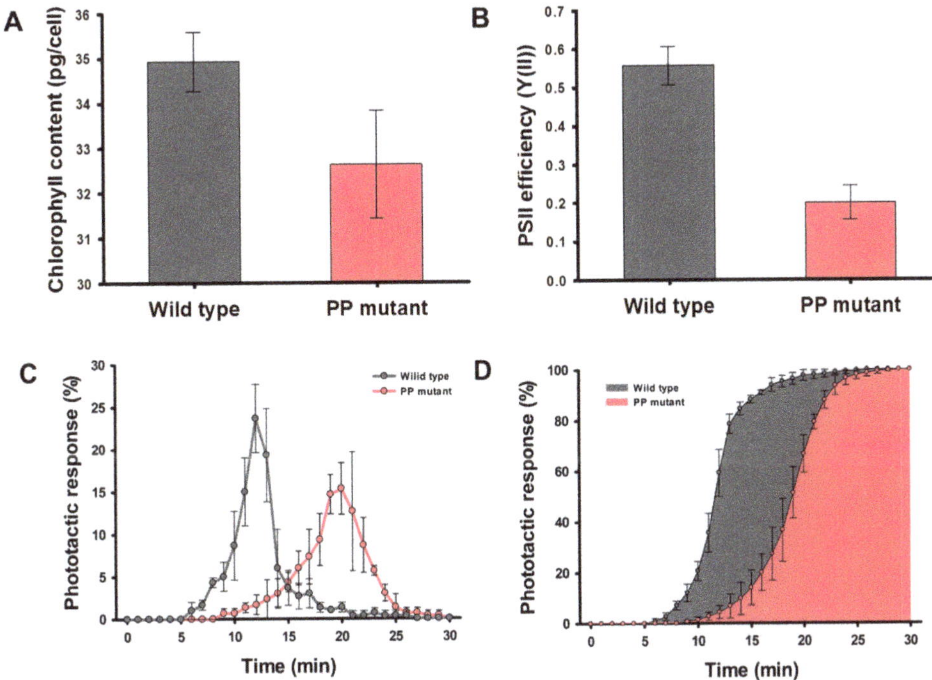

Figure 4. Optical characteristics of the wild type (grey) and PP mutant (red): (**A**) Chlorophyll content, and (**B**) Y(II) of the wild type and PP mutant. (**C**) The phototactic response, and (**D**) Cumulative histogram of phototactic response of the wild type and PP mutant for 30 min. The phototactic response was measured on 3000 cells per analysis. All data are the mean of three biological replicates.

2.3. Strain Selection Using the Microfluidic Device and Flask Cultivation

The microfluidic device was fabricated using polydimethylsiloxane (PDMS) to isolate mutated *H. pluvialis* with high photosynthetic efficiency and astaxanthin productivity based on the relationship between phototaxis and the photosynthetic efficiency. PDMS is useful for cell analysis using a microfluidic system because it is possible to efficiently monitor the cell movements and change in cell physiology at the single cell level due to its transparent property [38,39]. Ten thousand mutant libraries of the *H. pluvialis* were constructed by random mutagenesis using ultraviolet rays (Figure 1). After collecting cultures of all mutants, the screening process was repeated up to five times to increase the strains exhibiting a rapid phototactic response from a mixture of 10,000 mutants. After five screening cycles, a mixture of mutants showing a rapid reactivity to the blue light was obtained. The mixture was plated on an agar plate for isolating each strain as separated colonies. Next, 10 strains were selected from agar plates and cultured, and photoautotrophic growth was analyzed. The M1 showing the highest growth and astaxanthin production among 10 mutant strains were selected for further evaluation of their performances.

The wild-type strain and M1 mutant were inoculated into the flask for two stage cultivation (growth stage and induction stage) for 20 days (Figure 5A). In the growth stage, the M1 mutant showed faster and higher growth compared to the wild-type strain (Figure 5B). The maximum specific growth rate (μ_{max}) was calculated from the growth curve. The μ_{max} of M1 mutant was 0.702 h^{-1}, which was 1.17-fold higher than μ_{max} of the wild-type strain (0.598 h^{-1}). The Y(II) of the M1 mutant during the growth stage showed 0.63 ± 0.012, which was 13% higher than that of the wild type, suggesting that rapid phototactic response to the blue light can be helpful to improve Y(II) [30]. After 10 days of growth, the NIES-C medium was replaced with the nitrogen-deficient NIES-N medium because nitrogen deficiency enhances astaxanthin productivity of *H. pluvialis* [40]. Additionally, light intensity increased up to 200 µmol photon m^{-2} s^{-1} to induce the astaxanthin biosynthesis by oxidative stresses [41]. Under high light intensity, reactive oxygen species (ROS) are accumulated in microalgae by the over-reduction of the photosynthetic electron transport chain [42]. Microalgae produce carotenoids including astaxanthin as a defense mechanism against oxidative stress such as ROS [43]. Figure 5C shows the change in the ratio of carotenoid and chlorophyll according to the synthesis of the astaxanthin in the induction stage. It shows that the conversion to carotenoid occurs more efficiently in the M1 mutant than in the wild-type strain according to the environmental change. This is because the astaxanthin synthesis was rapidly conducted as a defense mechanism against oxidative stress caused by high light intensity. The productivity of astaxanthin of the M1 mutant showed 171.6 ± 4.16 mg L^{-1}, which was 1.26-fold higher than that of the wild-type strain.

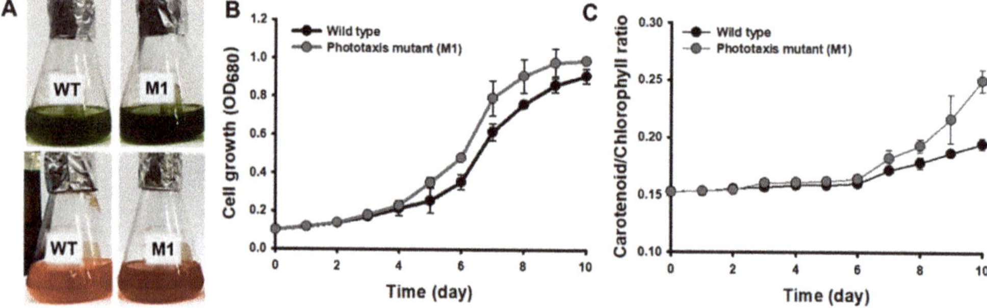

Figure 5. Flask cultivation of the wild type and M1 mutant screened by phototaxis: (**A**) Photo image of the wild type and M1 mutant under growth (up) and induction stage (down). (**B**) Absorbance in growth stage, and (**C**) Carotenoid/Chlorophyll ratio of in induction stage. The cells were grown in NIES-C medium and CO$_2$ concentration and light intensity were fixed at 5% and 40 µmol photons m^{-2} s^{-1}, respectively. Error bars indicate the standard deviations of the means for three experiments.

2.4. Validation of the M1 Mutant Performance in Large Scale Cultivation

In order to confirm the improved astaxanthin productivity of the M1 mutant, the flat panel photo-bioreactor of 100 L scale was manufactured using acrylic. The M1 mutant and wild-type strain pre-cultured in a 1 L flask for 10 days were inoculated and grown for 30 days (Figure 6A). In the growth phase, the M1 mutant showed a faster growth rate and 1.1-fold higher biomass production than the wild-type strain, which was similar to the results of the flask experiment (data not shown). This is because the M1 mutant has a higher photosynthetic efficiency than the wild type for the same light intensity condition (40 µmol photons m^{-2} s^{-1}). After the vegetative growth phase, cells were exposed to a continuous strong light at 200 µmol photons m^{-2} s^{-1} along with N-deprivation for the next 40 days in the red stage. The astaxanthin contents formed in the M1 mutant cells through an induction period of 40 days was 55.12 ± 4.12 mg g^{-1}, which was 1.26-fold higher than that of the wild-type strain (43.62 ± 3.98 mg g^{-1}) (Figure 6B). These results show that the M1 mutant has a 32% higher light reactivity and 18.5% faster light response rate than the wild-type strain, so it could rapidly sense the light faster than the wild-type strain against strong light irradiation during the induction period. Therefore, it is inferred that the M1 mutant promoted the synthesis of the astaxanthin to prevent oxidation of cells faster than the wild-type strain.

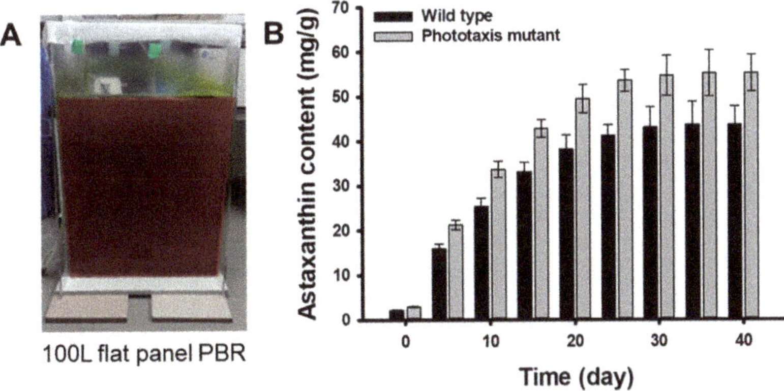

Figure 6. Large scale cultivation of the wild type and the M1 mutant: (**A**) Photo image of the 100 L flat panel photo-bioreactor. (**B**) Astaxanthin content of the wild type and M1 mutant. Cells suspensions were aerated at a flow rate of 0.1 vvm with 5% CO_2-enriched air. Error bars indicate the standard deviations of the means for three experiments.

Various mutagenesis and isolation strategies to enhance astaxanthin productivity from *H. pluvialis* are summarized in Table 1.

Table 1. Summary of strain development strategies for improved astaxanthin productivity from *H. pluvialis*.

Strain	Mutagenesis Strategy	Isolation Strategy	Astaxanthin Production		Growth Rate Improvement	Ref
			Unit	Improvement		
DPA12–1	UV, EMS [1]	DPA [2]	47.2 mg g^{-1}	1.7-fold	1.4-fold	[20]
EU3	UV, EMS	Nicotine	25.0 mg g^{-1}	1.3-fold	Similar	[44]
H$_2$–419–4	UV, EMS		37 pg cell^{-1}	1.3-fold	1.7-fold	[45]
M13	UV	Azide	174.7 mg L^{-1}	1.6-fold	Similar	[46]
Not specified	γ–ray		70.8 mg g^{-1}	2.4-fold	1.2-fold	[47]
M3	DBD [3] plasma	DPA	33.5 mg g^{-1}	1.5-fold	1.6 fold	[18]
B24	EMS	DPA	26.4 mg g^{-1}	1.3-fold	1.6-fold	[49]
M1	UV	Phototaxis	55.1 mg g^{-1}	1.3-fold	1.2-fold	This study

[1] EMS: ethyl methane sulfonate; [2] DPA: diphenylamine; [3] DBD: dielectric barrier discharge.

As a mutagenesis strategy, mainly UV irradiation and ethyl methane sulfonate (EMS) treatment were performed, and approaches such as γ–ray irradiation and dielectric barrier discharge (DBD) plasma irradiation were also carried out. As a strategy to isolate mutant strains, most studies have screened resistant strains using chemicals such as diphenylamine (DPA), nicotine, and azide as inhibitors. Wang et al. [20] developed strain *H. pluvialis* DPA12–1 with an astaxanthin content of 47.2 mg g^{-1}, 1.7-fold higher than that of the wild type, through UV irradiation and ethyl methane sulfonate (EMS) treatment. Chen et al. [44] reported that the astaxanthin content of *H. pluvialis* EU3, mutated by UV irradiation and EMS treatment and isolated by nicotine, improved to 25.0 mg g^{-1}, 1.3-fold higher than that of wild type. *H. pluvialis* H2–419–4 [47] mutated by UV irradiation and EMS treatment showed that astaxanthin production and growth rate were improved by 1.3-fold and 1.7-fold, respectively, compared to the wild type. Hong et al. [46] developed *H. pluvialis* M13, which showed 1.6-fold higher astaxanthin production than wild type by isolation using azide after UV irradiation. Cheng et al. [47] induced CO_2 stress in *H. pluvialis* mutated by γ–ray irradiation and astaxanthin production reached 70.8 mg g^{-1}, a 2.4-fold improvement compared to wild type. *H. pluvialis* M3 [48] was mutated by DBD plasma irradiation and isolated using DPA. Astaxanthin production of M3 mutant was 33.5 mg g^{-1}, which was

1.5-fold higher than that of wild type. Astaxanthin production of *H. pluvialis* B24 [49] isolated using DPA after mutation by EMS was found to be 26.4 mg g^{-1}. The growth rate of the mutant was slightly higher or comparable to that of the wild type. *H. pluvialis* M1 developed in this study showed that astaxanthin production (55.12 $\pm$ 4.12 mg g^{-1}) and growth rate were improved about 1.3-fold and 1.2-fold, respectively, compared to wild type. These results prove that the M1 mutant, selected using the microfluidic device, has similar or higher astaxanthin content than other mutants by UV irradiation. Although many studies have successfully developed H with improved astaxanthin productivity, chemicals that can cause environmental pollution, such as EMS, DPA, and nicotine, were used in the isolation process. Our strain selection strategy has the advantage of being able to efficiently screen strains with improved astaxanthin productivity in an eco-friendly approach.

3. Materials and Methods

3.1. Algal Strains and Culture Conditions

Haematococcus pluvialis NIES-144 (wild type) was obtained from the National Institute for Environmental Studies (Tsikuba, Japan). The PP mutant strain developed by chemical mutagenesis was used for determining whether differences in the photosynthetic efficiency affect the phototaxis [26]. The *H. pluvialis* was cultured by two-stage culture strategy; the 'green' stage for vegetative growth and the 'red' stage for the astaxanthin production. The cells were photo-autotrophically cultured in NIES-C medium (pH 7.5) with the vegetative growth at the low light intensity (40 μmol photons m^{-2} s^{-1} with a continuous light) for 10 days. After that, the astaxanthin accumulation was induced by the nutrient starvation with NIES-N medium (nitrogen deficient NIES-C medium, pH 7.5) at the strong light intensity (200 μmol photons m^{-2} s^{-1} with a continuous light) for 20 days in the red stage [50]. The light was provided by cool white LED lamps and the light intensities were measured by a LI-250 quantum photometer (Lambda Instrument, Blacksburg, VA, USA). The vegetative cells, with an initial cell density of 0.1 (OD$_{680}$), were inoculated into 100 mL Erlenmeyer flasks for the lab scale cultivation and 100 L flat panel photo-bioreactor, which was made of acrylic, for the large scale cultivation, respectively. All cell suspensions were aerated at a flow rate of 0.1 vvm with 5% CO$_2$-enriched air at 23 °C.

3.2. UV Irradiation Procedure for Mutagenesis

The wild-type cells (NIES-144) in the exponential phase experiencing the dark period were used for UV mutagenesis to improve transformation efficiency as described previously [51]. The cells diluted to 10,000 cells mL^{-1} were spread on an NIES-C agar plate, and were irradiated using 254 nm UVLS-225D Mineralight UV Display lamp (UVP, Upland, CA, USA), with the intensity of 0.02 mW cm^{-2}. UV intensities were measured by GT-510 UV meter (Giltron, Norwood, MA, USA). The survival rates of UV-treated cells depending on the UV energy were calculated by counting the colonies on solid agar plates under the different exposure times: 0 min (0 mJ cm^{-2}; control), 16 min (20 mJ cm^{-2}), 32 min (40 mJ cm^{-2}), 48 min (60 mJ cm^{-2}), and 60 min (80 mJ cm^{-2}). The mutant libraries were made under optimal conditions of 40 mJ cm^{-2} of UV energy (32 min of UV exposure time), representing about 2.6% of survival rates, and were kept for the regeneration during 24 h in the dark condition.

3.3. Fabrication of Microfluidic Device

A PDMS-based microfluidic device was fabricated using standard soft photolithography [52]. The design of the microstructure was generated using AutoCAD software and printed on transparent photomask film. A silicon mold master was fabricated by using SU-8 negative photoresist (SU-8 50, Microchem, Newton, MA, USA) on silicon wafers. The PDMS pre-polymer (10:1 mixture of 184 Sylgard base and curing agent, Dow Corning) was poured on the SU-8 mold and cured thermally at 80 °C. The PDMS layer containing the microchannel was bonded to a glass slide using oxygen plasma. The microfluidic device was composed of an inlet chamber (4 mm in diameter), outlet chamber (4 mm in diameter), and microchannel (30 mm in length, 50 μm in height) including observation zone (channel length 300 μm, width 100 μm, height 50 μm) near the outlet chamber. The width of the microchannel gradually decreased from 4 mm (at the inlet) to 150 μm (at the observation zone) to reduce hindrance of the phototactic movement by microchannel and facilitate the monitoring of the phototactic response at the single-cell resolution level (Figure 2).

3.4. Analysis of Phototactic Response in Microfluidic Device

Cells were grown for 10 days photo-autotrophically under continuous light at a light intensity of 40 μmol photons $m^{-2}\,s^{-1}$. The culture was then diluted to a density of 40,000 cells mL^{-1} and dark-adapted for 30 min to make cells sensitive to light stimulus. A 50 μL aliquot of the diluted culture was loaded into the inlet chamber of the microfluidic device, and an equal volume of NIES-C medium was loaded into the outlet chamber. Cells were collected into the inlet chamber via exposure to blue LED light at the end of the outlet chamber. After hydrostatic balance was established in the microchannel, a blue LED (70 μmol photons $m^{-2}\,s^{-1}$) was illuminated at the end of the inlet chamber to evoke negative phototaxis. The phototactic movements of cells were monitored and recorded under an inverted microscope (Olympus CKX41, Tokyo, Japan) equipped with a digital video camera (Canon EOS 700D, Tokyo, Japan). Cells arriving at the observation zone near the outlet chamber were counted, and their arrival times were automatically recorded for 30 min using custom software.

3.5. Phototaxis-Based Screening

Each colony on agar plates was transferred into 1 mL of NIES-C medium in each well of a 24-well microplate and cultured under low light condition (40 μmol photons $m^{-2}\,s^{-1}$) at 23 °C. All cultures of 10,000 transformants were then mixed, centrifuged, and resuspended in 6 mL of NIES-C medium to a cell density of 6.0×10^8 cells mL^{-1}. For screening of the mixture of 10,000 transformants, a microfluidic device with enlarged chambers (8 mm in diameter) and a microchannel (width from 8 mm to 0.4 mm, height 100 μm) was used. A 2.75 mL aliquot of dark-adapted cell mixture was loaded into the inlet chamber and exposed to blue LED light (70 μmol photons $m^{-2}\,s^{-1}$) for 10 min to isolate cells showing fast phototaxis. The cells arrived at the outlet chamber within 10 min and were collected by pipetting and recovered in NIES-C medium for 2 h. The cells were then used for the next cycle of phototaxis-assisted screening. After five cycles of screening, we obtained the cell mixture, and a 100 μL aliquot of the culture was plated onto NIES-C agar for the isolation of mutants as separated colonies. To compare the growth of mutant mixtures between each cycle of phototaxis-based screening, cells were inoculated to a density of 10,000 cells mL^{-1} and grown photo-autotrophically.

3.6. Analytical Methods

Chlorophyll fluorescence was measured with an FMS2 fluorometer (Hansatech, Norfolk, UK). Cells grown to an exponential phase (30 μg of chlorophyll a) were loaded onto a glass-fiber filter, and the filter was placed on the leaf clip. For determination of the PSII operating efficiency (Y(II)), cells (without dark-adaptation) were exposed to stepwise-increasing actinic light (from 1 to 900 μmol photons $m^{-2}\,s^{-1}$) for 20 s at each light intensity, and a saturating flash (3000 μmol photons $m^{-2}\,s^{-1}$, 0.7 s duration) was applied to measure Fm'. Y(II) was calculated as (Fm'–Fs)/Fm'. The cellular chlorophyll and carotenoid contents were determined spectrophotometrically as described [53,54].

The cell density was determined by measuring OD_{680} using UV-spectrophotometer (Shimadzu, Kyoto, Japan) or dry cell weight (DCW) every 24 h. DCW was only determined for large-scale cultivation by filtering aliquots of samples using pre-weighed filter paper. Then, the cell suspensions (10 mL) were filtered with GF/F glass microfiber filters (Whatman, Cambridge, UK) and dried at 105 °C overnight. DCW was determined by the difference between the mass of the biomass-containing filter paper and that of pre-weighed filter paper.

In order to identify the accurate amount of intracellular astaxanthin using a high performance liquid chromatography (HPLC) system, algal cells were collected by centrifuging the culture fluid at 3000 rpm for 10 min at 4 °C, and the astaxanthin in cell pellet was extracted using a homogenizer (TissueLyser II, Qiagen, Hilden, Germany) in the presence of methanol and glass beads [55]. Then, the homogenized lysates were saponified with 0.01 M KOH to convert the esterified astaxanthin into free forms.

After saponification, the astaxanthin concentration of each sample was determined by HPLC system equipped with two LC-10AD pumps and SPD-10 UV-Vis detector (Shimadzu, Kyoto, Japan). The extracts were separated using a 250 × 4.6 mm HS-303 hydrosphere C_{18} column (YMC, Kyoto, Japan). The mobile phase consisted of solvents A (dichloromethane: methanol: acetonitrile: water, 5.0: 85.0: 5.5: 4.5, *v/v*) and solvent B (dichloromethane: methanol: acetonitrile: water, 22.0: 28.0: 45.5: 4.5, *v/v*). For the effective separation of astaxanthin, a linear gradient system was used: 0% B for 8 min, a linear gradient from 0 to 100% B for 12 min, and 100% B for 50 min. The flow rate was set as 1.0 mL min^{-1} and the peaks were measured at 480 nm [50].

4. Conclusions

In this study, we screened mutant strains with photosensitivity using the negative phototaxis of the *H. pluvialis* in microfluidic device. It was confirmed that the more sensitive the mutant, the higher the astaxanthin productivity. In addition, large-scale cultivation in 100 L photo-bioreactor shows that the strains selected through phototaxis technology can be used in a more scaled-up culture for commercialization. We are planning follow-up studies to screen more cells faster and more efficiently. This screening strategy using the phototaxis of microalgae is advantageous in securement of *H. pluvialis* with improved growth rate and astaxanthin productivity.

Author Contributions: Conceptualization, K.H.L. and Y.C.; data curation, K.H.L., Y.C. and J.H.L.; funding acquisition, H.S.K.; methodology, K.H.L., Y.C. and J.H.L.; project administration, C.P., H.Y.Y. and H.S.K.; software, K.H.L., Y.C. and J.H.L.; supervision, C.P., H.Y.Y. and H.S.K.; writing—original draft, K.H.L. and Y.C.; writing—review and editing, C.P., H.Y.Y. and H.S.K. All authors have read and agreed to the published version of the manuscript.

Funding: This work was supported by the National Research Foundation of Korea (NRF) grant funded by the Korean Government (NRF-2019R1F1A1058873).

Institutional Review Board Statement: Not applicable.

Informed Consent Statement: Not applicable.

Data Availability Statement: The data presented in this study are available on request from the corresponding author.

Conflicts of Interest: The authors declare no conflict of interest.

References

1. Liu, X.; Osawa, T. *Cis* astaxanthin and especially 9-*cis* astaxanthin exhibits a higher antioxidant activity in vitro compared to the all-trans isomer. *Biochem. Biophys. Res. Commun.* **2007**, *357*, 187–193. [CrossRef] [PubMed]
2. Ranga Rao, A.; Harshvardhan Reddy, A.; Aradhya, S.M. Antibacterial properties of *Spirulina platensis, Haematococcus pluvialis, Botryococcus braunii* microalgal extracts. *Curr. Trends Biotechnol. Pharm.* **2010**, *4*, 809–819.
3. Pérez-López, P.; González-Garcia, S.; Jeffryes, C.; Agathos, S.N.; McHugh, E.; Walsh, D.; Murray, P.; Moane, S.; Feijoo, G.; Moreira, M.T. Life cycle assessment of the production of the red antioxidant carotenoid astaxanthin by microalgae: From lab to pilot scale. *J. Clea. Prod.* **2014**, *64*, 332–344. [CrossRef]
4. Shah, M.M.R.; Liang, Y.; Cheng, J.J.; Daroch, M. Astaxanthin-producing green microalga *Haematococcus pluvialis*: From single cell to high value commercial products. *Front. Plant Sci.* **2016**, *7*, 531. [PubMed]
5. Panis, G.; Rosales Carreon, J. Commercial astaxanthin production derived by green alga *Haematococcus pluvialis*: A microalgae process model and a techno-economic assessment all through production line. *Algal Res.* **2016**, *18*, 175–190. [CrossRef]
6. Li, J.; Zhu, D.L.; Niu, J.; Shen, S.D.; Wang, G. An economic assessment of astaxanthin production by large scale cultivation of *Haematococcus pluvialis*. *Biotechnol. Adv.* **2011**, *29*, 568–574. [CrossRef]

7.   Fassett, R.G.; Combes, J.S. Astaxanthin: A potential therapeutic agent in cardiovascular disease. *Mar. Drugs* **2011**, *9*, 447–465. [CrossRef]

8.   McWilliams, A. The Global Market for Carotenoids, BCC Research: Market Research Reports. 2018. FOD025F. Available online: https://www.bccresearch.com (accessed on 19 January 2021).

9.   Martínez-Álvarez, Ó.; Calvo, M.M.; Gómez-Estaca, J. Recent Advances in Astaxanthin Micro/Nanoencapsulation to Improve Its Stability and Functionality as a Food Ingredient. *Mar. Drugs* **2020**, *18*, 406. [CrossRef]

10.   Rammuni, M.N.; Ariyadasa, T.U.; Nimarshana, P.H.V.; Attalage, R.A. Comparative assessment on the extraction of carotenoids from microalgal sources: Astaxanthin from *H. pluvialis* and β-carotene from *D. salina*. *Food Chem.* **2019**, *277*, 128–134. [CrossRef]

11.   Khoo, K.S.; Lee, S.Y.; Ooi, C.W.; Fu, X.; Miao, X.; Ling, T.C.; Show, P.L. Recent advances in biorefinery of astaxanthin from *Haematococcus pluvialis*. *Bioresour. Technol.* **2019**, *288*, 121606. [CrossRef]

12.   Higuera-Ciapara, I.; Feliz-Valenzuela, L.; Goycoolea, F.M. Astaxanthin: A review of its chemistry and applications. *Crit. Rev. Food Sci. Nutr.* **2006**, *46*, 185–196. [CrossRef] [PubMed]

13.   Ranga Rao, A.; Siew Moi, P.; Ravi, S.; Aswathanarayana, R.G. Astaxanthin: Sources, extraction, stability, biological activities and its commercial applications—A review. *Mar. Drugs* **2014**, *12*, 128–152.

14.   Koller, M.; Muhr, A.; Braunegg, G. Microalgae as versatile cellular factories for valued products. *Algal Res.* **2014**, *6*, 52–63. [CrossRef]

15.   Lu, Q.; Li, H.; Zou, Y.; Liu, H.; Yang, L. Astaxanthin as a microalgal metabolite for aquaculture: A review on the synthetic mechanisms, production techniques, and practical application. *Algal Res.* **2021**, *54*, 102178. [CrossRef]

16.   Pham, H.M.; Kwak, H.S.; Hong, M.E.; Lee, J.W.; Chang, W.S.; Sim, S.J. Development of an X-shape airlift photobioreactor for increasing algal biomass and biodiesel production. *Bioresour. Technol.* **2017**, *239*, 211–218. [CrossRef] [PubMed]

17.   Silva, S.C.; Ferreira, I.C.F.R.; Dias, M.M.; Barreiro, M.F. Microalgae-Derived Pigments: A 10-Year Bibliometric Review and Industry and Market Trend Analysis. *Molecules* **2020**, *25*, 3406. [CrossRef] [PubMed]

18.   Kwak, H.S.; Kim, J.Y.H.; Sim, S.J. A microreactor system for cultivation of *Haematococcus pluvialis* and astaxanthin production. *J. Nanosci. Nanotechnol.* **2015**, *15*, 1618–1623. [CrossRef]

19.   Boussiba, S.; Bing, W.; Yuan, J.P.; Zarka, A.; Chen, F. Changes in pigments profile in the green alga *Haematococcus pluvialis* exposed to environmental stresses. *Biotechnol. Lett.* **1999**, *21*, 601–604. [CrossRef]

20.   Wang, N.; Guan, B.; Kong, Q.; Sun, H.; Geng, Z.; Duan, L. Enhancement of astaxanthin production from *Haematococcus pluvialis* mutants by three-stage mutagenesis breeding. *J. Biotechnol.* **2016**, *236*, 71–77. [CrossRef]

21.   Pereira, S.; Otero, A. *Haematococcus pluvialis* bioprocess optimization: Effect of light quality, temperature and irradiance on growth, pigment content and photosynthetic response. *Algal Res.* **2020**, *51*, 102027. [CrossRef]

22.   Abomohra, A.E.F.; El-Sheekh, M.; Hanelt, D. Protoplast fusion and genetic recombination between *Ochromonas danica* (chrysophyta) and *Haematococcus pluvialis* (chlorophyta). *Phycologia* **2016**, *55*, 65–71. [CrossRef]

23.   Nasri, N.; Keyhanfar, M.; Behbahani, M.; Dini, G. Enhancement of astaxanthin production in *Haematococcus pluvialis* using zinc oxide nanoparticles. *J. Biotechnol.* **2021**, *342*, 72–78. [CrossRef] [PubMed]

24.   Li, X.; Wang, X.; Duan, C.; Yi, S.; Gao, Z.; Xiao, C.; Agathos, S.N.; Wang, G.; Li, J. Biotechnological production of astaxanthin from the microalga *Haematococcus pluvialis*. *Biotechnol. Adv.* **2020**, *43*, 107602. [CrossRef] [PubMed]

25.   Yu, Q.; Li, Y.; Wu, B.; Hu, W.; He, M.; Hu, G. Novel mutagenesis and screening technologies for food microorganisms: Advances and prospects. *Appl. Microbiol. Biotechnol.* **2020**, *104*, 1517–1531. [CrossRef] [PubMed]

26.   Hong, M.E.; Choi, S.P.; Park, Y.I.; Kim, Y.K.; Chang, W.S.; Kim, B.W.; Sim, S.J. Astaxanthin production by a highly photosensitive *Haematococcus* mutant. *Process Biochem.* **2012**, *47*, 1972–1979. [CrossRef]

27.   Sandesh Kamath, B.; Vidhyavathi, R.; Sarada, R.; Ravishankar, G.A. Enhancement of carotenoids by mutation and stress induced carotenogenic genes in *Haematococcus pluvialis* mutants. *Bioresour. Technol.* **2008**, *99*, 8667–8673. [CrossRef]

28.   Hu, Z.; Li, Y.; Sommerfeld, M.; Chen, F.; Hu, Q. Enhanced protection against oxidative stress in an astaxanthin-overproduction *Haematococcus* mutant (Chlorophyceae). *Eur. J. Phycol.* **2008**, *43*, 365–376. [CrossRef]

29.   Sung, Y.J.; Kwak, H.S.; Hong, M.E.; Choi, H.I.; Sim, S.J. Two-dimensional microfluidic system for the simultaneous quantitative analysis of phototactic/chemotactic responses of microalgae. *Anal. Chem.* **2018**, *90*, 14029–14038. [CrossRef]

30.   Kim, J.Y.H.; Kwak, H.S.; Sung, Y.J.; Choi, H.I.; Hong, M.E.; Lim, H.S.; Lee, J.H.; Lee, S.Y.; Sim, S.J. Microfluidic high-throughput selection of microalgal strains with superior photosynthetic productivity using competitive phototaxis. *Sci. Rep.* **2016**, *6*, 21155. [CrossRef]

31.   Tamaki, S.; Tanno, Y.; Kato, S.; Ozasa, K.; Wakazaki, M.; Sato, M.; Toyooka, K.; Maoka, T.; Ishikawa, T.; Maeda, M.; et al. Carotenoid accumulation in the eyespot apparatus required for phototaxis is independent of chloroplast development in Euglena gracilis. *Plant Sci.* **2020**, *298*, 110564. [CrossRef]

32.   Ma, Z.E.; Helbling, E.W.; Li, W.; Villafañe, V.E.; Gao, K. Motility and photosynthetic responses of the green microalga *Tetraselmis subcordiformis* to visible and UV light levels. *J. Appl. Phycol.* **2012**, *24*, 1613–1621. [CrossRef]

33.   Choi, H.I.; Kim, J.Y.H.; Kwak, H.S.; Sung, Y.J.; Sim, S.J. Quantitative analysis of the chemotaxis of a green alga, *Chlamydomonas reinhardtii*, to bicarbonate using diffusion-based microfluidic device. *Biomicrofluidics* **2016**, *10*, 014121. [CrossRef] [PubMed]

34.   Ristori, T.; Rosati, G. The eyespot membranes of *Haematococcus pluvialis* flotow (Chlorophyceae): Their ultrastructure and possible significance in phototaxis. *Ital. J. Zool.* **1983**, *17*, 401–408.

35. Tamaki, S.; Mochida, K.; Suzuki, K. Diverse Biosynthetic Pathways and Protective Functions against Environmental Stress of Antioxidants in Microalgae. *Plants* **2021**, *10*, 1250. [CrossRef]
36. Ma, R.; Thomas-Hall, S.R.; Chua, E.T.; Eltanahy, E.; Netzel, M.E.; Netzel, G.; Lu, Y.; Schenk, P.M. Blue light enhances astaxanthin biosynthesis metabolism and extraction efficiency in *Haematococcus pluvialis* by inducing haematocyst germination. *Algal Res.* **2018**, *35*, 215–222. [CrossRef]
37. Wan, M.X.; Zhang, Z.; Wang, J.; Huang, J.K.; Fan, J.H.; Yu, A.Q.; Wang, W.L.; Li, Y.G. Sequential Heterotrophy–Dilution–Photoinduction Cultivation of *Haematococcus pluvialis* for efficient production of astaxanthin. *Bioresour. Technol.* **2015**, *198*, 557–563. [CrossRef]
38. Kwak, H.S.; Kim, J.Y.H.; Na, S.C.; Jeon, N.L.; Sim, S.J. Multiplex microfluidic system integrating sequential operations of microalgal lipid production. *Analyst* **2016**, *141*, 1218–1225. [CrossRef]
39. Meyvantsson, I.; Beebe, D.J. Cell culture models in microfluidic systems. *Annu. Rev. Anal. Chem.* **2008**, *1*, 423–449. [CrossRef]
40. Zhao, Y.; Yue, C.; Ding, W.; Li, T.; Xu, J.-W.; Zhao, P.; Ma, H.; Yu, X. Butylated hydroxytoluene induces astaxanthin and lipid production in *Haematococcus pluvialis* under high-light and nitrogen-deficiency conditions. *Bioresour. Technol.* **2018**, *266*, 315–321. [CrossRef]
41. Kobayashi, M.; Kakizono, T.; Nagai, S. Enhanced carotenoid biosynthesis by oxidative stress in acetate-induced cyst cells of a green unicellular alga, *Haematococcus pluvialis*. *Appl. Environ. Microbiol.* **1993**, *59*, 867–873. [CrossRef]
42. He, Q.; Yang, H.; Wu, L.; Hu, C. Effect of light intensity on physiological changes, carbon allocation and neutral lipid accumulation in oleaginous microalgae. *Bioresour. Technol.* **2015**, *191*, 219–228. [CrossRef] [PubMed]
43. Zuluaga, M.; Gueguen, V.; Letourneur, D.; Pavon-Djavid, G. Astaxanthin-antioxidant impact on excessive Reactive Oxygen Species generation induced by ischemia and reperfusion injury. *Chem. Biol. Interact.* **2018**, *279*, 145–158. [CrossRef] [PubMed]
44. Chen, Y.; Li, D.; Lu, W.; Xing, J.; Hui, B.; Han, Y. Screening and characterization of astaxanthin-hyperproducing mutants of *Haematococcus pluvialis*. *Biotechnol. Lett* **2003**, *25*, 527–529. [CrossRef] [PubMed]
45. Sun, Y.; Liu, J.; Zhang, X.; Lin, W. Strain H2-419-4 of *Haematococcus pluvialis* induced by ethyl methanesulphonate and ultraviolet radiation. *Chin. J. Oceanol. Limnol.* **2008**, *26*, 152–156. [CrossRef]
46. Hong, M.E.; Choi, H.; Kwak, S.; Hwang, S.W.; Sung, J.Y.; Chang, S.; Sim, S.J. Rapid selection of astaxanthin-hyperproducing *Haematococcus* mutant via azide-based colorimetric assay combined with oil-based astaxanthin extraction. *Bioresour. Technol.* **2018**, *267*, 175–181. [CrossRef] [PubMed]
47. Cheng, J.; Li, K.; Yang, Z.; Zhou, J.; Cen, K. Enhancing the growth rate and astaxanthin yield of *Haematococcus pluvialis* by nuclear irradiation and high concentration of carbon dioxide stress. *Bioresour. Technol.* **2016**, *204*, 49–54. [CrossRef]
48. Liu, J.; Chen, J.; Chen, Z.; Qin, S.; Huang, Q. Isolation and characterization of astaxanthin-hyperproducing mutants of *Haematococcus pluvialis* (Chlorophyceae) produced by dielectric barrier discharge plasma. *Phycologia* **2016**, *55*, 650–658. [CrossRef]
49. Gómez, P.I.; Inostroza, I.; Pizarro, M.; Pérez, J. From genetic improvement to commercial-scale mass culture of a chilean strain of the green microalga *Haematococcus pluvialis* with enhanced productivity of the red ketocarotenoid astaxanthin. *AoB Plants* **2013**, *5*, plt026. [CrossRef]
50. Kang, C.D.; Lee, J.S.; Park, T.H.; Sim, S.J. Comparison of heterotrophic and photoautotrophic induction on astaxanthin production by *Haematococcus pluvialis*. *Appl. Microbiol. Biotechnol.* **2005**, *68*, 237–241. [CrossRef]
51. Bišová, K.; Zachleder, V. Cell-cycle regulation in green algae dividing by multiple fission. *J. Exp. Bot.* **2014**, *65*, 2585–2602. [CrossRef]
52. Qin, D.; Xia, Y.; Whitesides, G.M. Soft lithography for micro- and nanoscale patterning. *Nat. Protoc.* **2010**, *5*, 491–502. [CrossRef] [PubMed]
53. Kirst, H.; Garcia-Cerdan, J.G.; Zurbriggen, A.; Melis, A. Assembly of the light-harvesting chlorophyll antenna in the green alga *Chlamydomonas reinhardtii* requires expression of the TLA2-CpFTSY gene. *Plant Physiol.* **2012**, *158*, 930–945. [CrossRef] [PubMed]
54. Baroli, I.; Do, A.D.; Yamane, T.; Niyogi, K.K. Zeaxanthin accumulation in the absence of a functional xanthophyll cycle protects *Chlamydomonas reinhardtii* from photooxidative stress. *Plant Cell* **2003**, *15*, 992–1008. [CrossRef]
55. Lamb, M.R.; Dutcher, S.K.; Worley, C.K.; Dieckmann, C.L. Eyespot-assembly mutants in *Chlamydomonas reinhardtii*. *Genetics* **1999**, *153*, 721–729. [CrossRef] [PubMed]

 *marine drugs*

Communication

Monocaprin Enhances Bioavailability of Fucoxanthin in Diabetic/Obese KK-A^y Mice

Kodai Nagata [1], Naoki Takatani [1], Fumiaki Beppu [1], Aya Abe [2], Etsuko Tominaga [2], Tomohisa Fukuhara [2], Makoto Ozeki [2] and Masashi Hosokawa [1,*]

[1] Faculty of Fisheries Sciences, Hokkaido University, 3-1-1 Minato, Hakodate 041-8611, Hokkaido, Japan; ko.d12n@gmail.com (K.N.); n-takatani@fish.hokudai.ac.jp (N.T.); fbeppu@fish.hokudai.ac.jp (F.B.)
[2] Taiyo Kagaku Co., Ltd., 1-3 Takaramachi, Yokkaichi 510-0844, Mie, Japan; aabe@taiyokagaku.co.jp (A.A.); etominaga@taiyokagaku.co.jp (E.T.); tfukuhara@taiyokagaku.co.jp (T.F.); mozeki@taiyokagaku.co.jp (M.O.)
* Correspondence: hoso@fish.hokudai.ac.jp

Abstract: Fucoxanthin is a marine carotenoid found in brown seaweeds and several microalgae. It has been reported that fucoxanthin has health benefits such as anti-obesity and anti-diabetic effects. To facilitate fucoxanthin applications in the food industry, it is important to improve its low bioavailability. We attempted the combined feeding of fucoxanthin-containing seaweed oil (SO) and monocaprin in a powder diet and analyzed the fucoxanthin metabolite contents in the liver, small intestine and serum of diabetic/obese KK-A^y mice. After 4 weeks of feeding with the experimental diets, the serum fucoxanthinol concentrations of the mice fed 0.2% SO and 0.5% monocaprin were higher than those of the 0.2% SO-fed mice. Furthermore, fucoxanthinol accumulation in the liver and small intestine tended to increase in a combination diet of 0.2% SO and 0.125–0.5% monocaprin compared with a diet of 0.2% SO alone, although amarouciaxanthin A accumulation was not different among the 0.2% SO-fed groups. These results suggest that a combination of monocaprin with fucoxanthin-containing SO is an effective treatment for improving the bioavailability of fucoxanthin.

Keywords: fucoxanthin; monocaprin; bioavailability; fucoxanthinol

Citation: Nagata, K.; Takatani, N.; Beppu, F.; Abe, A.; Tominaga, E.; Fukuhara, T.; Ozeki, M.; Hosokawa, M. Monocaprin Enhances Bioavailability of Fucoxanthin in Diabetic/Obese KK-A^y Mice. *Mar. Drugs* **2022**, *20*, 446. https://doi.org/10.3390/md20070446

Academic Editors: Ralph Urbatzka and Bill J. Baker

Received: 11 May 2022
Accepted: 5 July 2022
Published: 7 July 2022

Publisher's Note: MDPI stays neutral with regard to jurisdictional claims in published maps and institutional affiliations.

1. Introduction

Fucoxanthin is a marine carotenoid with unique structures such as an allene bond and epoxide in the molecule. It is an accessory pigment for photosynthesis contained in brown seaweeds and in several microalgae species. We have reported that dietary fucoxanthin exhibits beneficial functions to health such as anti-obesity and anti-diabetic effects in animal models [1,2]. Human studies have also shown a reduction in white adipose tissue weight and HbA1c level related to blood glucose regulation by fucoxanthin [3,4]. Therefore, there is interest towards fucoxanthin, for its utilization as a nutraceutical ingredient in the food industry. However, the uptake of fucoxanthin into intestinal cells is low compared with other carotenoids such as β-carotene and lutein [5]. Therefore, improving the low bioavailability of fucoxanthin is an important challenge.

In previous studies, lysophosphatidylcholine and lysoglycerogalactolipid has been reported to enhance carotenoid uptake into the intestinal Caco-2 cells [5,6]. These monoacyl-lipids are suggested to enhance the bioaccesibility of carotenoids, including fucoxanthin, by affecting the micellar state of digestive lipid components. In addition, fucoxanthin absorption was improved using an emulsion-based delivery system [7] and protein-based encapsulation [8]. Thus, improving the bioaccessibility and bioavailability of dietary fucoxanthin is important for enhancing its functionality.

Monoacylglycerol (MG) is an amphipathic lipid, and it is used as a nonionic emulsifier and as an additive in the food industry. The functional properties of MG depends on its fatty acid chain length and its unsaturation. Monocaprin, which is an MG-binding capric acid with a 10-carbon chain, is well known for its anti-bacterial activities [9]. A combination of monocaprin and doxycycline also showed effective anti-viral and wound-healing actions against herpes labialis [10]. Moreover, medium-chain fatty acids, including capric acid, are also reported to prevent the metabolic syndrome [11]. Thus, monocaprin is expected to be a functional lipid that affects the micellar state of carotenoids. Therefore, we paid attention to the ability of monocaprin to enhance the bioavailability of fucoxanthin. In addition, the mixing of fucoxanthin and monocaprin into a powder diet is a very easy-to-use method compared with delivery and encapsulation systems.

In this study, we investigated fucoxanthin bioavailability in diabetic/obese KK-A^y mice via the combined feeding of fucoxanthin-containing seaweed oil (SO) and monocaprin.

2. Results and Discussion

Dietary fucoxanthin is hydrolyzed to fucoxanthinol in the gastrointestinal tract and is converted to amarouciaxanthin A in the liver (Figure 1) [12]. We also reported that fucoxanthin metabolites are transported and accumulate in several tissues, such as the liver, white adipose tissue, and skeletal muscle (Figure 1) [13,14]. To enhance the bioavailability of fucoxanthin, we attempted a combined feeding of monocaprin and fucoxanthin mixed in a powder diet on diabetic/obese KK-A^y mice. In this study, we used SO containing 5% fucoxanthin and prepared the experimental diets as shown in Table 1. The 0.2% SO diet was prepared by replacing medium-chain triglyceride (MCT) in the control diet with SO (B, D, E, and F groups in Table 1). The fucoxanthin concentration was adjusted to 0.01% in the diet. Monocaprin was added into the diet at 0.125–0.5% by replacing soybean oil, as shown in Table 1.

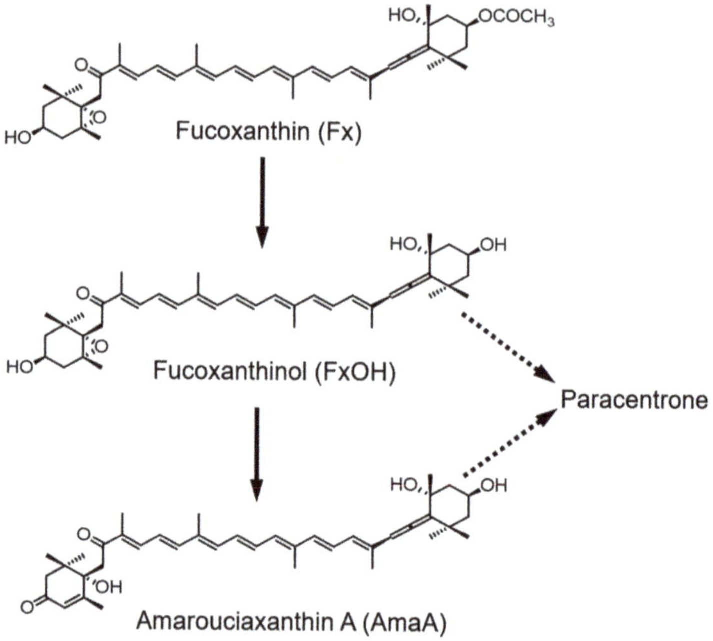

Figure 1. Fucoxanthin metabolites in the body.

After 4 weeks of feeding with the experimental diets and the control diet, there was no significant difference in the body weights among all of the groups (data not shown). SO-containing diets did not show an anti-obesity effect on KK-A^y mice because the fucoxanthin content in the diet was 0.01%, which is low compared with that in previous studies [1,2].

Table 1. Compositions of control and experimental diets used in the animal experiment. Fx: fucoxanthin; MCT: medium chain triglyceride; SO: seaweed oil.

Ingredient (g/kg Diet)	A: Control	B: SO 0.2%	C: Monocaprin 0.5%	D: SO 0.2% +Monocaprin 0.125%	E: SO 0.2% +Monocaprin 0.25%	F: SO 0.2% +Monocaprin 0.5%
Soybean oil	98.000	98.000	93.000	96.750	95.500	93.000
Seaweed oil (SO)	-	2.000	-	2.000	2.000	2.000
(Fx in diet)	-	(0.100)	-	(0.100)	(0.100)	(0.100)
MCT	2.000	-	2.000	-	-	-
Monocaprin	-	-	5.000	1.250	2.500	5.000
Corn starch	374.120	374.120	374.120	374.120	374.120	374.120
Dextrinized cornstarch	124.240	124.240	124.240	124.240	124.240	124.240
Casein	207.000	207.000	207.000	207.000	207.000	207.000
Sucrose	94.120	94.120	94.120	94.120	94.120	94.120
Cellulose	50.000	50.000	50.000	50.000	50.000	50.000
AIN-93 mineral mixture	35.000	35.000	35.000	35.000	35.000	35.000
AIN-93 vitamin mixture	10.000	10.000	10.000	10.000	10.000	10.000
L-Cystine	3.000	3.000	3.000	3.000	3.000	3.000
Choline bitartrate	2.500	2.500	2.500	2.500	2.500	2.500
tert-Butyl hydroquinone	0.014	0.014	0.014	0.014	0.014	0.014

Fucoxanthinol and amarouciaxanthin A, which are major metabolites of fucoxanthin, were analyzed, respectively, in the serum, and in the small intestines and livers of KK-A^y mice. Fucoxanthin was not detected in all of the mice, as in previous studies [13,14]. In the mice fed a 0.2% SO diet, the fucoxanthinol concentration was 0.28 µg/mL (Figure 2a). The addition of monocaprin in the 0.2% SO diet increased the serum fucoxanthinol concentration in a dose-dependent manner. A combination of 0.2% SO and 0.5% monocaprin significantly enhanced the fucoxanthinol concentration at 0.48 µg/mL, compared with the 0.2% SO diet group (Figure 2a). On the other hand, the amarouciaxanthin A concentrations in the serum were not different among the SO-fed groups (Figure 2b).

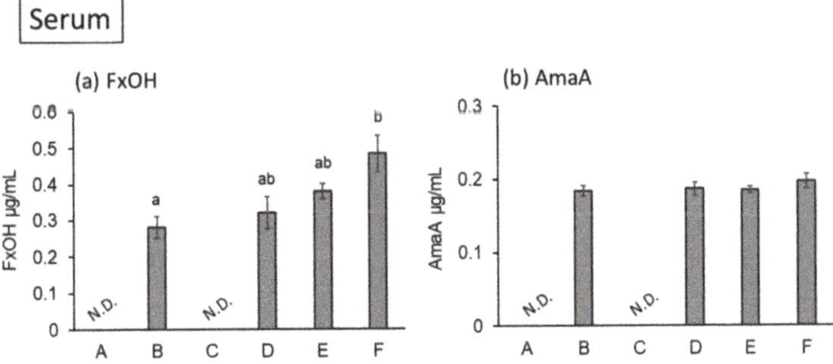

Figure 2. Carotenoid concentration in the serum of KK-A^y mice fed experimental diets for 4 weeks. (A) Control, (B) SO 0.2%, (C) monocaprin 0.5%, (D) SO 0.2% + monocaprin 0.125%, (E) SO 0.2% + monocaprin 0.25%, (F) SO 0.2%+monocaprin 0.5%. (a) FxOH concentration, (b) AmaA concentration. SO: seaweed oil; FxOH: fucoxanthinol; AmaA: amarouciaxanthin A; N.D.: not detected. Bars with different letters are significantly different; $p < 0.05$.

Fucoxanthinol accumulation was observed at 8.43 µg/g in the small intestine of the mice fed a 0.2% SO diet (Figure 3a). Monocaprin tended to increase fucoxanthinol accumulation in the small intestine but not in a dose-dependent manner. In the liver, fucoxanthinol accumulation also increased when using a combination diet of 0.2% SO and 0.125% or 0.25% monocaprin (Figure 4a). However, in the combination diet of 0.5% monocaprin and 0.2% SO, fucoxanthinol accumulation did not increase compared with that of the 0.2% SO and the 0.25% monocaprin diet. A high dose of monocaprin (0.5%) may activate the hepatic enzymes related to the degradation of fucoxanthin. This is required for further investigation.

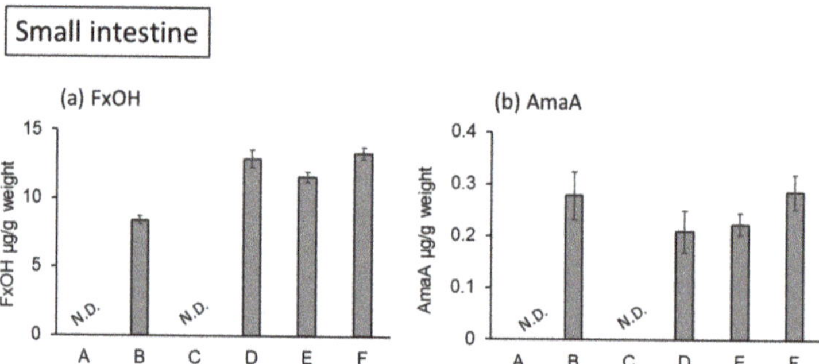

Figure 3. Carotenoid content in the small intestine of KK-A^y mice fed experimental diets for 4 weeks. (A) Control, (B) SO 0.2%, (C) monocaprin 0.5%, (D) SO 0.2% + monocaprin 0.125%, (E) SO 0.2% + monocaprin 0.25%, (F) SO 0.2% + monocaprin 0.5%. (**a**) FxOH content, (**b**) AmaA content. SO: seaweed oil; FxOH: fucoxanthinol; AmaA: amarouciaxanthin A; N.D.: not detected.

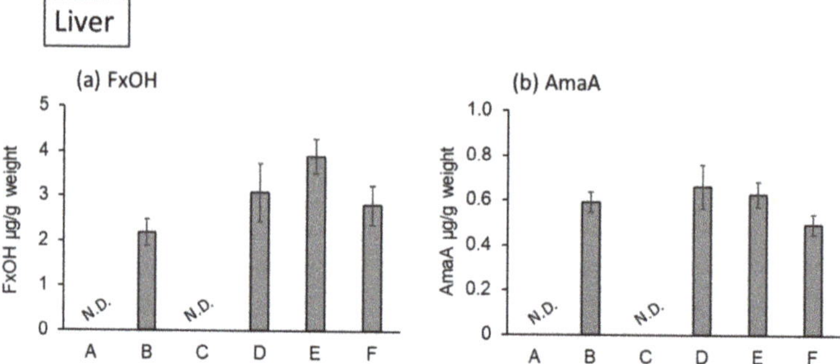

Figure 4. Carotenoid content in the liver of KK-A^y mice fed experimental diets for 4 weeks. (A) Control, (B) SO 0.2%, (C) monocaprin 0.5%, (D) SO 0.2%+monocaprin 0.125%, (E) SO 0.2%+monocaprin 0.25%, (F) SO 0.2%+monocaprin 0.5%. (**a**) FxOH content, (**b**) AmaA content. SO: seaweed oil; FxOH: fucoxanthinol; AmaA: amarouciaxanthin A; N.D.: Not detected.

On the other hand, amarouciaxanthin A accumulation was lower than fucoxanthinol accumulation in the small intestine and in the liver. Furthermore, amarouciaxanthin A accumulation in the small intestine and the liver was not different among SO-fed groups with/without monocaprin (Figures 3b and 4b).

Fucoxanthin-loaded particles composed of casein and chitosan, or of albumin and oleic acid have been reported to improve fucoxanthin bioavailability 4 h or 24 h after oral administration to mice [15,16]. Fucoxanthin–oleic acid–albumin complexes that have been dispersed in water also improved the bioavailability and antioxidant capacity in the eyes as well as in the serum of mice, compared with free fucoxanthin, after 15 days of administration [8]. Furthermore, an emulsifier of fucoxanthin with gum arabic and γ-cyclodextrin dissolved in water was reported to improve the bioavailability of fucoxanthin [17]. These experiments were conducted via the oral administration of fucoxanthin complexes in the water system. On the other hand, we demonstrated for the first time that monocaprin together with fucoxanthin in a powder diet, but not in a water system, increases fucoxanthinol accumulation in the serum of mice after 4 weeks of feeding. These results show that a combination diet of monocaprin with fucoxanthin-containing oil is effective for improving the bioavailability of fucoxanthin. It is suggested that the addition of monocaprin in a fucoxanthin-containing diet may have an effect on the micellar state of carotenoids in the digestion system and may improve fucoxanthin bioavailability. Mixing SO and monocaprin into a powder sample is a very easy process. From the current results, it is expected that fucoxanthin will be applied widely within the food industry fields.

We recently reported that dietary fucoxanthin inhibits hepatic oxidative stress and inflammation in non-alcoholic steatohepatitis model mice [18]. Therefore, our results provide beneficial information for the application of fucoxanthin and monocaprin in the food industry. On the other hand, the improvement of fucoxanthin stability is also important for applications. Sun et al. [19] reported that fucoxanthin microcapsules prepared using biopolymers are effective materials. Further examination is required to clarify the mechanism of a combined effect of fucoxanthin and monocaprin on stability as well as on bioavailability.

3. Materials and Methods

3.1. Chemicals

SO (fucoxanthin-5KM composed of 79% food oil, 20% brown seaweed lipid and 1% tocopherol) containing 5% fucoxanthin was purchased from Oriza Oil & Fat Chemical Co., Ltd. (Aich, Ichinomiya, Japan). Monocaprin was prepared by Taiyo Kagaku Co., Ltd. (Mie, Yokkaichi, Japan). Other chemicals were obtained from FUJIFILM Wako Chemical Co. (Osaka, Japan).

3.2. Animal Experiments

The diabetic/obese KK-A^y mice (4-week-old males) were obtained from CREA Japan Inc. (Tokyo, Japan). The mice were housed at 23 ± 2 °C and 50% humidity, with a 12 h light/12 h dark cycle and allowed free access to food and water. The control diet (A) containing 9.8% soybean oil and 0.2% MCT was prepared as shown in Table 1. MCT in the control diet was exchanged for 0.2% SO (B) in the experimental diets. Monocaprin was exchanged for soybean oil at 0.125–0.5% (C–F) in the experimental diets of each group, as shown in Table 1. After feeding the mice on a control diet for one week, six groups of seven mice were assigned so that there was no significant difference in body weight and blood glucose levels. Then, five experimental groups mice were acclimated by feeding a 0.5% monocaprin diet (C) for an additional one week. Control mice were fed a control diet (A), as shown in Table 1. After acclimation, each experimental diet and control diet (Table 1) were fed to the mice for 4 weeks. The mice were anatomized under anesthesia using isoflurane. The liver and small intestine were rapidly removed and stored at −30 °C. The serum was separated by centrifugation at 3000 rpm for 15 min. All procedures for animal care in this study were approved by the Ethical Committee of Experimental Animal Care of Hokkaido University (No 19-0083).

3.3. HPLC Analysis of Fucoxanthin Metabolites

The total lipid (TL) was extracted from mouse tissues and serum with chloroform:methanol (2:1, *v/v*) containing α-tocopherol (50 μg/mL) according to the Folch method [20]. The obtained TL was dissolved in *n*-hexane:acetone (7:3, *v/v*) and subjected to an HPLC system (LC-20AD and CBM-20A (Shimadzu, Kyoto, Japan), column: Mightysil Si 60 250 × 4.6 mm (Kanto Chemical Co., Inc., Tokyo, Japan), two columns were connected, column temperature: 25 °C, mobile phase: *n*-hexane:acetone (7:3, *v/v*), flow rate: 1.0 mL min^{-1}, detection: